Isotopes in Heterogeneous Catalysis

CATALYTIC SCIENCE SERIES

Series Editor: Graham J. Hutchings *(Cardiff University)*

Published

Vol. 1 Environmental Catalysis
edited by F. J. J. G. Janssen and R. A. van Santen

Vol. 2 Catalysis by Ceria and Related Materials
edited by A. Trovarelli

Vol. 3 Zeolites for Cleaner Technologies
edited by M. Guisnet and J.-P. Gilson

Vol. 5 Supported Metals in Catalysis
edited by J. A. Anderson and M. F. García

Forthcoming

Vol. 6 Catalysis by Gold
edited by G. C. Bond, C. Louis and D. Thompson

CATALYTIC SCIENCE SERIES — VOL. 4

Series Editor: Graham J. Hutchings

Isotopes in Heterogeneous Catalysis

edited by

Justin S J Hargreaves

S David Jackson

Geoff Webb

University of Glasgow, UK

ICP

Imperial College Press

Published by

Imperial College Press
57 Shelton Street
Covent Garden
London WC2H 9HE

Distributed by

World Scientific Publishing Co. Pte. Ltd.
5 Toh Tuck Link, Singapore 596224
USA office: 27 Warren Street, Suite 401-402, Hackensack, NJ 07601
UK office: 57 Shelton Street, Covent Garden, London WC2H 9HE

British Library Cataloguing-in-Publication Data
A catalogue record for this book is available from the British Library.

Catalytic Science Series — Vol. 4
ISOTOPES IN HETEROGENEOUS CATALYSIS

ISBN 1-86094-584-8

Printed in Singapore by World Scientific Printers (S) Pte Ltd

Contents

Contributors

D Duprez
Laboratoire de Catalyse
en Chimie Organique,
Université de Poitiers
and CNRS UMR 6503
40, Av. Recteur Pineau,
86022 Poitiers Cedex,
France

James G Goodwin, Jr
Department of Chemical
Engineering,
Clemson University,
Clemson, SC 29634-0909
USA

J S J Hargreaves
Department of Chemistry,
Joseph Black Building,
University of Glasgow,
Glasgow,
G12 8QQ, UK

E J M Hensen
Schuit Institute of Catalysis,
Faculty of Chemical Engineering,
Eindhoven University
of Technology,
P O Box 513,
5600MB Eindhoven,
The Netherlands

John R Jones
Chemistry, School of Biological
and Life Sciences,
University of Surrey,
Guildford,
Surrey GU2 7XH, UK

A M de Jong
Accelerator Laboratory,
Schuit Institute of Catalysis,
Faculty of Technical Physics,
Eindhoven University
of Technology,
P O Box 513,
5600MB Eindhoven,
The Netherlands

I M Mellor
Metalysis Ltd,
Unit 2, Farfield Park,
Manvers Way,
Wath upon Dearne,
Rotherham,
S63 5DB, UK

Zoltán Paál
Institute of Isotopes,
Chemical Research Center,
Hungarian Academy of Sciences,
Budapest P. O. Box 77,
H-1525, Hungary

Sourabh Pansare
Department of Chemical
Engineering,
Clemson University,
Clemson, SC 29634-0909,
USA

R A van Santen
Schuit Institute of Catalysis,
Faculty of Chemical Engineering,
Eindhoven University
of Technology, P O Box 513,
5600MB Eindhoven,
The Netherlands

Shui-Yu Lu
Molecular Imaging Branch,
National Institute of
Mental Health,
National Institutes of Health,
10 Center Drive,
MSC 1003,
Bethesda, MD 20892-1003,
USA

Amornmart Sirijaruphan
Department of Chemical
Engineering,
Clemson University,
Clemson, SC 29634-0909,
USA

J Sommer
Laboratoire de Physicochimie des
Hydrocarbures, UMR 7513,
Institute de Chimie,
Universite Louis Pasteur de
Strasbourg,
1 rue B. Pascal
67000-Strasbourg, France

Pál Tétényi
Institute of Isotopes,
Chemical Research Center,
Hungarian Academy of Sciences,
Budapest P. O. Box 77,
H-1525, Hungary

S Walspurger
Laboratoire de Physicochimie des
Hydrocarbures, UMR 7513,
Institute de Chimie,
Universite Louis Pasteur de
Strasbourg,
1 rue B. Pascal
67000-Strasbourg, France

Preface

Isotopes can be used as a way of detecting elements and their environments using a large range of techniques, including NMR, EPR, mass spectrometry, and radiation counters. Research effort using isotopes and the properties of isotopes has grown systematically since their discovery by JJ Thomson in 1913 and Aston's work in 1919 using the new technique of mass spectrometry. In 1943, George de Hevesy was awarded the Nobel Prize "for his work on the use of isotopes as tracers in the study of chemical processes", with Urey having already received the prize in 1934 for the "discovery of heavy hydrogen". After the Second World War, isotopes were more readily available, and by 1950 Taylor was using H_2/D_2 as a characterisation tool to probe the heterogeneity of catalyst surfaces, and Kemball was studying deuterium/alkane exchange over metal films. In a foretaste of things to come, in 1956 Eischens studied CO adsorption on platinum using mixtures of C-12 and C-13. Historically, deuterium and radioisotopes were pre-eminent with studies in the late 50s and early 60s on the mechanism of hydrogenation (Thomson, Bond, Wells, and Webb), Fischer-Tropsch synthesis (Emmett), and ammonia synthesis (Horiuti). Later studies took isotopic studies fully into the *in operando* regime with ICI scientists using $^{14}CO_2$ at 50 bar and 250°C to confirm that carbon dioxide was the source for methanol from a commercial $CO/CO_2/H_2$ feed, as proposed earlier. Isotope studies are the ultimate *in-situ/in operando* non-invasive characterisation method in heterogeneous catalysis. This volume comprises of a number of contributed chapters, bringing the use and application of isotopes in catalytic research up-to-date. Each chapter reviews the use of a particular isotope, isotopic reaction, or isotope based method. Both radioisotopes and stable isotope studies are covered.

In Chapter 1, Sommer and Walspurger give a detailed account of the use of deuterium in the determination of mechanism in acid-base catalysis. In such studies, deuterium is used both as a label to follow conversion pathways and also for kinetic isotope and equilibrium isotope measurements. In Chapter 2, Paal details the application of the ^{14}C radioisotope for the elucidation of reaction mechanism in hydrocarbon conversion and CO hydrogenation. Historically, a large volume of fundamental work has been performed with ^{14}C in these areas and this is comprehensively reviewed. The theme of radioisotopes is continued in Chapter 3, where Tétényi describes the application of the ^{35}S isotope in sulfide based hydrotreating catalysts. This is an especially powerful technique in such systems where there is high lability of sulfur species. Hargreaves and Mellor review isotopic oxygen exchange reactions with oxides in Chapter 4. The activity patterns of oxygen exchange are known to correlate well with those of oxidation activity for a number of substrates, although the exact identity of the exchange sites remains elusive in most cases. In Chapter 5, Duprez gives a comprehensive review of the use of H/D and $^{18}O/^{16}O$ exchange kinetics in the elucidation of diffusivity in heterogeneous catalysis, illustrating the dynamic nature of species under reaction conditions. This is followed in Chapter 6, by a review of the application of the steady state transient isotopic kinetic method contributed by Pansare, Sirijaruphan and Goodwin. In this powerful technique, a pulse of labelled reactant is admitted to a working catalytic reaction and the resultant transient response can be analysed in terms of pathway, and number and activity distribution of active sites. In the concluding chapter, de Jong, Hensen and van Santen review the positron emission profiling technique and present a case study of its application in ammonia oxidation. It is an effective technique for tracing the movement of labelled atoms through a catalytic bed and for the elucidation of reaction mechanism.

We would like to take this opportunity to express our appreciation to those who have written chapters and to thank them for their diligence in responding to tight deadlines so effectively. We would also like to thank the staff of Imperial College Press for their kind assistance in the production of this book and to Professor Graham Hutchings (Series Editor) for his kind invitation to contribute a volume to this series.

J S J Hargreaves,
S D Jackson,
G Webb.

Glasgow, 2005

Chapter 1

Deuterium-Labelling in Mechanistic Studies on Heterogeneous Acid-Base Catalysts

J Sommer and S Walspurger

1. Introduction

Among the tools used to study reaction mechanisms, the replacement of an atom by one of its isotopes has proven unique in its efficiency. In organic chemistry, which is mostly concerned with carbon and hydrogen containing compounds, replacing protium with deuterium has frequently been used, especially in tracer studies, to follow reaction paths and in kinetic studies and to determine isotope effects on reaction rates. In nature, hydrogen is essentially composed of atoms in which the nucleus is a single proton. However, it contains 0.0156% of deuterium, in which the nucleus also contains a neutron. A major source of deuterium is heavy water, D_2O, which is prepared on industrial scale by the electrolytic enrichment of normal water.

The stability of this nuclide as well as its nuclear properties, which are favourable for NMR observation, contributed to the development of its use in mechanistic studies. In comparison, tritium, which is another heavy hydrogen nuclide containing 2 neutrons and 1 proton, has more limited applications as a consequence of its radioactivity. The properties of the three nuclides are given in Tables 1 and 2.

The use of deuterium in heterogeneous catalysis for mechanistic studies can be divided into two different categories. The largest application is in tracer studies, involving the labelling of one or more selected atoms in the starting compound and localising the isotope in the products of rearrangement or isomerisation reactions. In this way, migration of atoms

Table 1. Some atomic properties of protium, deuterium and tritium.

Property	H	D	T
Relative Atomic Mass	1.007 825	2.014 102	3.016 049
Nuclear Spin Quantum Number	1/2	1	1/2
NMR Frequency (at 2.5 tesla)/MHz	100.56	15.360	104.68
NMR Relative Sensitivity (constant field)	1.000	0.009 64	1.21
Radioactive Stability	Stable	Stable	β^- $t_{1/2}$ 12.35 y

Table 2. Some physical properties of $H_2D_2T_2$.

Property	H_2	D_2	T_2
Melting Point/K	13.957	18.73	20.62
Boiling Point/K	20.39	23.67	25.04
Heat of Dissociation/kJ mol^{-1} (at 298.2 K)	435.88	443.35	446.9
Zero Point Energy/kJ mol^{-1}	25.9	18.5	15.1
Internuclear Distance/pm	74.14	74.14	(74.14)

via either an intra- or inter-molecular process can be monitored. The second application is the use of deuterium in the study of the isotope effect on reaction rates (i.e., KIE: the kinetic isotope effect) and/or on equilibria (i.e., EIE: the equilibrium isotope effect). In this application, the effect of replacing protium (H) by deuterium (D) is related to the difference of the zero point energy of the C-H or C-D bond respectively in the ground state. Substituting protium with deuterium lowers the ground state of a molecule, providing important information on the transition state (i.e., geometry and the nature of the bond which is broken in the reaction).[1–3]

In relation to the economic importance of acid catalysed hydrocarbon transformations such as cracking and isomerisation, the largest number of isotopic studies published thus far concern the exchange or migration of protium, deuterium and ^{13}C in alkanes.

2. H/D Exchange in the Strongest Liquid Superacids

In order to understand the mechanistic implications of these acid-catalysed processes and to distinguish between carbenium and carbonium intermediates, it is important to recall the results obtained, when the reactions are conducted under superacidic conditions.[4]

When small alkanes come into contact with the strongest superacids such as HF-SbF_5, even at temperatures below 0°C, rapid protium exchange occurs between the acidic protons and the hydrocarbon. This is due to the fact that, in the strongest acid media, hydrocarbons behave as sigma bases[5] which may accept a proton on the various C-H or C-C bonds, leading to 3 center-2 electron transition states (or carbonium ions), as shown in Scheme 1 for iso-pentane.

The rapid protonation-deprotonation process can be monitored by using either a deuterated alkane or a deuterated superacid. When the reaction is carried out in the presence of carbon monoxide, it is also possible to compare the isotope exchange process with the protolysis reaction which produces a smaller alkane and a smaller carbenium ion which is trapped by CO, forming stable oxocarbenium ions that are easily converted into esters for analysis[6] (e.g., Scheme 2).

The reaction has been carried out by our group on all small alkanes from C_1 to C_5.[6,7] A comparison between exchange and ionisation rates shows that reversible protonation is much faster than ionisation and thus is directly related to the relative basicity of the various sigma-bonds, as expected from inductive effects (Table 3). Extensive reversible-protonation is not accompanied by intramolecular atom scrambling in protonated alkanes, as shown in a study involving ^{13}C labelled propane and propane 1,1,1,3,3,3 d_6 in the HF-SbF_5 system.[9] This demonstrates that skeletal rearrangement does not occur via carbonium ions and necessitates carbenium intermediates.

Schemes 3(a) and 3(b) illustrate pathways in protonated iso-butane and protonated propane respectively. The hypothesis of alkane protonation by solid acids such as zeolites, heteropolyacids, sulfated zirconias, chlorinated aluminas, or supported acids, used in the petrochemical industry, has been a controversial issue since its proposal by Haag and Dessau (1984).[10]

A large number of papers related to this subject has been published since then, which can be classified into three types of mechanistic approaches: (i) studies of initial product distribution, (ii) hydrogen deuterium exchange studies and (iii) theoretical calculations. In this review we will limit ourselves essentially to H/D isotope exchange studies between the acidic catalyst and small alkanes. In most cases, it is possible to observe intermolecular H/D exchange in the absence of side reactions such as intramolecular atom scrambling, isomerisation or cracking.

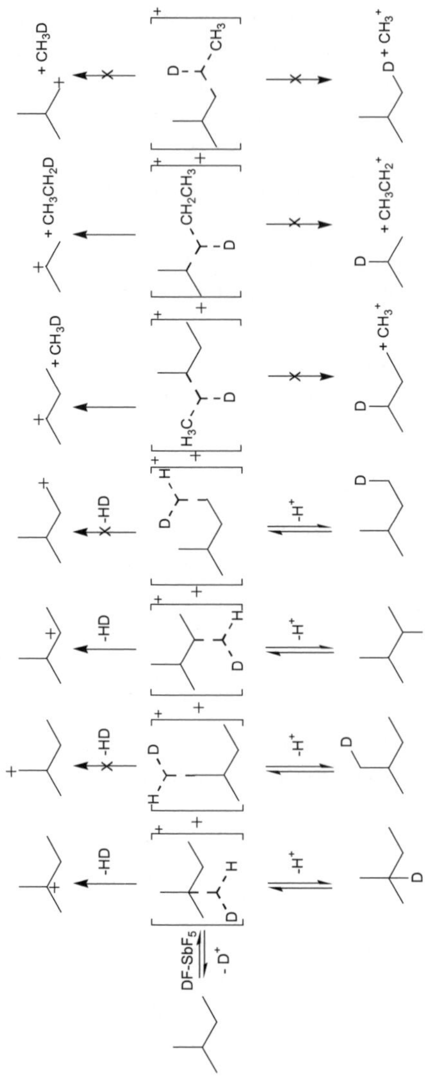

Scheme 1. H/D exchange and ionisation of isopentane in DF-SbF_5.

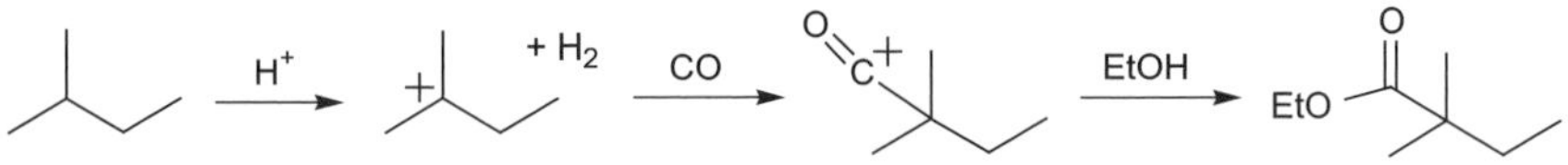

Scheme 2. Quenching of protolysis products by carbon monoxide.

Table 3. H/D exchange observed in small alkanes reacted with DF/SbF_5.[8]

	SbF_5 (mol % in DF)[a]	Exchange (atom %)[b] on		
		Primary C-H	Secondary C-H	Tertiary C-H
Propane	12	12.0 *1*	16.5 *1.38*	
n-butane	12.7	28.1 *1*	38.9 *1.38*	
Isobutane	12.7	7.7 *1*		17.5 *2.27*
Isopentane	12.4	12.1 *1*	16.2 *1.34*	19.3 *1.60*

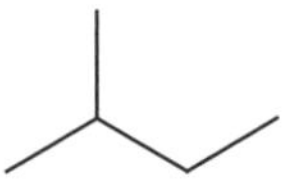

All the numbers in italic are normalized to the exchange in the primary position.
[a]Mol % determined by weight, ±1% of the indicated value.
[b]Determined by ^1H, ^2H NMR, ±3% of the indicated value.

3. H/D Exchange between Alkanes and Solid Acid Catalysts

3.1. *The number of exchangeable acid sites on the catalyst*

In order to follow both quantitatively and kinetically the isotope exchange between catalyst and alkanes, it is necessary to know the number of exchangeable sites on the catalyst. Numerous techniques are available to assess the acidity of solid acids and a special issue of *Topics in Catalysis*

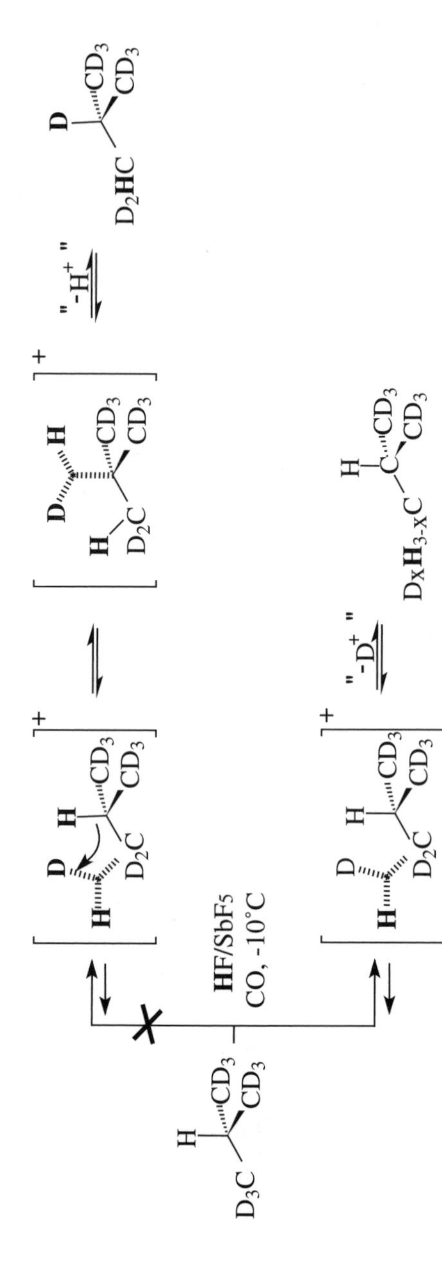

Scheme 3(a). Absence of an intra-molecular rearrangement in protonated iso-butane.

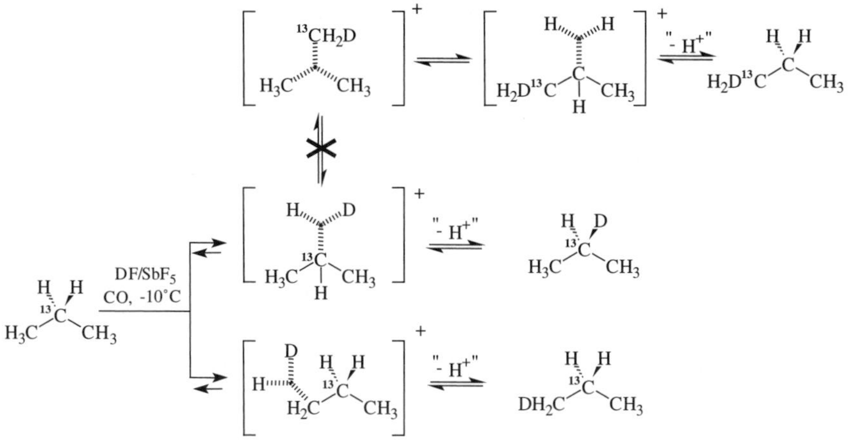

Scheme 3(b). Absence of an intra-molecular rearrangement in protonated propane.

has been devoted to this subject.[11] The most reliable experimental procedure for the assessment of the number of surface acid sites for the H/D exchange process was proposed by Olindo *et al.*[12] It is based on a double exchange process as described by Eqs. (1–3) (Scheme 4).

$$\text{Zeol-OH} \xrightarrow[\text{excess}]{D_2O} \text{Zeol-OD} + H_2O + HDO + D_2O \quad (1)$$

$$\text{Zeol-OD} \xrightarrow[\text{excess}]{H_2O} \text{Zeol-OH} + H_2O + HDO + D_2O \quad (2)$$

$$H_xOD_y \xrightarrow[\text{excess}]{(CF_3CO)_2O} \mathbf{x}\, CF_3COOH + \mathbf{y}\, CF_3COOD \quad (3)$$

Scheme 4. Double exchange process for zeolites.

Initially, a very large excess of gaseous D_2O is used to exchange all the Brönsted acid sites present on the catalyst (Eq. 1). The titration of the O-D sites, thus produced is performed via back-exchanging the deuterium present on the solid surface (Eq. 2) with an excess of distilled water (a 3% H_2O in nitrogen stream). The partially exchanged water H_xOD_y, composed of H_2O, HDO and D_2O, is collected in a U-tube cooled at −117°C (the melting point of ethanol), weighed and chemically trapped by an excess of trifluoroacetic anhydride. This reaction yields a mixture of deuterated and non-deuterated trifluoroacetic acid, the ratio of which can be easily estimated by $^1H/^2H$ NMR. This double exchange procedure is used to prevent the participation of atmospheric water in the measurement. Table 4 presents some results of the acid site density, measured for promoted and non-promoted sulfated zirconia (SZ) catalysts.

The anhydride method is proven to be extremely reliable, whereas the use of iso-butane as the exchange template may lead to secondary reactions. Recently, Louis *et al.*[13] have extended this method to the titration of the total amount of the Brönsted acid sites of zeolites. Figure 1 shows the comparison between the amount of Brönsted acid sites determined by this method and the theoretical amount estimated by means of the Si/Al ratio.

The Brönsted acid site density measured on all zeolite samples such as MOR, MFI, SAPO, EMT, BEA and HUSY, is in good agreement with the theoretical amount estimated by the Si/Al ratio. Prior to the measurement, the samples were calcined at 450°C. By knowing the crystalline structure of such materials, one is able to determine the number (n) of H atoms, thereby estimating the Si/Al ratio, which is a key parameter for the study of the catalysis by zeolites. Indeed, the chemical formula of a protonic zeolite is given by $H_nAl_nSi_{x-n}O_y \cdot zH_2O$, relating the number of protons to the chemical composition. Figure 1 also shows the results obtained for the heteropolyacid $H_3PW_{12}O_{40}$, which normally possesses 0.89 ± 0.05 mmol H/g.[14] By applying the D_2O/H_2O isotope exchange method, approximately 0.91 ± 0.05 mmol H/g are measured, demonstrating further the reliability and the versatility of this method to other crystalline solid acids.

Figure 2 shows a comparison between the Brönsted acid site density measured with the anhydride method and a classical propylamine adsorption

Table 4. Effect of alumina on Brönsted acid sites density in SZA.

Catalyst	Al_2O_3 (% mol)	Brönsted acid site density (mmolH$^+$/g)			
		Iso-butane Method		Anhydride Method	
SZ	0	0.061	0.064		
SZA1	0.7	—	0.078		
SZA3	2.7	—	0.131	0.129	0.133
SZA5	4.5	—	0.210	0.204	
SZA9	7.3	—	0.167	0.176	
SZA15	12.4	—	0.251	0.200	

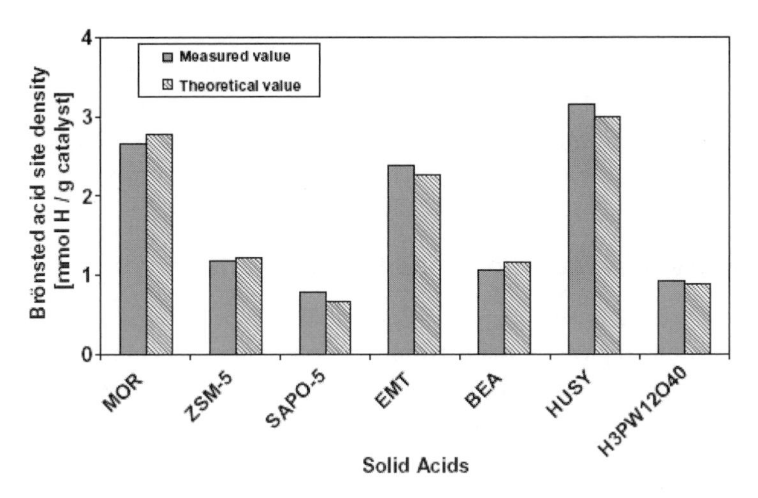

Fig. 1. Theoretical and measured Brönsted acid sites on solid acids.

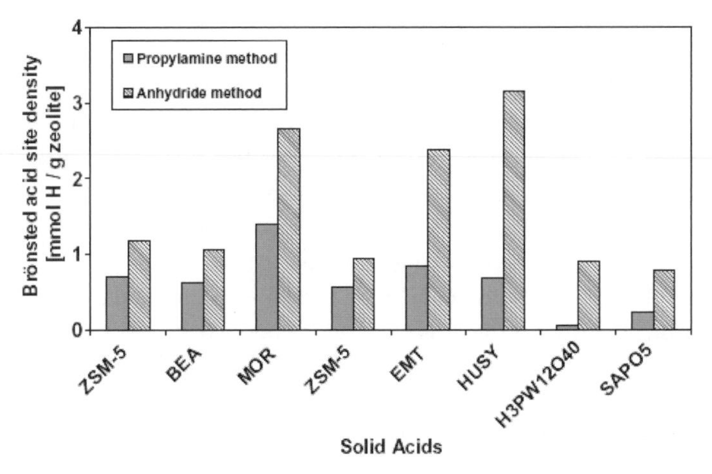

Fig. 2. Comparison between titration with an amine and with isotopic exchange using water.

technique, based on Hofmann elimination.[15] The general trend observed is a lower quantity of Brönsted acid sites titrated with the propylamine method, via measuring the propene released from the solid, compared with the anhydride method. While the former is based on the protonation of an alkylamine, probably involving acid sites of sufficient strength, the latter does not discriminate between the hydroxyl groups. Unlike other methods which can be limited by mass transfer constraints, the shape of the catalytic bed, the need to reach a thermodynamic equilibrium between a basic probe and the solid surface, and which seek to discriminate between acidic and non-acidic protons, this D_2O/H_2O isotope exchange technique can be easily used to count the total number of Brönsted acid sites in solid acids.

3.2. *H/D exchange with methane and ethane*

C_1 and C_2 alkanes are a special case for this type of study, as they are very resistant to forming the corresponding primary carbenium ions. Thus, the acid catalysed exchange can only proceed via a concerted mechanism in which, only in the transition state, a positive charge may develop. CH_5^+, the methonium ion, was shown experimentally as well as theoretically to be stabilised by solvation, even at extremely low concentration in liquid superacids.[16] It was suggested that this ion is the probable transition state of the hydron exchange between acid and methane, as methane exchanges readily its hydrons with acidic OH groups on zeolites and sulfated zirconia albeit at temperatures above 500 K.[17]

Several authors suggested using the rate of H/D exchange between solid acids and methane as an indicator of the relative acidity of these catalysts.[18–20] The first theoretical and experimental work in this field was published by van Santen and coworkers in Nature.[21] For the theoretical part, the zeolite structure was represented by a small $H_3Si\text{-}OH\text{-}Al(OH)_2\text{-}SiH_3$ cluster, and for the experimental part, proton forms of Faujasite and MFI were chosen. The study concluded that there was a concerted exchange mechanism, in which no positive charge developed in the transition state (Fig. 3) with an activation energy of $150 \pm 20\,\mathrm{kJ\,mol^{-1}}$.

A more sophisticated theoretical approach was subsequently adopted by Vollmer and Truong,[22] using an embedded-cluster model including the Madelung potential which resulted in a carbonium ion like transition state, closer in both geometry and charge distribution with the model calculated in a superacidic environment. However, this proposal is in contrast with many other contributions. Recently, Engelhard and Valyon reported H/D exchange between CD_4 and the OH-groups of various H-zeolites and γ-alumina.[23,24] As the activation energy was found to be independent of Brönsted acidity, the authors suggested that dissociative adsorption over a Lewis acid-base pair occurs. This conclusion is similar to that reached for the H/D exchange occurring between methane-d_4 and OH sites, catalysed by basic oxides. In 1965, Larson and Hall[25] reported that such exchange, which could be observed even at room temperature, was catalysed only

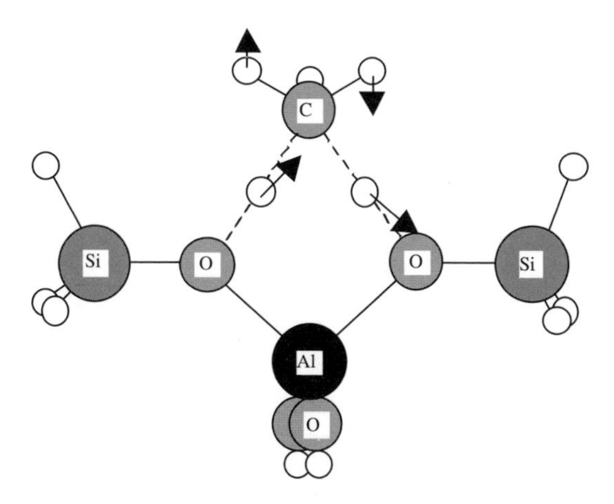

Fig. 3. Transition state proposed by van Santen *et al.* for H/D exchange between zeolite and methane.[21]

by a small number of active sites ($\leq$1% of the hydroxyl groups) and had a very low activation energy of approximately 17 kJ mol^{-1}. As methane had the highest rate in comparison with other alkanes, a carbanionic type of intermediate was suggested. Moreover, in the mid 1970s, Kemball and coworkers[26,27] studied the H/D exchange between methane and D_2 in the presence of γ-alumina, again proposing a carbanionic intermediate.

In the late 1990s, the kinetics of H/D exchange between methane, deuterated FAU and MFI zeolites was studied by Schoofs *et al.*[20] in the temperature range of 450–550°C, in order to study structural (MFI vs FAU) and chemical (Si/Al ratio) effects on the reaction rate. The formation of mono-, bis-, ter- and per-deuterated isotopomers of methane was monitored with an on-line mass spectrometer. It was found that for all catalysts and temperatures, the exchange occurred via a consecutive reaction, i.e., $CH_4 \rightarrow CH_3D \rightarrow CH_2D_2 \rightarrow$ and so forth, in the absence of side reactions (Fig. 4). The activation energies of the exchange process were of the order of 150 $\pm$ 10 kJ mol^{-1}. H/D exchange was also observed at 300°C on sulfated zircona (SZ) and sulfated zirconia promoted with alumina (SZA).[17] These catalysts were more reactive and at 400°C the presence of ethylene was detected as a consequence of methane activation. Scheme 5 was suggested. The apparent activation energy of the exchange process was found to be substantially lower for SZ catalysts ($\sim$95 kJ mol^{-1}) than for HZSM-5 (130 kJ mol^{-1}). In comparison with theoretical calculations concerning the exchange with sulfuric acid,[28] a cyclic transition state was suggested as shown in Scheme 6.

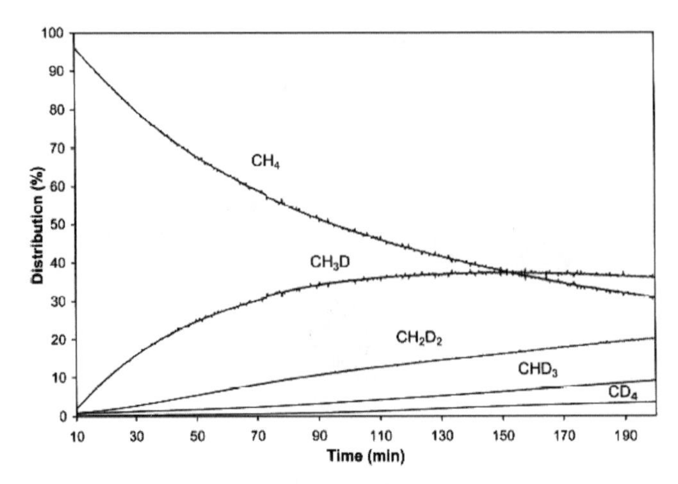

Fig. 4. Distribution of the isotopomers as a function of time for the H-D exchange of 0.5% CH_4 in He at 500°C over MFI.

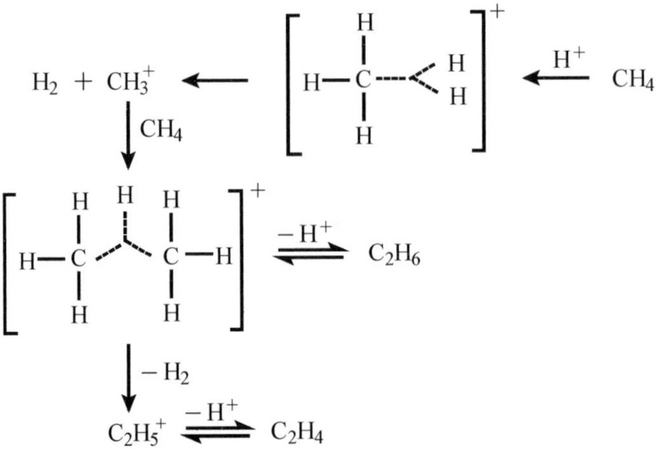

$$H_2 + CH_3^+ \longleftarrow [H_3C\cdots H_2]^+ \xleftarrow{H^+} CH_4$$

$$H_2 + CH_3^+ \xrightarrow{CH_4} [H_3C\cdots H\cdots CH_3]^+ \underset{}{\overset{-H^+}{\rightleftharpoons}} C_2H_6$$

$$[H_3C\cdots H\cdots CH_3]^+ \xrightarrow{-H_2} C_2H_5^+ \overset{-H^+}{\rightleftharpoons} C_2H_4$$

Scheme 5. Reaction mechanism of methane conversion over sulfated zirconia catalysts.

Comparing the exchange rates per acid site for FAU and MFI (Fig. 5),[29] Schoofs found that for the same Al/Si + Al ratio, the exchange rate was higher for MFI than for FAU-type zeolites in accordance with an important structural influence. A kinetic isotope effect ($k_H/k_D = 1.7$) was measured for MFI type zeolite, suggesting a faster exchange between CD_4 and

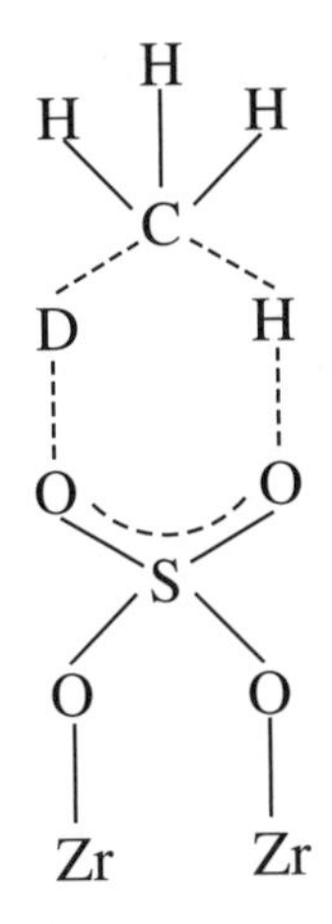

Scheme 6. Transition state for the H/D exchange of methane over SZ-based catalysts.

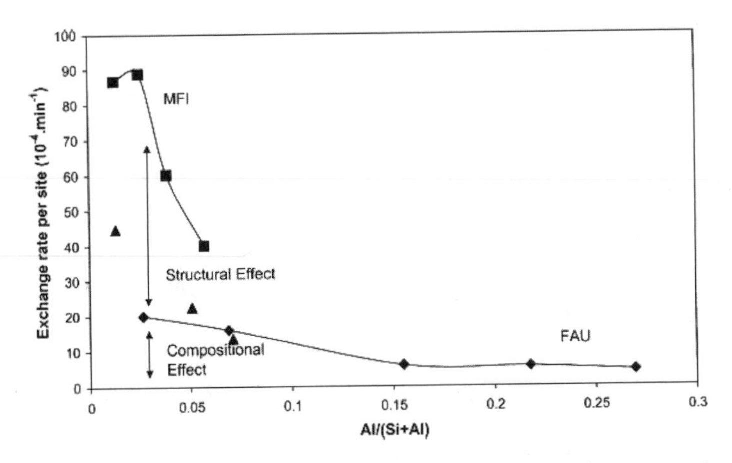

Fig. 5. Structural and compositional effects on H/D exchange rate between methane and deuterated MFI and FAU.

H-zeolite, than between CH_4 and D-Zeolite. This finding indicated that the breaking of O-H/O-D bond was involved in the limiting step, and the acid proton moves towards the C atom of methane, before the H atom of methane moves toward the O-atom. This was in line with the findings of Truong and Vollmer (see above). It was also supported by the fact that for ethane, the apparent activation energies were lower due to charge stabilisation by inductive effect from the additional methyl group.

3.3. *H/D exchange with alkanes having more than 2 carbon atoms*

The reactivity of alkanes having more than two carbon atoms is much higher than methane and ethane, as they are prone to giving secondary or tertiary carbenium ions. Iso-butane exchanges its primary protons even at room temperature on zeolitic material, albeit very slowly.[30] The suggestions that these small alkanes exchange hydrons with the catalyst following the same scheme as C_1 or C_2 alkanes (carbonium ion type intermediates), can be repeatedly found in the literature, and were shown to be erroneous on various grounds. The initial proposal by Mota and colleagues[31] that H/D exchange between 3-methylpentane and HUSY occurred via carbonium ion intermediates, was made on the basis of incorrect interpretation of ^{2}D NMR spectra.[32] More recently, Stepanov[33] and Ivanova[34,35] made the same proposition for the exchange taking place between propane and HZSM-5. However, their conclusion could also be disproved simply because the temperature range $T > 473\,K$ was too high to observe the intermediate stages.[36,37]

The mechanism of the exchange process between small alkanes C_nH_{2n+2}, $n \geq 3$ and solid acids has been thoroughly studied in our group in the last decade. The results were all in agreement with the intermediate formation of an alkene which, by reprotonation, yielded regioselectively deuterated alkanes as shown in Scheme 7.

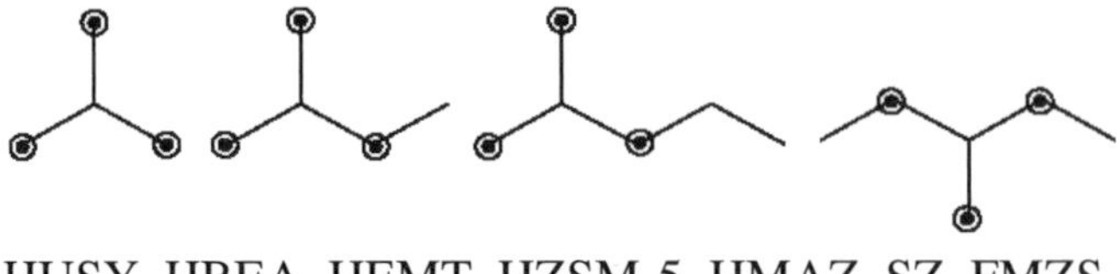

Acids : HUSY, HBEA, HEMT, HZSM-5, HMAZ, SZ, FMZS, HPA...

Scheme 7. Regioselectivity of H/D exchange between solid acids and various alkanes.

Deuterium is introduced following Markovnikov's rule,[38] that is, corresponding to the most stable carbenium ion intermediate. As desorption only occurs via hydride transfer from a fresh alkane, no D is found in the branching position, (or C_2 in propane). The mechanistic scheme suggested is shown in Scheme 8.

On solid acids, even more so than in liquid media, the presence of free carbenium ion intermediates should not be considered. These intermediates are adsorbed (or solvated) on the surface, in the form of relatively polarised alkoxy species which undergo reactions possibly with nucleophilic assistance from neighbouring oxygen-lone pairs.

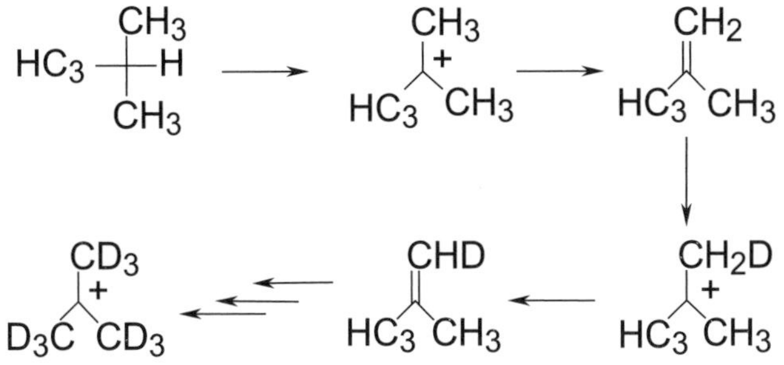

Last Step : Hydride Transfer

$$(CD_3)_3C^+ + (CH_3)_3CH \longrightarrow (CH_3)_3C^+ + (CD_3)_3CH$$

Scheme 8. Selective deuteration in iso-butane and hydride transfer.

The rates of exchange between iso-butane and HBEA have been measured in the temperature range of 130–160°C and an apparent activation energy of $65\,\mathrm{kJ\,mol^{-1}}$ was found, substantially higher than that for H-ZSM-5 ($50\,\mathrm{kJ\,mol^{-1}}$). The rates of exchange at a given temperature are determined not only by the number of acid sites, but more by the activity of the zeolite (or the structure of the reactant).

Scheme 9 represents a theoretical catalytic cycle:

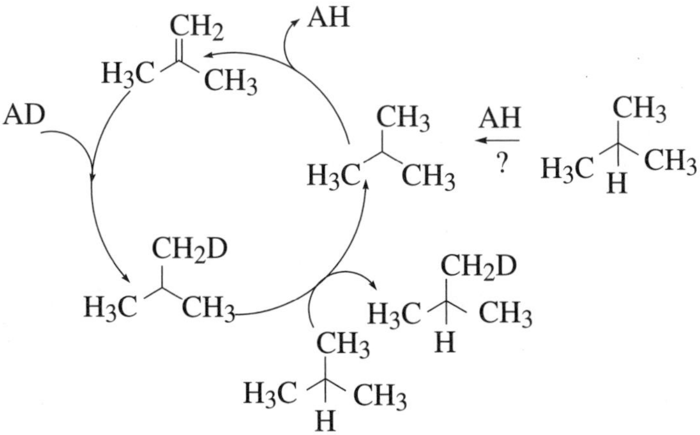

Scheme 9. Theoretical catalytic cycle for the H/D exchange between deuterated solid acid and iso-butane.

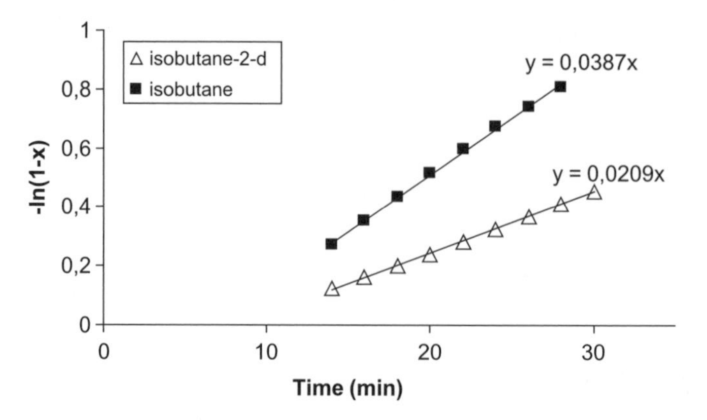

Fig. 6. H/D exchange rates for isobutane and deuterated isobutane in methine position on D-BEA at 150°C – Primary Kinetic Isotopic Effect.

In order to distinguish between two possible slow steps in the catalytic cycle, a kinetic isotope study was performed by Sassi,[39] comparing the exchange rates of mono-deuterated iso-butane (1-d-2-methylpropane) and non-deuterated isobutane. A KIE of 1.85 can be ascribed to C-H (or C-D) tertiary bond rupture in the slow step of the catalytic cycle. Two steps in this mechanism can be considered as the slowest: (1) the activation step of iso-butane to generate the t-butyl cation (generation of the carbenium ion from the alkane), and (2) the hydride transfer from the incoming iso-butane to the t-butyl cation, which creates a new undeuterated cation on the surface and desorbs the deuterated product. In order to distinguish between these options, small amounts of iso-butene were added to the feed stream. Protonation of iso-butene is a very facile step which generated the adsorbed t-butyl cation. The results obtained in the presence of alkenes show no change in the isotope effect, indicating that the slow step in the catalytic cycle is the breaking of the tertiary C-H bond of the iso-alkane during the hydride transfer step.

This conclusion was also reached by Schoofs on the basis of the presence of perdeuterated olefins among the products.[40] Linear alkanes, known to be less reactive, also undergo H/D exchange with the solid acid by the same mechanistic scheme, at slower rates and above 150°C. This exchange reaction occurs in a very clean way, since no side products from cracking and isomerization are observed. The cations, which are adsorbed on the surface are prone to de-protonation, but the alkenes formed are rapidly re-protonated before substantial oligomerization can take place.[41] However,

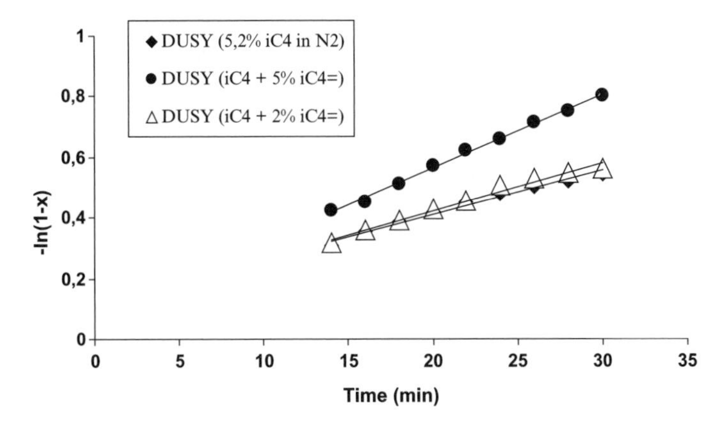

Fig. 7. Effect of the addition of iso-butene on the rates of H/D exchange reaction between isobutane and D-USY at 150°C.

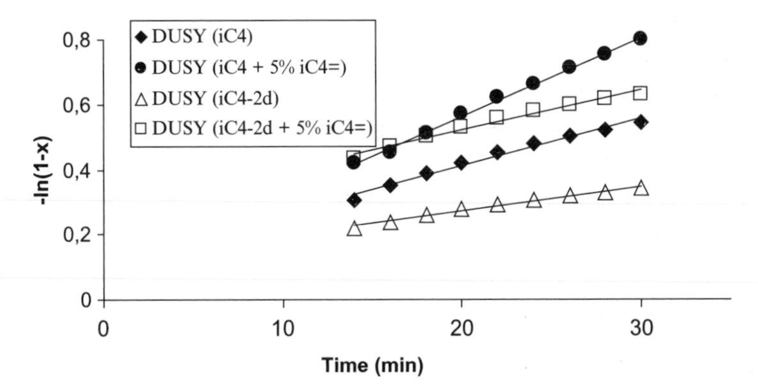

Fig. 8. Effect of the addition of iso-butene on the primary kinetic isotope effect on D-USY at 150°C.

in contrast with branched alkanes, the regiospecificity of H/D exchange could not be demonstrated since the olefinic bond can be generated at any carbon along the chain, leading to non regioselective deuteration as shown in Scheme 10.

The occurrence of carbenium ions as reaction intermediates is strongly supported by the observation that the isotopic exchange can be totally suppressed in the presence of carbon monoxide.[42] Furthermore, trapping of the intermediate carbenium ions by CO and water has been observed by *in situ* NMR spectroscopy, when iso-butane, water, and CO reacted on HZSM-5 zeolite to form pivalic acid.[43] Regarding the low conversion,

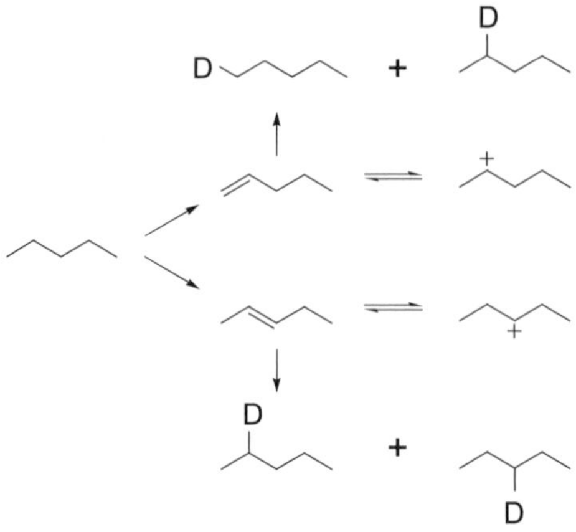

Scheme 10. Distribution of deuterium in the case of H/D exchange on n-hexane.

only a limited number of acid sites are suggested to be strong enough for the initial protolytic activation to take place. Once carbenium ions are generated, they become the active catalyst.

In all these cases, the carbonium ion mechanism (direct H/D exchange via penta-coordinated carbon) can be rejected primarily on the basis of: 1) lower activation energies observed, and 2) the observation of rearrangement products due to carbenium ion intermediates.

3.4. *The case of propane*

Propane activation on solid acids has been the subject of various mechanistic studies, especially in relation with aromatisation.[44,45] On the basis of extensive H/D and ^{13}C scrambling observed by *in situ* solid-state NMR in the 200–300°C range, several authors[33,35,46] suggested that the reaction was governed by penta-coordinated carbonium ion intermediates, by analogy with superacid conditions. Monitoring hydron redistribution directly between catalyst and substrate, as well as via internal isotope scrambling, is a very powerful tool in studying a rather complex mechanism. However, at 200–300°C, the energy provided is sufficient to overcome the activation energy of some of the less demanding pathways, which will stay unnoticed under these conditions. For this reason, we have reinvestigated in detail the exchange and scrambling reactions with selectively labelled propane on aluminium doped sulfated zirconia (SZA). This is active at the lower temperature range which is more favourable in distinguishing between

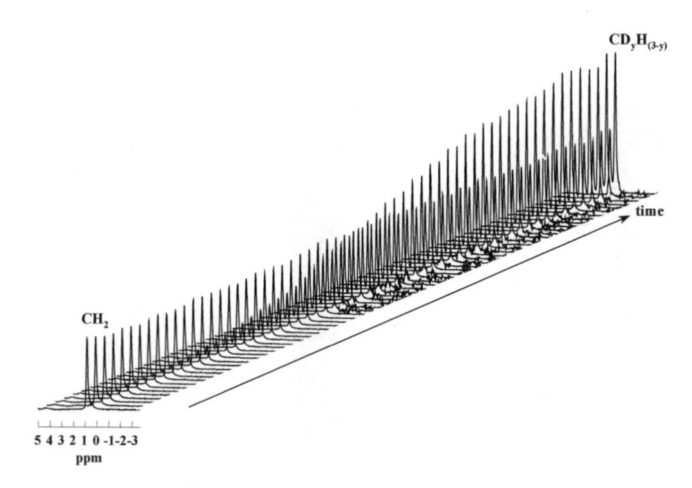

Fig. 9. Stack plot of the ^1H MAS NMR spectra during the H/D exchange reaction between propane-1,1,1,3,3,3-d_6 and Al_2O_3-promoted sulfated zirconia catalyst (SZA) at 48°C.

the various competitive pathways.[36] Under very mild conditions (30°C to 120°C), three competitive and consecutive pathways could be observed in the course of the reaction:

(i) *A regioselective H/D exchange* between the acidic hydrons of the catalyst and the primary deuterons of 1,1,1,3,3,3 d_6 propane. This process could be directly followed by ^1H MAS NMR (Fig. 9). The formation of the active species is not completely clear, but with the sulfated zirconia which has oxidising properties,[47,48] one can easily imagine a dehydrogenative oxidation leading to propene-like species adsorbed on the surface as a 2-alkoxy intermediate. The following scheme emphasizes the occurrence of a Markovnikov distribution of the label in propane (Scheme 11). The carbenium ion is immediately trapped by a lone pair yielding the alkoxy species which may be deprotonated with the assistance of a neighbouring oxygen. The alkene so formed is then redeuterated, generating a new carbenium or deuterated alkoxy species desorbed by hydride transfer, yielding the product 1-*d* propane.

(ii) *An intra-molecular H/D scrambling* between the methyl and methylene proton of propane, independent of the inter-molecular exchange, was monitored by ^2H MAS NMR. The 2-alkoxy species is in equilibrium with the 1-alkoxy propane via a hydride shift from the methyl to methylene position (Scheme 12). To avoid the generation of a primary propyl ion, this shift can be facilitated by nucleophilic assistance from a neighbouring oxygen of the surface, generating the 1-alkoxy species known to be a stable

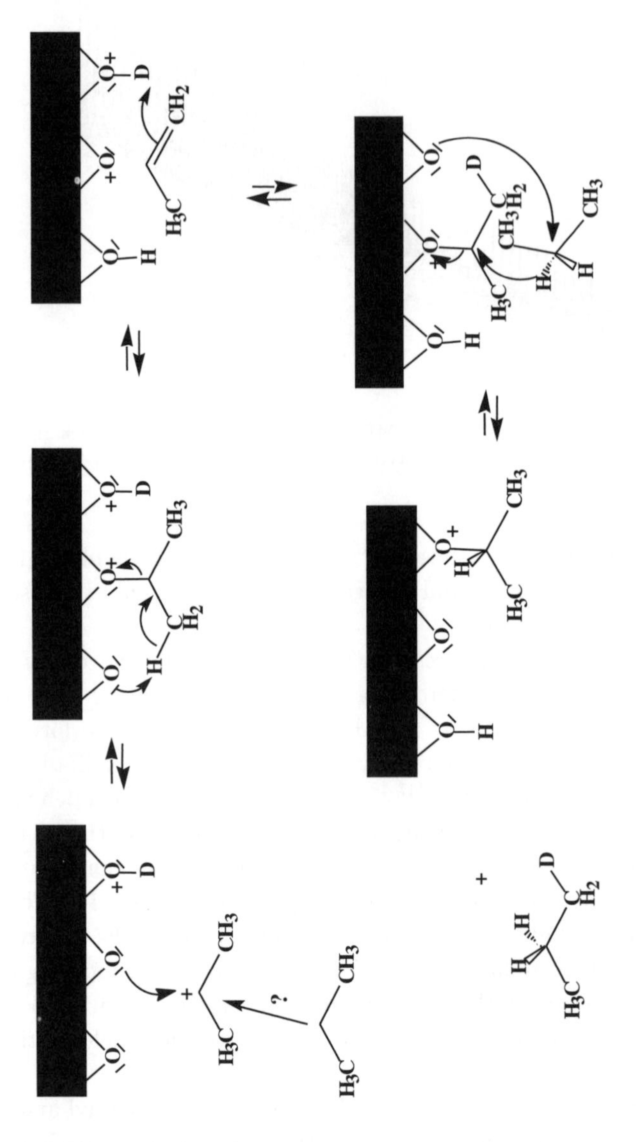

Scheme 11. Proposed mechanism for the regiospecific exchange of hydrons on methyl position.

intermediate on solid acids.[49–52] This process is slower than the previous process at low temperature, but its participation in the label redistribution increases dramatically with temperature. Thus, at 100°C, both processes have almost equivalent rates and the distinction can no longer be made.

(iii) *An intramolecular ^{13}C scrambling*, indicative of a skeletal rearrangement, intervenes at higher temperatures. This process has been found to be much slower than redistribution of hydrons in the methylene position. However, from the determination of the apparent activation energy of each of these processes, [approximately 60 kJ mol^{-1} for the regioselective exchange and about 80 kJ mol^{-1} for processes (ii) and (iii)], it appears that redistribution of hydrons and carbon scrambling have a common transition state: the 1-alkoxy species. In fact, the carbon scrambling implies the intermediacy of protonated cyclopropane (PCP, Scheme 13), a well-known intermediate in carbenium ion chemistry.

Despite the fact that "free" carbenium ions are too improbable on the solid surface, but should be considered as a more or less ionic alkoxy species, all these results are in line with classical carbenium ion chemistry. The general mechanistic scheme shown in Fig. 10 summarises the exchange between propane and sulfated zirconia.

The measurement of the activation energy for each process suggests that processes 2 and 3 undergo a common transition state responsible for both vicinal hydride migration and protonated propane formation. However, a partial contribution of H/D exchange via carbonium ion intermediates may occur at much higher temperature. In this instance, the resulting isotope distribution will most probably be statistical, as the differentiation between the slowest and the fastest steps will disappear owing to the high rates. In order to prove the implication of such a carbonium type mechanism, the experiment should demonstrate that the most basic C-H bonds exchange preferentially. This has not been done until now.

3.5. *Competitive reactions evidenced by H/D exchange*

3.5.1. *Molecular rearrangement*

Contrary to small hydrocarbons, alkanes with a number of carbons higher than 6 do not show a regiospecific H/D exchange either on zeolites or in D_2SO_4. This is clearly evidenced in the case of the 3-ethylpentane (3EP), in which only the 6 methylene hydrons next to the branching should be involved in the exchange according to the general scheme (Scheme 9,

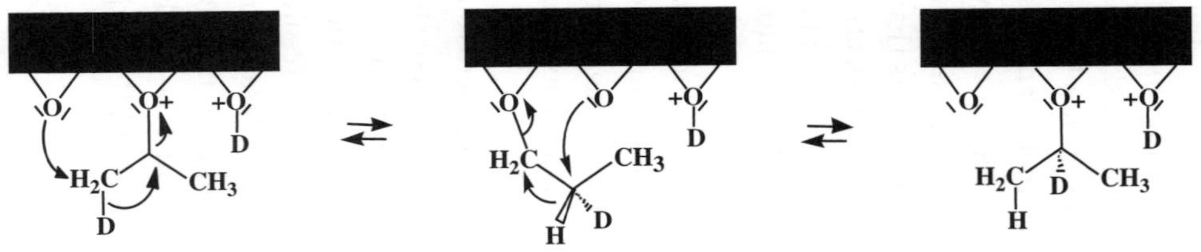

Scheme 12. Mechanism for the internal scrambling of hydron in propane.

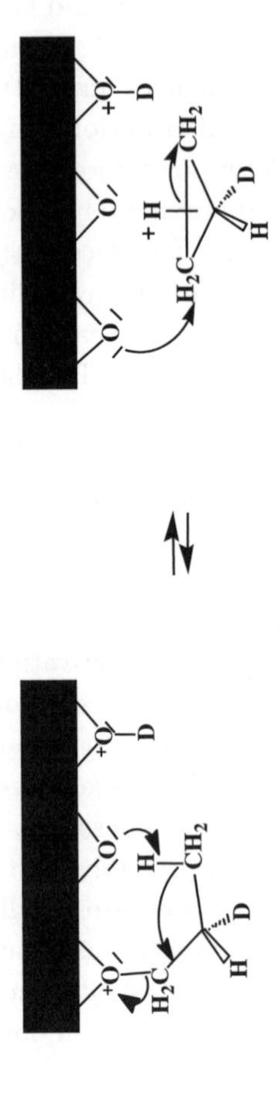

Scheme 13. Complete redistribution of labels via the protonated cyclopropane.

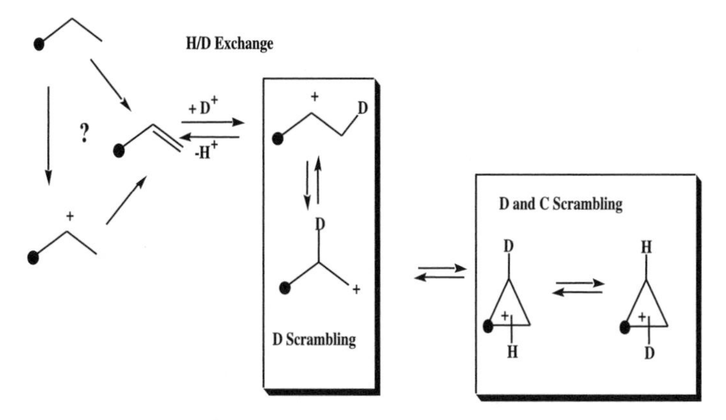

Fig. 10. Initial products and intermediates resulting from H/D exchange of propane 1-^{13}C over SZA.

iso-butane being the smallest branched alkane). In fact, all three types of hydrons were exchanged during the reaction in both media.[53] The triethylcarbenium ion (I) once formed, rearranges very rapidly, via an ethyl shift, to the 3-methyl-3-hexylium ion (II) which is in equilibrium with the corresponding alkenes (Scheme 14). As the exchange takes place, the recovered hydrocarbons consist of a large proportion of undeuterated starting material and a small fraction of highly deuterated isomers. It is likely that for steric reasons, the H/D exchange is much faster in the isomers (methylhexane derivatives) than for the starting 3EP.

It is worth noting that in the case of sulfuric acid, this competition between isomerization and deprotonation/reprotonation is greatly enhanced. Indeed, in this medium, the final hydride transfer can only take place at the interface; consequently, repetitive deprotonation/reprotonation steps occur, yielding only highly deuterated products, whereas at the surface of the solid where the hydride transfer is faster, giving a statistical distribution of isotopologues.

3.5.2. *Deactivation of acidic catalysts*

When the contact time between iso-butane and deuterated zeolites (or D_2SO_4) was prolonged to obtain extensive deuteration of the alkane (>90%), deuterium also appeared gradually in the tertiary position.[54] The appearance of deuterium in this position cannot be explained by a simple hydride transfer from iso-butane to the tertio-butyl cation. Alkenyl and polyenyl ions that were previously identified in sulfuric acid as a product

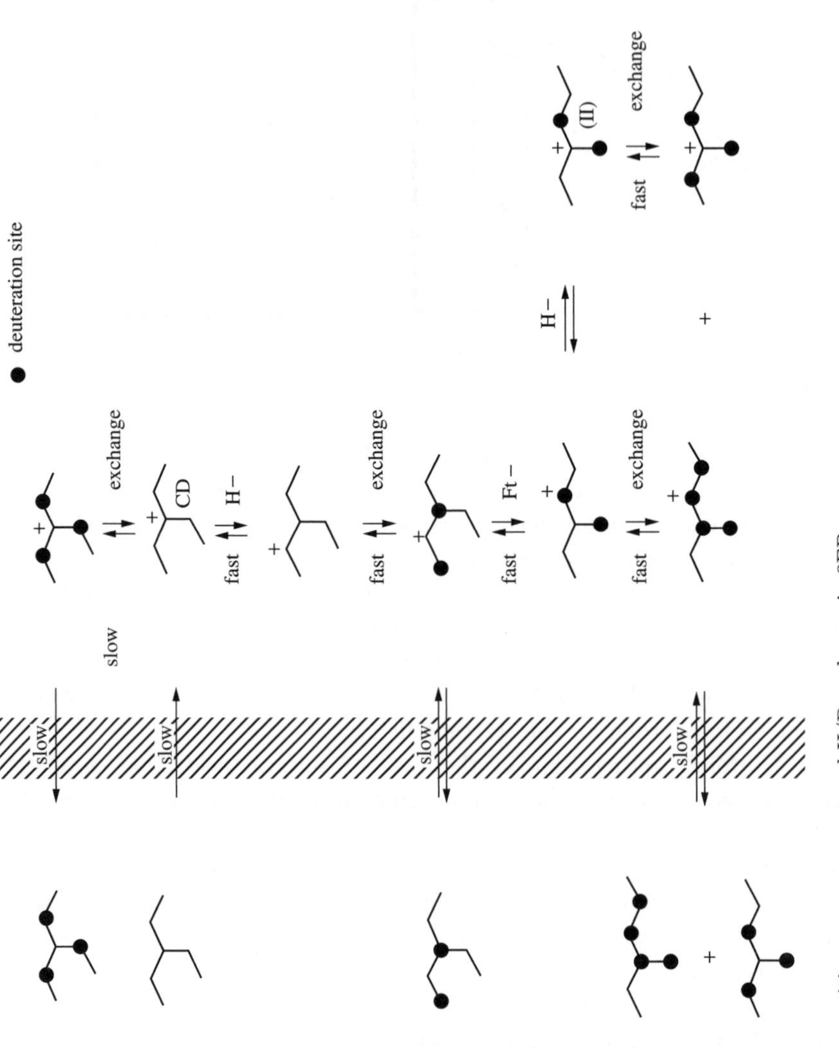

Scheme 14. Competitive rearrangement and H/D exchange in 3EP.

of oligomerisation,[55] are able to release a deuteride ("D^-"). These species have been observed at the surface of the sulfated zirconia by UV-Visible spectroscopy,[56] with a characteristic strong adsorption band at 292 nm. The involvement of such ions has been shown more recently in the reaction of n-butane with sulfated zirconia using UV-Visible spectroscopy[57] and ^{13}C NMR.[58] The mechanism of the formation of polyenyl species, accompanied by the appearance of deuterium in the methine hydron of isobutane (steps 3 and 6), is described in Scheme 15.

The formation of such hydrogen-deficient species becomes significant only after long contact times between the alkane and either D_2SO_4 or D_2O-exchanged solid acids. It leads to further transformations to aromatic and polyaromatic species known to poison the surface of solid acids, leading to coke formation and deactivation of the catalyst.[59,60] For the solid acid, the catalyst can be regenerated by burning off the coke at high temperature (500–550°C with zeolites).

3.5.3. *The inhibiting effect of CO*

Carbon monoxide has been widely used to characterise the hydroxyl groups of solid acids, since its interaction with acidic protons modifies the strength of the C-O bond.[61,62] Analysis by IR spectroscopy is a tool of choice to obtain information on the surface of the catalyst. CO acts as a basic probe and according to previous studies, its complexation with Lewis acid sites (LAS) decreases the acidity and consequently, the activity of the catalyst in n-butane isomerization.[63] However, it is uncertain if the interaction of CO with LAS is the main origin of catalyst deactivation, since n-butane isomerisation is also inhibited by CO in the absence of LAS.

4. Conclusion

In the last decade, deuterium labelling of alkanes or solid acid catalysts has proved to be a preferred tool for mechanistic studies. The availability of high resolution multinuclear NMR spectroscopy and mass spectrometry has facilitated the interpretation of the experimental results. It appears that despite the differences in media between liquid and solid acids, all results in acid-catalysed hydrocarbon chemistry can be rationalised by mechanisms involving solvated carbenium ion intermediates, as known from classical organic chemistry. Nevertheless, the determination of the real nature of the

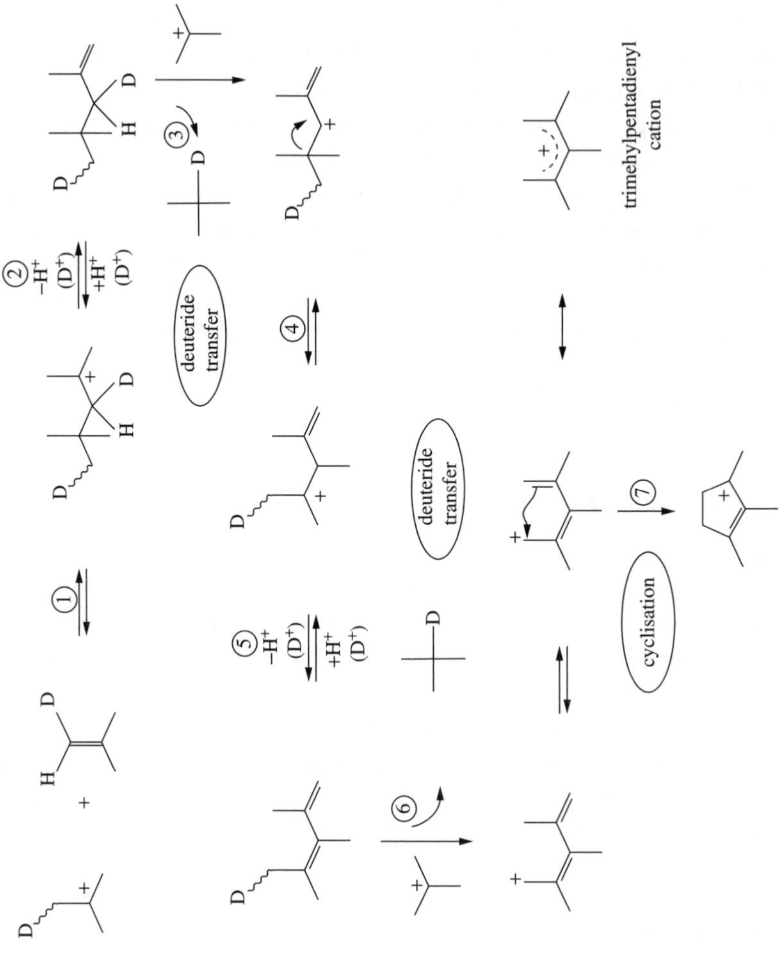

Scheme 15. Formation of cycloalkenyl ions from isobutane and incorporation of deuterium in the tertiary position.

solvation of the intermediates or transition states on solid acids necessitates further investigation.

References

[1] Melander L, Saunders WH, *Reaction rates of Isotopic molecules*, Wiley-Interscience, New York, 1980.
[2] O'Ferral RAM, *J Chem Soc* **13**: 785–790, 1970.
[3] Mattson O, Westaway KC, *Adv Phys Org Chem* **31**: 143, 1998.
[4] Olah GA, Prakash GKS, Sommer J, *Superacids*, John Wiley & Sons Inc. New York, 1985.
[5] Olah GA, *Angew Chem Int Edit* **12**: 173, 1973.
[6] Sommer J, Bukala J, *Acc Chem Res* **26**: 370–376, 1993.
[7] Sommer J, Jost R, *Pure Appl Chem* **12**: 2309–2318, 2000.
[8] Goeppert A, Sommer J, *New J Chem* 26: 1335–1339, 2002.
[9] Goeppert A, Sassi A, Sommer J, Esteves PM, Mota CJA, Karlsson A, Ahlberg P, *J Am Chem Soc* **121**: 10628–10629, 1999.
[10] Haag WO, Dessau RM, *Proceedings of the 8th International Congress on Catalysis, Berlin*, 305–316, 1984.
[11] Fripiat JJ, Dumesic JA, *Top Catal* **4**: 1997.
[12] Olindo R, Goeppert A, Habermacher D, Sommer J, Pinna F, *J Catal* **197**: 344–349, 2001.
[13] Louis B, Walspurger S, Sommer J, *Catal Lett* **93**: 81–84, 2004.
[14] Essayem N, Coudurier G, Vedrine JC, Habermacher D, Sommer J, *J Catal* **183**: 292–299, 1999.
[15] Kresnawahjuesa O, Gorte RJ, de Oliveira D, Lau LY, *Catal Lett* **82**: 155–161, 2002.
[16] Ahlberg P, Karlsson A, Goeppert A, Nilsson Lill SO, Diner P, Sommer J, *Chem-Eur J* **7**: 1936–1943, 2001.
[17] Hua WM, Goeppert A, Sommer J, *Appl Catal A-Gen* **219**: 201–207, 2001.
[18] Walspurger S, Goeppert A, Haouas M, Sommer J, *New J Chem* **28**: 266–269, 2004.
[19] Jentoft RE, Gates BC, *Catal Lett* **72**: 129–133, 2001.
[20] Schoofs B, Martens JA, Jacobs PA, Schoonheydt RA, *J Catal* **183**: 355–367, 1999.
[21] Kramer GJ, van Santen RA, Emeis CA, Nowak AK, *Nature* **363**: 529–531, 1993.
[22] Vollmer JM, Truong TN, *J Phys Chem B* **104**: 6308–6312, 2000.
[23] Engelhardt J, Valyon J, *React Kinet Catal Lett* **74**: 217–224, 2001.
[24] Valyon J, Engelhardt J, Kallo D, Hegedus M, *Catal Lett* **82**: 29–35, 2002.
[25] Larson JG, Hall K, *J Phys Chem* **69**: 3080–3089, 1965.
[26] Robertson PJ, Scurrell MS, Kemball C, *J Chem Soc* 903–912, 1975.
[27] John CS, Kemball C, Pearce EA, Pearman AJ, *J Chem Research* 400–401, 1979.
[28] Goeppert A, Diner P, Ahlberg P, Sommer J, *Chem-Eur J* **8**: 3277–3283, 2002.

[29] Schoofs B, *PhD Thesis*, Leuven, 2000.
[30] Mota CJ, Jost R, Hachoumy M, Sommer J, *J Catal* **172**: 194–202, 1997.
[31] Mota CJA, Martins RL, *J Chem Soc Chem Comm* 171–173, 1991.
[32] Sommer J, Hachoumy M, Garin F, Barthomeuf D, Vedrine J, *J Am Chem Soc* **117**: 1135–1136, 1995.
[33] Stepanov AG, Ernst H, Freude D, *Catal Lett* **54**: 1–4, 1998.
[34] Ivanova II, Rebrov AI, Pomakhina EB, Derouane EG, *J Mol Cat A* **141**: 107–116, 1999.
[35] Ivanova II, Pomakhina EB, Rebrov AI, Derouane EG, *Top Catal* **6**: 49–59, 1998.
[36] Haouas M, Walspurger S, Sommer J, *J Catal* **215**: 122–128, 2003.
[37] Haouas M, Walspurger S, Taulelle F, Sommer J, *J Am Chem Soc* **126**: 599–606, 2004.
[38] Vollhardt KPC, *Organic Chemistry*, WH Freeman & Co., New York, 1987.
[39] Hua WM, Sassi A, Goeppert A, Taulelle F, Lorentz C, Sommer J, *J Catal* **204**: 460–465, 2001.
[40] Schoofs B, Schuermans J, Schoonheydt RA, *Micropor Mesopor Mat* **35**(6): 99–111, 2000.
[41] Xu BQ, Sachtler WMH, *J Catal* **165**: 231–240, 1997.
[42] Sommer J, Jost R, Hachoumy M, *Catal Today* **38**: 309–319, 1997.
[43] Luzgin MV, Stepanov AG, Sassi A, Sommer J, *Chem-Eur J* **6**: 2368–2376, 2000.
[44] Biscardi JA, Iglesia E, *J Phys Chem B* **102**: 9284–9289, 1998.
[45] Biscardi JA, Iglesia E, *Phys Chem Chem Phys* **1**: 5753–5759, 1999.
[46] Ivanova II, Pomakhina EB, Rebrov AI, Derouane EG, *Top Catal* **6**: 49–59, 1998.
[47] Farcasiu D, Ghenciu A, Li JQ, *J Catal* **158**: 116–127, 1996.
[48] Babou F, Coudurier G, Vedrine JC, *J Chim Phys et de Phys Chim Biol* **92**: 1457–1471, 1995.
[49] Benco L, Hafner J, Hutschka F, Toulhoat H, *J Phys Chem B* **107**: 9756–9762, 2003.
[50] Correa RJ, Mota CJA, *Phys Chem Chem Phys* **4**: 375–380, 2002.
[51] Hunger M, Seiler M, Horvath T, *Catal Lett* **57**: 199–204, 1999.
[52] Kazansky VB, Senchenya IN, *J Catal* **119**: 108–120, 1989.
[53] Sommer J, Habermacher D, Jost R, Sassi A, *Appl Catal A-Gen* **181**: 257–265, 1999.
[54] Sommer J, Sassi A, Hachoumy M, Jost R, Karlsson A, Ahlberg P, *J Catal* **171**: 391–397, 1997.
[55] Deno NC, Carboniums Ions, Wiley Interscience, New York, Vol. II, 1970.
[56] Chen FR, Coudurier, Joly JF, Vedrine J, *J Catal* **143**: 616, 1993.
[57] Spielbauer D, Mekhemer GAH, Bosch E, Knozinger H, *Catal Lett* **36**: 59–68, 1996.
[58] Luzgin MV, Arzumanov SS, Shmachkova VP, Kotsarenko NS, Rogov VA, Stepanov AG, *J Catal* **220**: 233–239, 2003.
[59] Haw JF, Nicholas JB, Song WG, Deng F, Wang ZK, Xu T, Heneghan CS, *J Am Chem Soc* **122**: 4763–4775, 2000.

[60] Chua YT, Stair PC, Nicholas JB, Song WG, Haw JF, *J Am Chem Soc* **125**: 866–867, 2003.
[61] Neyman KM, Strodel P, Ruzankin SP, Schlensog N, Knozinger H, Rosch N, *Catal Lett* **31**: 2–3, 1995.
[62] Morterra C, Bolis V, Cerrato G, Magnacca G, *Surf Sci* **309**: 1206–1213, 1994.
[63] Pinna F, Signoretto M, Strukul G, Cerrato G, Morterra C, *Catal Lett* **26**: 3–4, 1994.

Chapter 2

Application of ^{14}C Radiotracer for the Study of Heterogeneous Catalytic Reactions

Z Paál

1. Introduction

Radiotracers can be very useful in elucidating reaction pathways, kinetic parameters and mechanisms. The application of the ^{14}C isotope has been evident for reactions of organic compounds. The heydays of radiotracers were in the second half of the 20th century, mainly in the 1960's and 1970's, when the complicated separation techniques of the 1950's could be replaced by radio-gas chromatography. These studies were concentrated in a few laboratories by P H Emmett in Baltimore, his student Joe Hightower, W K Hall in Pittsburgh, H Pines in Evanston, J Happel in New York, B H Davis in Lexington, S J Thomson and G Webb in Glasgow; as well as several groups in Moscow (G V Isagulyants, S Z Roginskii and others), H Schulz in Karlsruhe, J Dermietzel in Leipzig and P Tétényi in Budapest. There have been very few publications on the use of ^{14}C tracer in the 21st century.

The "classical" reviews on the use of isotopes including ^{14}C in the study of the mechanism of heterogeneous catalytic reactions were published between 1969 and 1986.[1–4] A recent book on isotopic tracers includes also a chapter on heterogeneous catalysis.[5] The present review will discuss case studies, concentrating mainly on the results of the above-mentioned research groups, including our own. Some results published in journals of low circulation or in books, sometimes in German, Russian or Hungarian, are also discussed.

2. Investigation of Reaction Pathways by Using Mixtures Containing Labelled Reactant

The rates of formation and further reaction of possible intermediates in multi-step reactions may be very different. If we consider a triangular reaction and know that both A $\rightarrow$ C *and* B $\rightarrow$ C takes place, the question if (*a*) a direct A $\rightarrow$ C reaction or the A $\rightarrow$ B $\rightarrow$ C stepwise process prevails, or (*b*) possibly both processes take place, can be clarified by reacting a mixture of A + B, with one component labelled (denoted by *). If A* + B is reacted to obtain C, the appearance of radioactivity in B gives evidence that the stepwise process is possible. This will be illustrated in the next example (Scheme 1A and 1B).

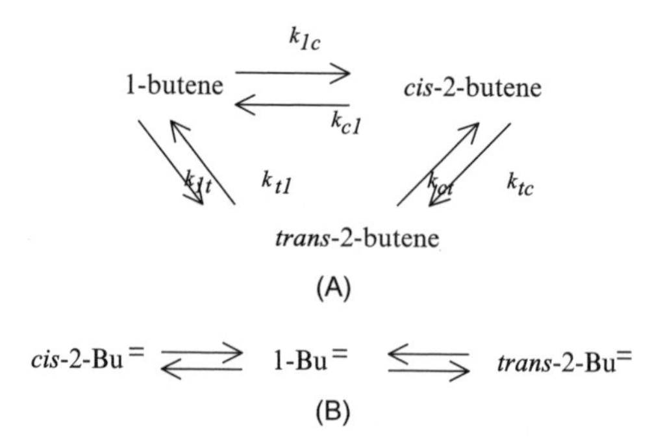

Scheme 1. Two pathways of acid catalysed isomerisation of butenes.

2.1. *Hydrocarbon cracking and isomerisation over acidic catalysts*

Hightower *et al.* (1965) studied the cracking of hexadecane by using ^{14}C tracer.[6] Smaller hypothetical cracking products labelled with ^{14}C were added to the reactant. The appearance of radioactivity in various fractions gave evidence on the possible role of the primary cracking product, identical with the labelled molecule in the formation of end products. Ethene played practically no role in the cracking process. Alkenes with 3 or more C atoms were, in turn, very important in product distribution. Alkene isomers with the same carbon number were in thermodynamical equilibrium with each other. These isomerisation processes were much more rapid than hydrogen transfer at 545 K. These alkenes were true intermediates of alkane cracking products with the same C number. All unsaturated additives (including ethene) preferentially produced coke. Their relative probability of incorporation into the coke (if hexadecane $\equiv$ 1) is e.g., 0.07 for propane, 2.65 for propene and ~8.6 for pentene-1.

The possible reactions of isomerisation of butenes ($Bu^=$) involve either a triangular (Scheme 1A) or a consecutive scheme (Scheme 1B). The mixture of traces of [^{14}C] labelled 1-$Bu^=$ plus *cis*-2-$Bu^=$ or the inverse mixture, traces of [^{14}C] labelled *cis*-2-$Bu^=$ plus 1-$Bu^=$, as well as the respective 1:1 mixtures of the above butenes, were reacted in pulse-microcatalytic experiments on alumina and silica-alumina catalysts.[7] A detailed analysis of the six rate coefficients of the assumed triangular reaction showed that all constants were higher than zero, thus Scheme 1B was valid. The ratios of rate coefficients k_{1c}/k_{1t} and k_{c1}/k_{ct} were much higher than expected from thermodynamic equilibrium, while the ratio k_{t1}/k_{tc} was lower or just higher than that value. The specific radioactivity of coke was close to that of *trans*-butene, when its mixture with ~0.2% labelled $1Bu^=$ was reacted.[7] By plotting the ratio of the instantaneous specific radioactivity for the $2Bu^=$ product to the initial specific radioactivity of the labelled isomer in the feed ($1Bu^=$) in a very wide conversion range (from ~5 to 80%), the triangular mechanism (Scheme 1A) also proved to be true. The points were rather far from the curves calculated, assuming the validity of Scheme 1B.[8]

Cyclohexane dehydrogenation represents another "classical" example for isotopic studies. Balandin's "sextet mechanism"[9] predicted direct dehydrogenation of cyclohexane over several metals, assuming a planar reactive chemisorption of the reactant. Cyclohexene is also readily dehydrogenated to benzene. The use of hydrocarbons labelled with ^{14}C established the true reaction pathway. Tétényi and co-workers[10,11] reacted a mixture of [^{14}C]-cyclohexane and inactive cyclohexene on different metals and measured the specific radioactivity of the fractions (cyclohexane, cH, cyclohexene, $cH^=$ and benzene, Bz) in the product at low conversion values (Table 1).[12] The specific radioactivity values are expressed as "**r%/m%**", i.e., the distribution of the radioactivity in the corresponding component (**r%**) is divided by the distribution of products in mole % (**m%**), as described in Ref. 13. The feed (50% cH plus 50% $cH^=$) contained all radioactivity in the cH, its **r%/m%** (denoted as α) being 2.0, while that of $cH^=$ was 0. Dilution of the cyclohexane fraction was caused by the hydrogenation of inactive $cH^=$. The radioactivity of Bz was usually less than that of cH, indicating that it was produced from both components, $cH^=$ reacting with higher rates. The appearance of radioactivity in $cH^=$ indicated that the cH $\rightarrow$ $cH^=$ reaction indeed took place. The higher the **r%/m%** (β) of $cH^=$, the closer was its equilibration with cH. This was approached at 563 K on Pt, where

Table 1. Relative molar radioactivities (**r%/m%**) in the products after reacting a mixture of [^{14}C]-Cyclohexane and Cyclohexene.[a]

Catalyst	T, K	The Values of **r%/m%** in		
		Cyclohexane	Cyclohexene	Benzene
Feed		2.1	0	—
Pt	533	1.52	0.08	0.45
Pd	573	1.95	0.02	0.33
Rh	578	1.60	0.03	0.40
Ir	573	2.03	0.02	0.07
Ni	578	1.57	1.31	0.56
Ni	608	1.67	0.90	0.65
Co	573	1.87	0.025	0.021

[a]Single flow reactor, no carrier gas added, after Ref. 15.

the effluent contained only 7% cH$^{=}$ and 28% Bz. The situation on Ni was similar, whereas Rh represented the opposite extreme. Derbentsev *et al.*[14] used a mixture of cyclohexane and [^{14}C]-cyclohexene, and found that the "direct" dehydrogenation process cH $\rightarrow$ Bz was much more rapid on Re/C catalysts than the stepwise route. A detailed calculation of the relative rates for the dehydrogenation of cH, and cH$^{=}$ from the ratios of specific radioactivities of the cH$^{=}$ (i.e., β) and Bz (γ), revealed that cH$^{=}$ dehydrogenation was 5–8 times more rapid on Pd and Pt, whereas this ratio was between 17 and 100 over the other metals.[15] The ratio of specific radioactivities (γ/β) showed a continuous function of the atomic diameter, large metal atoms (Pt, Pd) facilitate the planar adsorption of the six-membered ring and favor "direct" dehydrogenation (cH $\rightarrow$ Bz), while the stepwise process prevailed on smaller metal atoms such as Ni and Co, where "edgewise" adsorption would be more important. The data published for Re[14] are in good agreement with those measured on other metals (Fig. 1).

Dehydrocyclisation of alkanes to aromatic compounds is one of the basic reactions of naphtha reforming, which is one of the most important industrial catalytic processes.[16] The nature of the ring closure step is one of the main questions of understanding the chemistry of dehydrocyclisation. Sharan reviewed earlier work on ^{14}C tracer studies of this reaction.[17] An excellent discussion on the development of ideas and the role of [^{14}C] isotope in elucidating reaction pathways has been published by Davis.[18] Two basic ideas have competed in Scheme 2. One assumed stepwise dehydrogenation of open-chain alkanes then cyclisation as one of

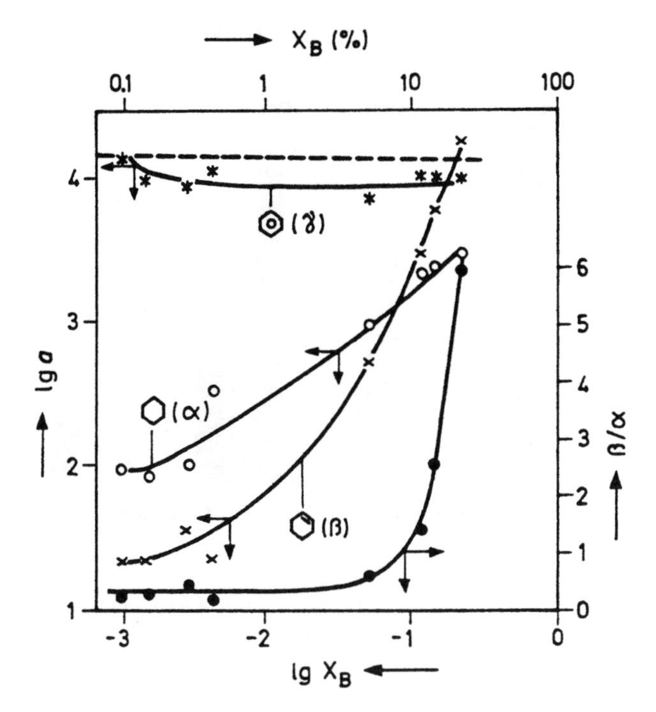

Fig. 1. Specific radioactivities of individual fractions and the ratio of the corresponding values for $cH^{=}/cH$, as a function of benzene conversion. Adapted from Ref. 13.

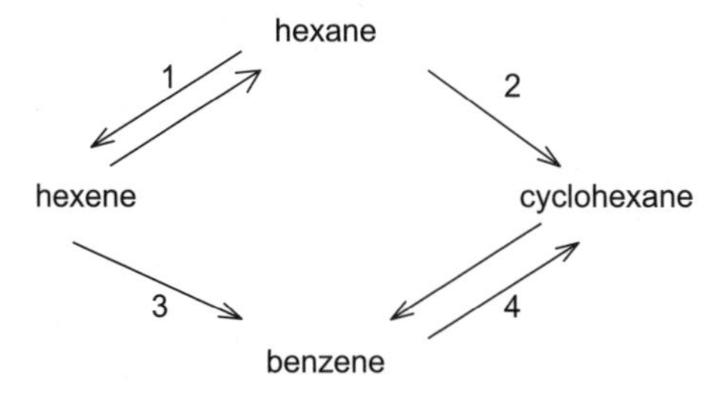

Scheme 2. Hypothetical starting process of C_6 ring formation from hexane.

the closing steps of the sequence, the other school assumed that the six-membered ring would form via direct alkane → cycloalkane reaction, followed by the dehydrogenation of the cyclohexane ring. The former route was proven for oxide catalysts and the latter route was regarded to be valid for Pt/C catalysts.[18,19]

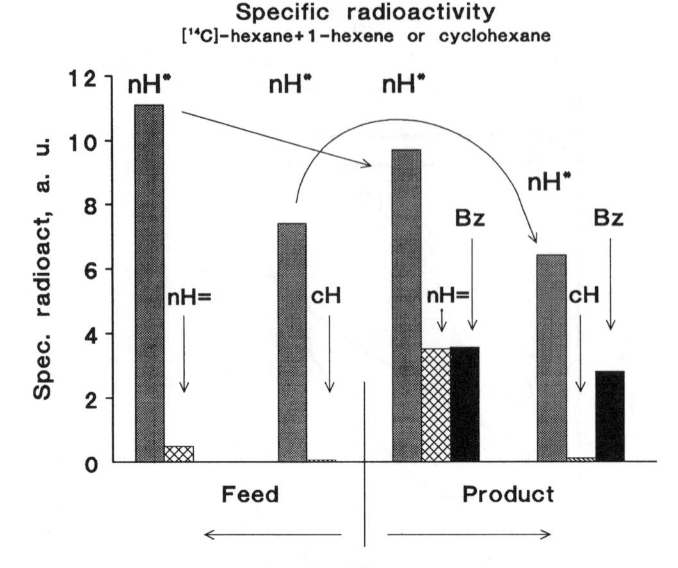

Fig. 2. Specific molar radioactivities in the components of feed and product when reacting a mixture of [^{14}C]-n-hexane (65%) plus inactive 1-hexene (35%) and [^{14}C]-n-hexane (62%) plus inactive cyclohexane (38%). Pulse system, catalyst: Pt black, carrier gas: helium.

Reacting mixtures with one component labelled, facilitated the determination of which route prevailed on given catalysts and under given conditions. Figure 2 depicts results obtained on Pt with mixtures of [^{14}C]-hexane plus 1-hexene or [^{14}C]-hexane plus cyclohexane.[19,20] The appearance of radioactivity in hexene and the identical specific radioactivity of hexene and benzene indicates that the alkene was an exclusive intermediate of aromatisation. This exclusivity was confirmed by the absence of radioactivity in the cyclohexane fraction. Similar results were obtained when [^{14}C]-hexane and hexene were reacted in the presence of hydrogen at 753 K, close to the temperature of industrial reforming.[19] Kinetic calculations with the mixture of [^{14}C]-1-hexene and inactive hexane showed that up to 90% of benzene was formed *via* hexene, over Pt/C at 573 K.[22] Comparing the radioactivities of benzene to that of the [^{14}C]-cyclohexane (mixed with inactive hexane), indicated that the rate of cycloalkane aromatisation was about 1.5 times more rapid than dehydrocyclisation, up to ~55% cyclohexane conversion.[23,24] The absence of deuterium exchange between n-octane and methylcyclohexane[14] also supported the proposal that cyclohexane dehydrogenation and dehydrocyclisation were independent processes.

The reaction hexenes → hexadienes was demonstrated without using radiotracers both on oxide[26,59] and metal catalysts, Ni[27] and Pt.[28] Mixtures containing [^{14}C]-hexene contributed to the clarification of the further reaction pathway.[20,29] These studies showed that neither the hexene → cyclohexane nor the hexene → cyclohexene ring closure pathway took place.[20] Table 2 indicates that radioactivity appeared in both the hexatriene and 1,3-cyclohexadiene fractions when their inactive form was admixed to radioactive hexene. The aromatisation of both inactive components was much more rapid than that of hexene, therefore their specific radioactivities showed very low absolute values, however, these were still higher than that of benzene produced mainly from these non-radioactive precursors. The true precursor of ring closure should have been *cis-cis*-1,3,5-hexatriene. Its ring closure takes place without any catalyst from ~513 K.[30] The stepwise dehydrogenation of open-chain hydrocarbons produces *cis*- and *trans*-isomers of alkenes and alkadienes. Any *cis-cis*-triene

Table 2. Relative molar radioactivities (**r%/m%**) in the products after reacting various mixtures containing [^{14}C]-hexene.[a]

A. Mixtures of [^{14}C]-hexene + Cyclohexadiene[b]			
	The Values of **r%/m%** in		
	Hexane + hexenes	1,3-cyclo-hexadiene[c]	Benzene
Feed	3.9	0.013	Absent
Products	4.2	0.074	0.030
	4.1	0.051	0.030

B. Mixtures of [^{14}C]-hexene + *trans-trans*-1,3,5-hexatriene[c]				
	Hexane + hexenes	Hexadienes-	1,3,5-hexatriene	Benzene
Feed	4.75	Absent	0.004	Absent
Products	3.55	0.12	0.012	0.04
	3.55	0.25	0.022	0.05

[a]Pulse system, carrier gas: helium, catalyst: Pt black, T = 633 K. Radioactivity was determined by trapping GC fractions in a cool trap and monitoring them by a liquid scintillator [20] This resulted in their incomplete separation.
[b]24% [^{14}C]-hexene, 1% cyclohexene, 75% cyclohexadiene. Due to the incomplete separation of hexenes and cyclohexene resulted in the appearance of radioactivity as an artifact in the cyclohexene impurity. After Ref. 20.
[c]21% [^{14}C]-hexene, 79% 1,3,5-hexatriene. After Ref. 29.

formed from the mixture in Table 2B must have reacted very rapidly to benzene, thus its detection in the product was highly unlikely. The *trans* → *cis* geometric isomerisation takes place reportedly via half-hydrogenated intermediates[31] and requires surface hydrogen. With less hydrogen present, surface unsaturated intermediates with *trans*-structure may "freeze" on the catalyst, causing coke accumulation.[19,32]

The above results are valid if the equilibration between the adsorbed and gas phase of corresponding compounds (feed, intermediates and/or products) is more rapid than their further reaction.[33] In the absence of sufficient hydrogen, this may not be the situation. Davis[18] did not reject the triene intermediates but criticised the "stepwise dehydrocyclisation" as described above, since he doubted that the requirement of gas–surface equilibration is fulfilled under every experimental condition. The desorption of surface intermediates with more double bonds is hindered, *compared with their further reaction*. Davis's argument that "the aromatic is as unsaturated as the triene" (Ref. 18) is true; and so is the second half of the sentence, "and readily desorbs as a product", but the two treatments should be separated. Benzene is an end product, thus it will not react further before its hindered desorption takes place. In fact, the degree of dehydrogenation of the surface benzene precursor can be too deep, thus their removal to the gas phase is a hydrogenative process, for example

$$C_6H_4(ads) + 2H(ads) \rightarrow C_6H_6(gas) \tag{4}$$

The reality of such processes has been confirmed experimentally by at least two other independent methods, Temperature Programmed Reaction — TPR[34] and Transient Response Method.[35]

As far as the reactivity of the unsaturated cyclic benzene precursor (assumedly adsorbed cyclohexadiene) is concerned, its opposite process, i.e., hydrogenation to cyclohexane is also possible. Reacting a mixture of [^{14}C]-cyclohexane and hexane on 0.8% Pt/KL catalyst in a closed loop reactor, showed unchanged specific radioactivity of the cyclohexane fraction in 12-fold hydrogen excess, but this value dropped to ca. one-half of the original value at ~15% cyclohexane conversion in 48-fold hydrogen excess.[36] Two hypotheses were put forward to explain this phenomenon. One hypothesis was that on this rather disperse catalyst with special site geometry, a "direct" cyclohexane formation was possible (although not typical, as described above). The other hypothesis assumed that hydrogenation of the "surface unsaturated C_6 ring" competed successfully with aromatisation in such a high H_2 excess. The latter hypothesis was suggested for cyclohexane formation on a Pt-Ge catalyst with apparently small free Pt islands.[37] This catalyst had a very low activity in benzene hydrogenation, thus the reactive chemisorption of the six-membered ring must have been hindered.

2.2. Formation and rupture of C_5-cyclic hydrocarbons

Metal catalysed skeletal isomerisation of alkanes was first reported in the 1960's. Surface cyclopentanes were regarded as one of the intermediates[38] competing with another mechanism "bond shift"[38,39] or "C_3-cyclic" isomerisation.[40] The latter is the only possibility for isomerising substituted butanes, whereas the two pathways can compete with alkanes having five of more C atoms in their main chain. Reacting 3-[^{14}C]-methylcyclopentane (3MP*, ~76%) plus inactive methylcyclopentane (MCP, ~24%) was used to clarify the role of the "C_5-cyclic route" on Pt black, although the desorption of the *surface* C_5-cyclic intermediate competes with its further reaction to isomers. The **r%/m%** values of this mixture were monitored between 573 and 663 K in the conversion range of ~8 to 25% (Table 3).[41] The relative molar radioactivity of MCP was much less than that of 3MP*, and lower than the corresponding values of 2MP isomer. Both values increased at higher temperatures in H_2. The absolute **r%/m%** values in 5% hydrogen were lower than those observed in pure H_2,

Table 3. Relative molar radioactivities (**r%/m%**) in the products after reacting a mixture of 3-[^{14}C]-methylpentane (3MP) and methylcyclopentane (MCP).

Carrier Gas	T, K	The Values of **r%/m%** in			
		2MP	3MP	MCP	2MP/MCP
Feed		0	1.31	0	—
H_2	573	—	1.28	0.01	0
	603	0.20	1.26	0.12	1.67
	618	0.29	1.27	0.22	1.32
	633	0.44	1.28	0.31	1.42
	648	0.51	1.24	0.36	1.42
	663	0.63	1.20	0.41	1.53
5% H_2+95% He	573	0.18	1.30	0.08	2.2
	603	0.43	1.29	0.12	3.5
	633	0.38	1.28	0.10	3.8

Pulse system, catalyst: Pt black. Adapted after Ref. 41.

but those in 2MP were higher than the **r%/m%** of MCP. This means that ring opening of MCP took place to give 2MP and 3MP, diluting the original labelled 3MP* in this manner. The conversion of 3MP* to both MCP and 2MP was larger at higher temperature but smaller in lower H_2 excess. The higher **r%/m%** values of 2MP indicate the parallel occurrence of bond shift isomerisation, marked at higher temperature. The **r%/m%** values of the isopentane fragment were close to that of 3MP*, while n-pentane was almost inactive, in good agreement with the loss of the labelled methyl group in the latter, as opposed to the splitting of an unlabelled terminal C atom to form isopentane.[41]

2.3. *Aromatisation of branched alkanes*

Branched alkanes with less than 6 C atoms in their main chain also produce benzene.[19,21,32] They must, however, undergo skeletal rearrangement before forming the aromatic ring. Sárkány[42] illustrated this with the example of 3-methylpentane, distinguishing three possible pathways: (a) the formation of methylcyclopentane and its subsequent ring enlargement, (b) vinyl shift involving the breaking of the secondary–tertiary C–C bond, (c) CH_x addition-abstraction producing toluene as the primary product, followed by its demethylation. Route (a) is important in industrial reforming and was attributed to the acidic function of bifunctional catalysts (see above). Route (c) involves alkane homologation, producing larger molecules from the starting hydrocarbons (see below) and a subsequent loss of the methyl group. Reacting mixtures of an aromatic hydrocarbon and methylpentane (MP) isomers (one component labelled with ^{14}C) would produce benzene with unchanged specific radioactivity by the first two pathways, but demethylation could lead to the loss of the labelled methyl group. Experiments were carried out with mixtures of (I) 2MP and [^{14}C]-toluene,[43] and (II) with benzene and methyl labelled [^{14}C]-3MP.[42] The catalysts were Co or Ni of various preparations, both metals exhibiting quite high fragmentation activity up to >90%.[43] In case I, the residual radioactivity of toluene between 6–36% conversion dropped to 0.55–0.85% of the original value (ρ_0) due to demethylation, while the specific radioactivity of benzene was $(0.12–0.35)^*\rho_0$, indicating incorporation of the original methyl label into the ring. Analogous results were obtained in case II. The specific radioactivity of benzene was lower, with that of toluene higher than the value of 3MP in the starting mixture, indicating that route (c) actually took place on Ni and also on Pt.[42] In the latter, the radioactivity of benzene from the mixture of 3MP* and MCP (see above) was very close to that of the 3MP* feed,[41] indicating that route (a) of aromatisation (ring enlargement) was negligible on Pt black.

Homologation reactions were described in several metal catalysed processes as side reactions at high temperatures, and under hydrogen-deficient conditions.[44,45] The latter condition favoured products that appeared as aromatics. Metals such as W, Rh, Ni active in multiple fragmentation were found to be active, while Pt showed moderate activity. On metal films, one or two-step chain growth of alkanes (i.e., adding one or two C atoms to the original chain) was assumed, with branched chains showing lower reactivity.[44,46] This reaction was studied with one-component feeds of [1-^{14}C]-hexane or [3-^{14}C]-methylpentane, and the relative molar radioactivity (**r%/m%**) was monitored in the effluent of pulse reactions (Table 4). The molar radioactivities of C_6 products from hexane are identical to that of the feed, that of fragments and toluene are higher. If a surface CH_x is added to the feed chain, the **r%/m%** of toluene can give information on its origin. Pt catalyses "single rupture", with some superimposed "multiple rupture" producing methane.[47] The combination of the two reactions resulted in ~3.3 fragments per fragmented molecule. The multiple rupture of labelled hexane takes place the following way, denoting the C_1 fragments as surface entities

$$^{14}CH_3\text{–}CH_2\text{–}CH_2\text{–}CH_2\text{–}CH_2\text{–}CH_3 \rightarrow 5CH_x +^{14} CH_x+ \qquad (5)$$

Table 4. Relative specific radioactivities in the products of formed from radioactive alkanes.[a]

Component	Spec. Radioactivity, Impulse s^{-1} mg^{-1}	
	Hexane	3-methylpentane
$<C_6$	1.11	1.09
C_6 isomers	0.98	1.01
MCP and/or benzene[b]	0.95	1.06
C_7 saturated	—	0.65
Toluene	1.16	0.87

[a]Pt black, pulse system, carrier gas H_2, T = 673 K. Averages of 4 or 5 runs, scattering: ±1 to ±8%. After Ref. 46.
[b]Methylcyclopentane (MCP) and benzene were incompletely separated. MCP prevailed from 3-methylpentane and benzene from hexane.

i.e., 1/6 of the C_1 fragments is radioactive. The molar radioactivity of toluene was 1.16, in good agreement with the above assumption. In this case, if "end demethylation" prevailed,[48] 50% of the initially produced CH_x fragments would be radioactive and the above value for toluene should have approximated 1.5. With 3-methylpentane, the splitting of the secondary–tertiary C–C bond is preferred, as opposed to the primary–tertiary bond.[60]

$$\begin{array}{ccc} \overset{{}^{14}CH_3}{\overset{|}{CH_3\text{–}CH_2\text{–}CH_2\text{–}CH_2\text{–}CH_3}} & \rightarrow & \overset{{}^{14}CH_3}{\overset{|}{CH_3\text{–}CH_2\text{–}CH_3}} + 2CH_x \end{array} \tag{6}$$

One molecule fragmented on Pt produced ~2.2 fragments,[50] i.e., just a slight methane excess.

If the alkylating C_1 unit is not radioactive, the molar radioactivity of the C_7 product from the labelled feed should be $6/7 = 0.83$, which is in good agreement with the result of Table 5. Comparing the results from [^{14}C]-hexane and inactive benzene,[46] as well as the reverse mixture,[51] indicates that alkylation of the "ready" aromatic ring[52] as well as degradation of larger products (e.g., dimers)[53] were less likely pathways.

Dehydrogenation of cyclohexanol to phenol can start with the formation of cyclohexanone intermediate (the stable tautomer of "cyclohexenone").[54] The loss of the hydroxyl group from either the starting cyclohexanol (to give cyclohexane), or from phenol (to give benzene) can also take place. Dehydration of cyclohexanol to cyclohexene is also possible[55] as summarised in Scheme 3. These reactions were studied over various metal catalysts, by using different mixtures containing one labelled component. Typical results obtained on Cu, Ni and Pt catalysts are summarised in Table 5.[56–58] The specific radioactivities decreased in the sequence: cH-ol > cH-one > phenol on Cu and Ni catalysts, while a different order cH-ol > phenol > cH-one was observed on Pt, as well as on Pd.[59] Thus, the sequential reaction $1 \rightarrow 2$ leads to phenol in the former two catalysts and the direct route to phenol 1A is possible on Pt and Pd.[57,59] The following relative rates were determined for Ni catalyst:

$$1 > 2 \gg 5 > 4.$$

Comparing 10 metals of Group 8–10 (plus Re).[59] revealed that Rh and Ir were also rather active catalysts of aromatisation, but only in hydrogen excess. Cyclohexanone was the preferred product on Os, Co, Fe, Ru and Re. Concerning the formation of hydrocarbons, reaction (3) does not occur. Hydrocarbons are formed either by dehydrogenation to cyclohexene,

Table 5. Relative molar radioactivities (**r%/m%**) obtained in various mixtures of [^{14}C]-cyclohexanol.[a]

A. Mixtures of [^{14}C]-cyclohexanol + Cyclohexanone				
Catalyst	T, K	The Values of **r%/m%** in		
		Cyclohexanol	Cyclohexanone[a]	Phenol
Feed I		1.58	0.12	—
Cu	573	0.97	1.01	0.87
Ni	523	1.12	1.05	0.77
	573	1.10	0.98	0.96
	573	1.21	0.95	0.86
Feed II		1.85	0.13	—
Pt	533	1.72	0.36	0.66
	533	1.67	0.28	0.54

Flow reactor. Feed: ~1:1 cyclohexanol and cyclohexanone. Conversion of cyclohexanol: 30–60%. Adapted after Refs. 56, 57.

B. Mixtures of [^{14}C]-cyclohexanol + Hydrocarbons						
Catalyst	The Values of **r%/m%** in					
	C_6H_{12}	C_6H_{10}	C_6H_6	$C_6H_{10}O$[a]	$C_6H_{11}OH$	C_6H_5OH
Feed I[b]	0	—	—	0.12	1.38	—
Pt, 543 K	0	0	1.11	0.54	0.94	1.15
Feed II[c]	—	0	—	*	3.7	—
Pt, 543 K	0	0	0.25	4.8	2.5	2.7
Feed III[d]	—	—	—	0.17	1.74	—
Pt, 543 K	—	—	0.63	0.28	1.77	0.83

Pulse system., carrier gas: N_2. Adapted after Ref. 58.
[a]The apparent radioactivity of cyclohexanone is due to its incomplete separation from the labelled cyclohexanol. It was rather high in the case of Feed II in Table 2B.
[b]Feed: 72% [^{14}C]-cyclohexanol, 28% cyclohexane.
[c]Feed: 27% [^{14}C]-cyclohexanol, 73% cyclohexene.
[d]Feed: 53% [^{14}C]-cyclohexanol, 47% cyclohexanone.

or by hydrogenative dehydroxylation of phenol to benzene. Re was the only metal producing radioactive cyclohexane from [^{14}C]-cyclohexanol; [^{14}C]-cyclohexene was produced on Co, Re, Ru, and perhaps Ir, whereas Os was the only metal catalyst which did not produce radioactive benzene.[59] The C_6 cyclic hydrocarbons undergo hydrogenation — dehydrogenation reactions as discussed earlier. The rate of benzene formation from cyclohexane

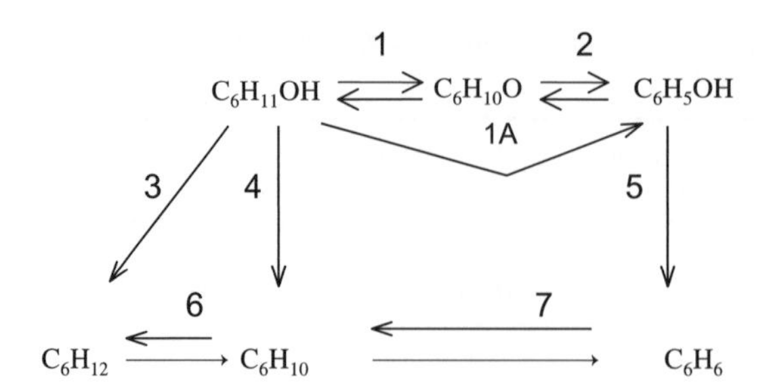

Scheme 3. Reactions of cyclohexanol on metal catalysts.

on the different metals showed differences up to 1 000. Such differences were not observed in the reaction of cyclohexanol → phenol.

Methanol synthesis
Methanol is synthesised from "synthesis gas" (i.e., a mixture of CO plus H_2). Along with their simple combination reaction

$$CO + 2H_2 \rightarrow CH_3OH \tag{7}$$

another process, the hydrogenation of CO_2 can also take place

$$CO_2 + 3H_2 \rightarrow CH_3OH + H_2O \tag{8}$$

Mixtures of $CO + H_2$ and $CO_2 + H_2$ were reacted with one component, either CO or CO_2, containing ^{14}C label.[60] Table 6 shows that on a mixed oxide catalyst, hardly any label appeared in the methanol end product from ^{14}CO, but the use of $^{14}CO_2$ resulted in radioactive methanol. A side reaction of CO and water diluted considerably the radioactivity of the CO_2 fraction after reaction. Reaction (8) was found valid on iron catalysts. Related studies (mainly those carried out in Moscow) have been summarised by Rozovskii.[61]

3. Study of Reaction Kinetics by Isotope Tracer

So far, the *mere appearance* of radioactivity in the assumed intermediate was used to obtain information on the likely route of the catalytic process. Following the changes at various levels of conversion would give information on the reaction kinetics, too. Before discussing the quantitative method, a semi-quantitative study will be considered.

Table 6. Relative specific radioactivities in the products of methanol synthesis.[a]

Component	Spec. Radioactivity, Impulse s^{-1} mg^{-1}			
	Mixture 1		Mixture 2	
	Feed	Product	Feed	Product
CO	400	190	0	490
CO_2	0	9	5900	790
CH_3OH	—	5	—	1450

[a]Expressed as pulses per mg $BaCO_3$ after oxidizing the products.
Catalyst: $CuO/ZnO/Al_2O_3$. After Ref. 60.

3.1. *Benzene hydrogenation*

Discovering the real intermediates would produce some experimental difficulties due to their high reactivities. The presumptive intermediate cyclohexene ($cH^=$) hydrogenated much more rapidly than benzene, thus, even if radioactive $cH^=$ was formed from [^{14}C]-benzene, its concentration in the gas phase could be much lower than on the surface. At higher conversions, $cH^=$ conversion approached 100% and disappeared from the gas phase. The consumption of the original form of the key component (no matter if it was present in the original feed or formed *in situ*) represented the applicability limit of multicomponent mixtures with one labelled component. These difficulties were overcome by using the mixture of [^{14}C]-benzene and 1,3-cyclohexadiene.[13] The diene component was hydrogenated very rapidly, being often absent in the effluent, but cyclohexene indicated the importance of "stepwise" hydrogenation with the appearance of radioactivity in its primary product. Hydrogenation involves the addition of H atoms, formed by dissociating H_2 on the metal. The probability is 0.4 that a C_6H_{10} surface intermediate, produced by sequential "random" addition of H atoms to any position of the six-membered ring, had the structure of a chemisorbed cyclohexene.[15] The ratio of relative molar radioactivities $cH{=}/cH$ was between 0.01 and 0.27 at low benzene conversions (0.4–2.7%),[13] thus a preference of 1,3,5-hydrogen addition can be assumed (Scheme 4).

The β/α ratio was rather constant in the benzene conversion range between 0.1 and ~8%. Its value increased sharply at higher conversions. This increase was attributed to the consumption of inactive cyclohexene, representing the primary product of diene hydrogenation. The results could

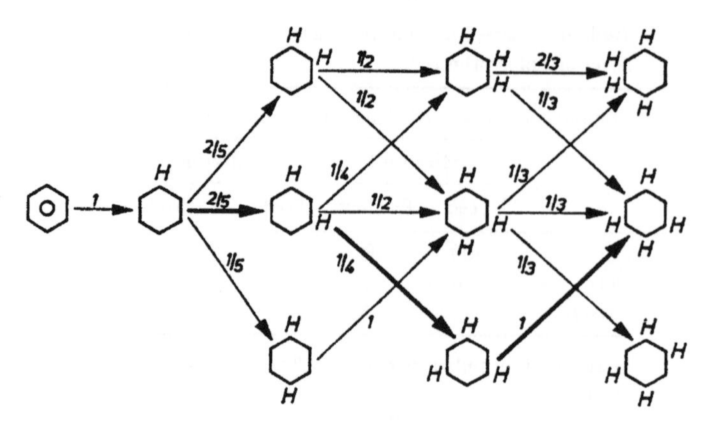

Scheme 4. Statistical probabilities of subsequent random addition of H atoms to the benzene ring. The probabilities of elementary steps has been indicated on the arrows. The thicker arrow indicates the 1,3,5-addition avoiding cyclohexene formation. Adapted after Ref. 15.

be interpreted in terms of a triangular mechanism of benzene hydrogenation, including a "stepwise" and a "direct" route. The latter may take place by random addition of hydrogen atoms. There is a case (1,3,5-addition) when cyclohexene with one double bond was not an intermediate.[62] The stepwise hydrogenation may involve "edgewise" adsorption of the benzene ring and unsaturated intermediates. The role of the latter surface species was pointed out recently by using *in situ* surface chemical methods during reaction.[63]

Cyclohexadiene was converted totally on a Ni catalyst when the benzene conversion was ~5–6%. From this point onwards, the concentration of $cH^{=}$ dropped dramatically[13] with a simultaneous increase in its specific radioactivity exceeding that of cH, when the Bz conversion reached ~10% as shown in Fig. 3. The formation of radioactive cyclohexene from $[^{14}C]$-benzene supplied further evidence of the existence of stepwise hydrogenation, even if it is not the exclusive route.

3.2. *The kinetic isotope method*

This technique has been developed for studying sequential reactions by Gál and co-workers.[33] Theory and earlier results were summarised in a book,[64] describing it as a dynamic method that utilised the temporal changes of specific (or molar) radioactivities of assumed intermediates.

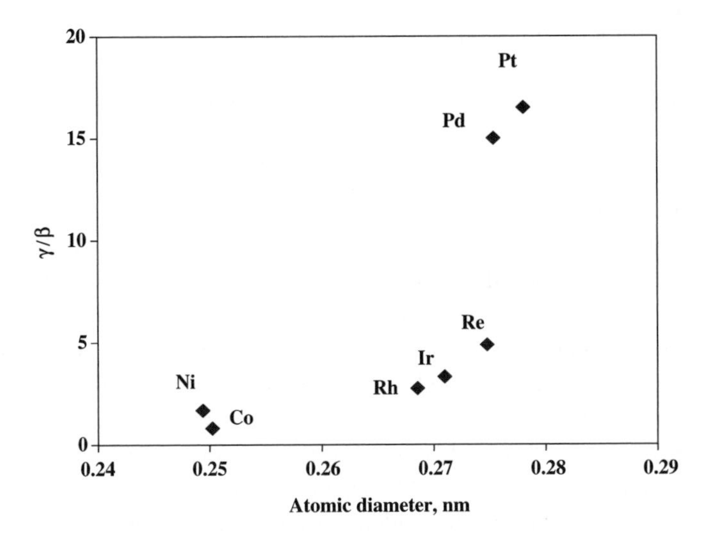

Fig. 3. The ratio of the two pathways of benzene formation (as characterised by the ratio of the specific radioactivities of cH$^=$, β and Bz, γ from a mixture of [^{14}C]-cyclohexane and inactive cyclohexene) as a function of the atomic diameter of the metal catalyst. The atomic distances in the six-membered ring are as follows (considering also possible valence angles in various conformations) are: C_1–C_2: 0.153 nm; C_1–C_3: 0.2506–0.2632 nm. Redrawn after Ref. 15; the point for Re taken from Ref. 14.

Case (**a**).

Consider a reaction sequence

$$\mathrm{A} \rightarrow \mathrm{B} \rightarrow \mathrm{X} \rightarrow \mathrm{Y} \rightarrow \mathrm{C} \rightarrow \text{etc.} \tag{9}$$

When a small portion of labelled X is introduced, the changes of the specific activity of X(a) and that of Y(b) as a function of time, can be described by the following equations

$$da/dt = -\mathrm{aw}_1/[\mathrm{X}] \quad \text{and} \quad db/dt = (\mathrm{a}-\mathrm{b})\mathrm{w}_2/[\mathrm{Y}] \tag{10}$$

at $t = 0$, $b = 0$. As X transforms into Y, the value of b will increase in time. When b reaches its maximum, $db/dt = 0$ and $a = b$. Thus, the value of b gives a maximum curve and the curves of a and b intersect at the maximum value of b.

In the case of another sequence (**b**)

$$\begin{array}{c} \mathrm{A} \rightarrow \mathrm{B} \rightarrow \mathrm{X} \rightarrow \ \mathrm{Y} \rightarrow \mathrm{C} \rightarrow \text{etc} \\ \nearrow \\ \mathrm{K} \end{array} \tag{11}$$

the value of b as a function of time has also a maximum, but $b < a$, i.e., there is no intersection.

In case **c** (i.e., when another intermediate enters between X and Y):

$$A \rightarrow B \rightarrow X \rightarrow Z \rightarrow Y \rightarrow C \rightarrow \text{etc} \quad (12)$$

a and b will intersect but not at the maximum of b.

The principles of using this method for heterogeneous catalytic reactions and the corresponding kinetic equations were reviewed by Bauer and Dermietzel.[65] Adding a tracer to a component of a reversible consecutive reaction of components A, B and C would correspond to case **a** as described above, while case **c** will be valid for a triangular reaction (cf. Scheme 1). This will be illustrated using the example of xylene isomerisation on an industrial Pt/Al_2O_3 catalyst.[66] A feed containing 89% *p*-xylene, 10% [^{14}C]-*m*-xylene and 1% *o*-xylene was reacted. The specific relative activities of the effluent, plotted as a function of residence time (Fig. 4), indicate two rather different behaviours. Figure 4A shows that the results in the presence of *hydrogen* corresponded to case **a**. The intersection of the curves for *o*-xylene and *m*-xylene are at the maximum specific radioactivities of the former feed component. The same was found with 10% labelled m-xylene added to 89% *o*-xylene and 1% *p*-xylene.[65] A different picture is obtained in the presence of nitrogen (Figure 4B). The specific radioactivity of *m*- and *o*-xylene intersect but not at the maximum of the latter, as in case **c**. A new product class also appeared (equimolar amounts of toluene and trimethylbenzene, TMB). Their specific radioactivity shows a maximum, but never intersects with that of *m*-xylene (→ case **b**). Thus, the direct isomerisation pathway observed in hydrogen transforms into a more complicated one, involving a disproportionation step of the *o*- and *m*-xylene, with TMB serving as an intermediate between these two isomers. This represents a new pathway for the formation of *m*-xylene.

The kinetic isotope method was also applied to study whether the "stepwise" or "direct" route is valid for the dehydrogenation of the cyclohexane ring.[1,14,67] As mentioned before, plotting specific radioactivities as a function of the contact time confirmed the predominance of "direct pathway" on Re and the "stepwise" dehydrogenation on chromia.

4. Reaction Pathway Studies by the Determination of the Position of the Label

The alternative method to the application of tracers is by reacting one labelled feed and determining the position of the label in the product. This

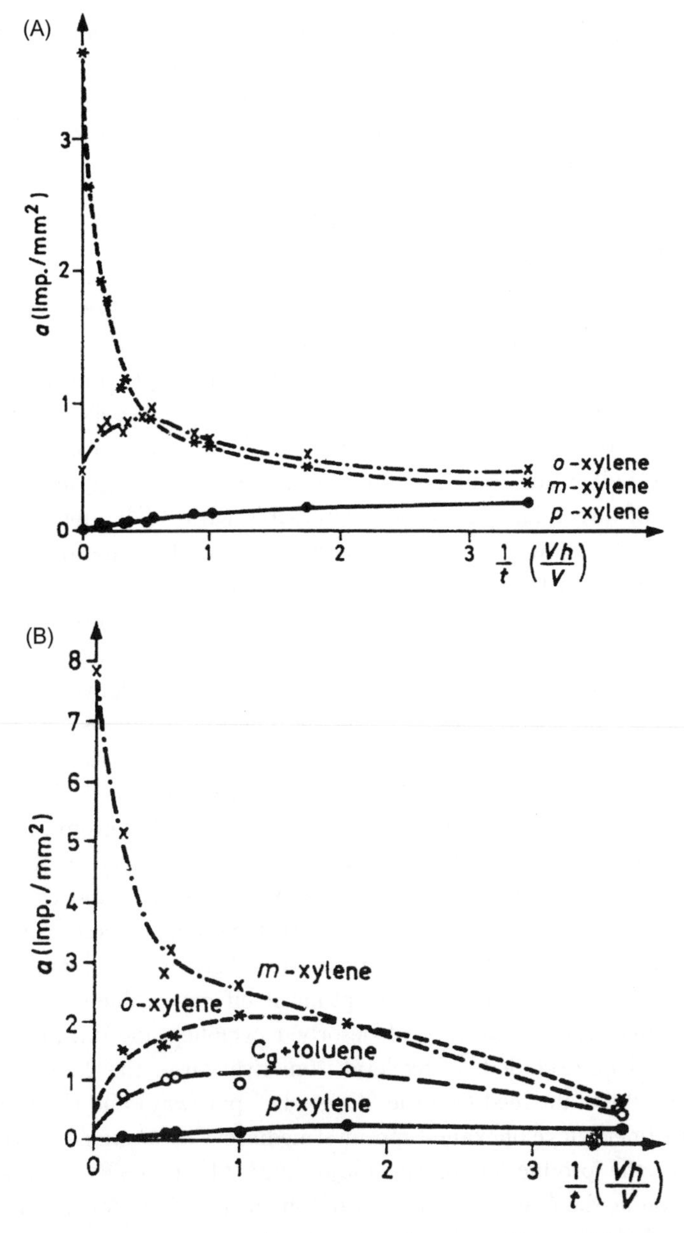

Fig. 4. Xylene isomerisation on Pt/Al_2O_3. The relative specific activities are plotted as a function of the time of residence in a single-flow reactor Adapted after Refs. 65, 66. A: in the presence of H_2; B: in the presence of N_2.

method was called the "labelled molecule" approach[17] and it required rather large amounts of product(s), their subsequent chemical processing and involved, as a rule, laborious and sophisticated processes. Most of the publications using this method appeared in the "classical golden age" of radioisotope application.

4.1. *Dehydrocyclisation*

In his work as early as 1958, Mitchell[68] observed 24–29% methyl label in toluene formed from 1-[^{14}C]-n-heptane on Cr_2O_3/Al_2O_3 catalyst, containing K_2O and CeO_2 additive both at low and high conversions. Formation of transannular CH_2 bridge in the surface C_6-cyclic intermediate was suggested. The same catalyst was used to study aromatisation of 3-[^{14}C]-methylheptane and 2,2-dimethyl-4-[^{14}C]-methylpentane.[69] Three pathways of direct C_6 cyclisation are possible with 3-[^{14}C]-methylheptane (C_1–C_6; C_2–C_7; side methyl–C_7). The first two produce methyl labelled *o*- and *p*-xylene, and the third route ring labelled ethylbenzene. The distribution of [^{14}C] in the aromatic products confirmed that more than 90% of them was formed via C_6 ring closure.

Pines and co-workers published several studies on the dehydrocyclisation of various C_7 and C_8 alkanes on oxide and metallic catalysts.[70] Oxidising the alkylaromatic product with $KMnO_4$ produced CO_2. Its radioactivity indicated the fraction of label in the side alkyl group(s). 1-[^{14}C]-n-heptane should form toluene with 50% of the label in the methyl group, if a direct C_6 ring closure occurred. 40% or more label was reported in the methyl position[71] in the overall product on unsupported Cr_2O_3 and Cr_2O_3/non-acidic Al_2O_3. The authors assumed a surface 7-membered ring and a subsequent methyl insertion to explain this distribution, mainly on the "non-acidic" catalyst. Since cycloheptane itself gave toluene much faster than heptane, the appearance of desorbed cycloheptane during aromatisation could be excluded.[71] Skeletal isomerisation of the 2,2-dimethyl-3-[^{14}C]-methylpentane feed (*via* the "C_3-cyclic" pathway) and the formation of a 1,1-dimethylcyclohexane-type intermediate was assumed as the step preceding the production of the major part (41 to 68%) of xylene on Cr_2O_3/non-acidic Al_2O_3.[72] The rest of toluene may have formed by a combination of C_4 fragments. Note that alkanes with a quaternary C atom are not able to form polyene precursors; thus, this assumption does not contradict the "triene" mechanism as discussed above. Aromatisation of labelled octanes[73] suggested the formation of *o*-xylene *via* direct formation of a

C_6 ring ("2,7-cyclisation" — producing unlabelled methyl groups), since the amount of methyl label was 2–4%. C_7- or C_8-cyclic intermediates were suggested for the other two xylene isomers. The results obtained on oxide catalysts has been summarised by Pines and Goetschel.[74] A later study of the Pines group[75] indicated that "direct C_6 ring closure" prevailed also on Pt/Al_2O_3.

Feighan and Davis developed a six-step degradation process including oxidation of the methyl group, its exchange to an amino group, formation of a quinoline ring and stepwise oxidation of its carbocyclic ring containing the ring C atoms of toluene.[76] This permitted direct monitoring of the [^{14}C] label in the ring, too. For example, if toluene was formed from [4-^{14}C]-n-heptane by direct C_6 ring closure (independently of the degree of unsaturation of the intermediate undergoing cyclisation), the label should be in the *meta*-position (C_3) with respect to the methyl group. The *meta*- and *ortho*- C atoms of the ring (C_2 and C_3 positions) produced CO_2 in the fifth and sixth reaction step respectively. In one selected example, ~92% carbon was found in *meta* (C_3) position with respect to the methyl group, together with ~4.5% label in C_2, ~2.5% in C_1 position and ~1.5% in the CH_3 group. Detailed experiments reacting [1-^{14}C]-n-heptane and [4-^{14}C]-n-heptane on various chromia-alumina,[77] Pt/Al_2O_3,[78] $Pt\text{-}Re/Al_2O_3$ and $Pt\text{-}Sn/Al_2O_3$[79] catalysts indicated that at least 80%, if not more (up to 96%) of toluene was produced by direct C_6 ring closure. The comprehensive review by Davis[18] discussed in detail almost every available result and the possible role of various possible C_5-cyclic intermediates.

4.2. *Skeletal transformations of cyclic hydrocarbons*

On typical industrial bifunctional catalysts, ring closure supposedly occurred *via* formation of an alkylcyclopentane ring from the alkene intermediate, followed by acid-catalyzed $C_5 \rightarrow C_6$ ring expansion. Csicsery and Burnett[80] added [^{14}C]-1,3-dimethylcyclopentane tracer to heptane, which was then reacted over a non-acidic Pt/SiO_2 catalyst. About 9% of radioactivity in the C_7 products was present as toluene, while the rest representing iso-heptanes was formed by random ring opening of the cyclopentane ring. Thus, $C_5 \rightarrow C_6$ ring enlargement took place, but it was not crucial, especially if we consider that toluene could be formed by aromatisation of methylhexane primary products.

Huang *et al.* reacted the mixture of 97% n-octane plus 3% propyl-[1-ring^{14}C]-cyclopentane on Pt/KL zeolite.[81] The radioactive label appeared in the methyl- and ethyl-cyclopentane products, indicating that they were originated from the splitting of C–C bonds in the side chain of propylcyclopentane, rather than from C_5-cyclisation of octane fragments. The probability cyclopentane formation corresponded to that expected from random splitting but the breaking of other side alkyl C–C bonds was less probable than those expected values.

Isomerisation of ethylcyclohexane to various dimethylcyclohexane isomers can involve theoretically two pathways: A: transfer of the methyl group of the side chain to the ring or B: $C_6 \rightarrow C_5 \rightarrow C_6$ ring contraction–expansion (Scheme 5). If a [^{14}C] label is introduced into the side alkyl group, mechanism A would produce side alkyl labelled products, whereas the label would appear in the ring with mechanism B. The composition of the product, after reacting ethylcyclohexanes over Ni/SiO_2–Al_2O_3 at 633 K, was close to the thermodynamic equilibrium.[82] They were dehydrogenated to aromatics, the xylenes oxidised first to benzoic, phthalic, terephtalic and isophthalic acids, followed by catalytic decarboxylation to form benzene and CO_2. In this way, [^{14}C] in the ring and in the side chain could be distinguished. The pronounced radioactivity (~50–80%) in the benzene fraction points to the existence and even predominance of mechanism B. These may involve several subsequent ring contraction–expansion steps (see below).

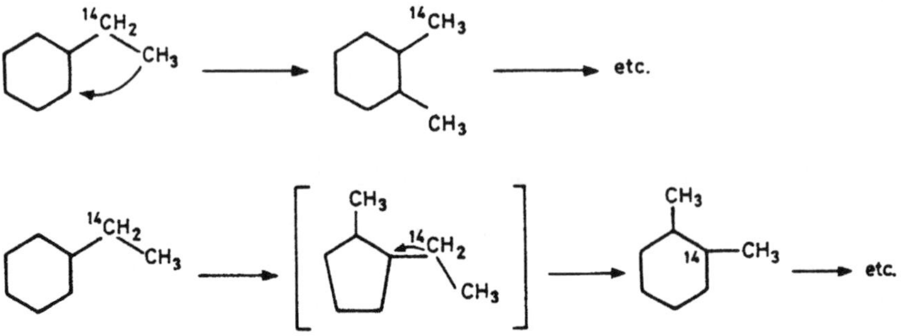

Scheme 5. Reaction pathways of isomerisation of ethylcyclohexane. Re-drawn after Ref. 82.

Reacting ring-labelled and side chain labelled ethylbenzene and *m*-xylene on industrial Pt/Al_2O_3 gave further information[83] on the possible pathways of xylene isomerisation discussed in the previous section. The reaction of labelled ethylbenzene gave a considerable amount of [^{14}C] label in the aromatic ring of xylene products, indicating the importance of ring

Table 7. Distribution of ^{14}C label in the xylene isomers formed from [^{14}C] labelled ethylbenzene and [^{14}C]-methyl-*m*-xylene.

Initial Label Position	H_2 Pressure, MPa	Distribution of the ^{14}C Label	
		Methyl Group	Ring
Ethylbenzene ring-[^{14}C]	1.8	21	78
Ethylbenzene 7-[^{14}C]	1.8	10	90
Ethylbenzene 8-[^{14}C]	1.8	71.5	28.5
m-xylene-methyl-[^{14}C]	0.1	99	1
m-xylene-methyl-[^{14}C]	1.1	84	16
m-xylene-methyl-[^{14}C]	1.6	80	20
m-xylene-methyl-[^{14}C]	2.1	72.5	27.5

Single-flow reactor, T = 723 K. After Ref. 83.

contraction–expansion steps in xylene formation, supported by the appearance of alkylcyclopentenes and even trace amounts (<0.1%) of methylcycloheptene. The authors assumed that partly or fully hydrogenated C_6-cyclic compounds were present in this process. Methyl migration was the main route of xylene isomerisation, but the ring radioactivity increased in higher hydrogen excess (Table 7), promoting ring rearrangement reactions on the bifunctional catalyst.

All the studies mentioned determined the position of the label as CO_2 after proper oxidation processes. Another possible method can be pyrolysis of alkanes.[84] Pulses of various [^{14}C] labelled hydrocarbons (butadiene, pentene, cyclopentadiene, alkanes, alkylbenzenes) were introduced into an inert gas stream, then passed through a heated reactor, and the pyrolysis products (methane, >C_1 fragments and benzene) were analysed by a gas chromatograph equipped by a mass and a radioactivity detector. The temperature was kept between 873 and 1050 K, while the decomposition of aromatics required higher temperatures. With conversions between 5 and 90%, an accuracy of ±5% could be reached.

4.3. *Fischer-Tropsch synthesis*

This is a catalytic process leading to the formation of higher hydrocarbons (and/or alcohols) from CO and H_2 closely related to methanol synthesis. Earlier isotopic studies were reviewed by Eidus.[85] A very detailed paper by Raje and Davis summarised ^{14}C tracer studies up to 1995.[86] Some important milestones will only be surveyed here and a few more recent results will be added.

The possible steps of Fischer-Tropsch (FT) reaction and its catalysts (Fe, Co, Ru, Ni) represent a very complicated system.[87] The catalysts usually need a "formation" or "self-organisation", meaning that the full activity will only be reached after a certain period. This means that for Fe-based catalysts, a part of the initial Fe oxide is transformed into iron carbide. This was investigated as early as 1948 by the ^{14}C tracer method.[88] A fused iron catalyst was carbided with $[^{14}C]$. The synthesis product from $CO/H_2 = 1{:}1$ reactant contained 10–15% labelled molecules, almost independently of the reaction conditions, even in repeated runs, indicating the minor role of carbide incorporation into hydrocarbons. The "formation" of a Fe-Al-Cu catalyst at 523 K and various H_2/CO ratios required 100 to 2000 minutes. The yield of "retained carbon" decreased gradually, while the FT yield increased more abruptly after this period.[87]

Techniques described in the previous sections should be combined with the ^{14}C tracer studies of Fischer-Tropsch reaction. One has to add a labelled compound (e.g., alkenes or alcohols) suspected to participate in one of the above steps, and the products have to be separated and the individual components analysed for their radioactivity. No chemical processing is necessary for each fraction since their chromatographic separation is possible. As far as the role of added ^{14}C tracer is concerned, Davis[86,89] distinguished four possibilities. If the added radioactive component is a chain initiator, the specific activity is constant as a function of the carbon number. If the added component is a chain propagator (random addition of CO or the tracer component to the growing chain), or if there is an equilibration between it and the CO component, the specific activity should increase linearly with increasing C number. If the added tracer leads to chain growth via polymerisation, the radioactivity depends on the carbon number of the FT product. The polymerisation of $[^{14}C]$-ethene would lead to radioactivity incorporation into products with even C numbers, while products with odd numbers of C atoms would be inactive[90]

1. Initiation: $[^{14}C]\text{-}C_2 + C_1 \rightarrow [^{14}C]\text{-}C_2\text{–}C\text{–}C\text{–}C\ldots \quad (13)$

2. Polymerisation: $[^{14}C]\text{-}C_2 \rightarrow [^{14}C]\text{-}C_2\text{–}[^{14}C]\text{-}C_2\text{–}[^{14}C]\text{-}C_2\ldots \quad (14)$

3. Propagation: $C_1 + [^{14}C]\text{-}C_2 \rightarrow C_1\text{–}C_1\text{–}[^{14}C]\text{-}C_2\text{–}C^{14}C_1\ldots C_1\ldots C_1 - [^{14}C]\text{-}C_2\text{–}C_1\ldots \quad (15)$

4. Initiation and propagation:

$$C_1 + [^{14}C]\text{-}C_2 \rightarrow [^{14}C]\text{-}C_2\text{–}C_1\text{–}C_1\ldots C1\ldots[^{14}C]\text{-}C_2\text{–}C_1\text{–}C_1\text{–}C_1 - [^{14}C]\text{-}C_2\ldots \quad (16)$$

More detailed schemes have been presented in Refs. 86, 91 and 92.

Apart from labelled ethene,[90,93] FT synthesis was carried out in the presence of other labelled compounds such as ethanol,[94,95] larger alkenes[86] and higher alcohols.[89,96,97] Most results indicate that more than one mechanism is responsible for the distribution of radioactivity in the products. Figure 5[91] illustrates that polymerisation of labelled ethene on Co produced more label in even C-number alkanes up to C_{10}. Almost constant radioactivities were observed in the $>C_6$ products with added [^{14}C] propanol. Smaller products showed more incorporated ^{14}C and the monomethyl-alkanes contained more radioactivity. The authors concluded that these alkenes participate both in chain initiation and chain propagation. Alcohols, in turn, initiated chain growth but did not participate in chain propagation on an industrial Fe catalyst.[96] Neither ethene nor ethanol (or ethene formed by its dehydration) participated in the chain termination step.[98]

Radiotracers contributed to clarify that the chain initiation is different on Fe and Co or Ni catalysts.[86] This question was first studied by using

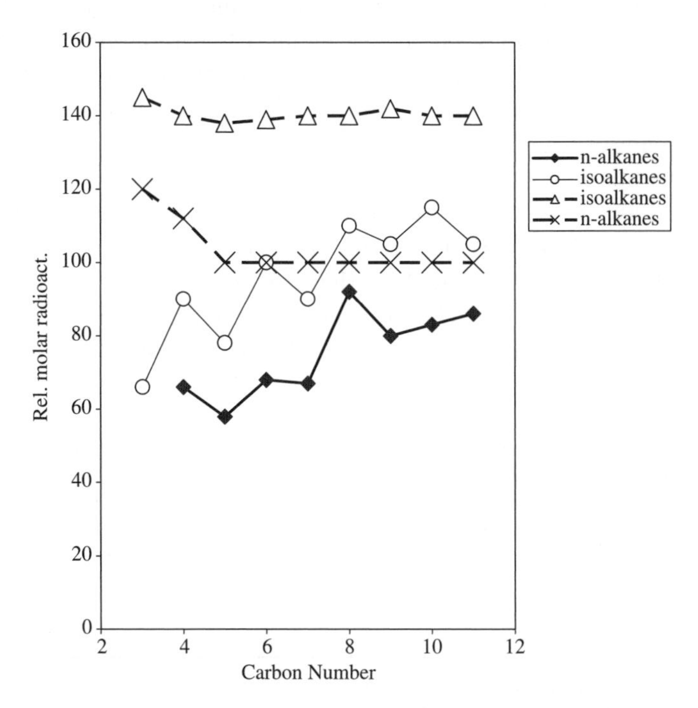

Fig. 5. Relative molar radioactivities of n-alkanes (♦) and monomethylalkanes (○) in the presence of [^{14}C]-ethene tracer as well as n-alkanes (X) and monomethylalkanes (Δ) produced in the presence of [^{14}C]-propene during Fischer-Tropsch synthesis on Co catalyst. Re-drawn after Refs. 91, 92.

ketene tracer, labelled with [^{14}C] either in the methylene or in the carbonyl group.[99] The radioactivity in hydrocarbon products increased linearly (up to C_{10}) when $CH_2={}^{14}CO$ was added to the synthesis gas mixture on both Co and Ni catalysts. The same was observed with $^{14}CH_2=CO$ on an iron catalyst, but the radioactivity of hydrocarbons (up to C_8) was constant in this case and was about 1/3 of the initial radioactivity of labelled ketene. This corresponds to case 1 as discussed above.[86,89] Schulz[87] distinguished 4 major reactions (CH_2 formation for the "carbide" chain initiation; chain growth and chain termination) along with at least 7 minor processes on Co and Ni catalysts.

The incorporation of ^{14}C in products with longer chains is not complete. One molecule per 18 C atoms was found to be radioactive in the C_6–C_{29} fraction, formed in the presence of [^{14}C]-ethanol tracer on a Co catalyst,[95] as opposed to one in 60 on Fe.[86] With 1-[^{14}C]-hexadecene added to a synthesis gas on Co/thoria catalyst, at ~70% conversion, the relative molar activity of the C_{16} alkane (normalised to C_6 = 100) was 291 000, that of the C_{17}–C_{20} fraction 6 000 (constant), but the analogous values for C_2–C_{14} were as small as ~50 to ~500, increasing linearly with larger alkanes.[87,91] Table 8[91] indicates that only a fraction of the ^{14}C label from ethene was incorporated into chain growth products on Co and Fe catalysts. About the same amount of ^{14}C (31%) was found in FT products formed with 1[^{14}C]-1-propene, but only 18% when 1-[^{14}C]-1-hexadecene was applied.[87] About 50% of radioactive ethene gave methane, and 50% chain growth products on a Co catalyst,[100] the specific activity of higher products being practically constant between C_4 and C_{32}. Less than 10% of ^{14}C from labelled ethene was incorporated into C_{10}–C_{17} products, but the incorporation from ^{14}C-ethanol was 60–80 times higher on a fused iron catalyst.[86] The difference

Table 8. Transformation of [^{14}C]-ethene in various reactions under the conditions of Fischer-Tropsch synthesis.

Catalyst	Conv,. %	Hydrogenation.	Chain Growth	Fragmentation	To CH_4
Co[a]	~100	66.7	29	4.3	4.3
Fe[b]	76.6	67.4	9.1	0.1	0.1
Fe[c]	77.1	65.0	11.5	0.6	0.6

After Ref. 90. Flow system, $CO:H_2$ = 1:2.
[a]Co + 18% ThO_2 catalyst, T = 463 K, 1 atm.
[b]Precipitated Fe catalyst, T = 493 K, 20 atm.
[c]Fused Fe catalyst, T = 593 K, 20 atm.

was even higher with labelled propene and propanol. 10–20% of added [^{14}C]-1-propanol was incorporated into longer alkanes, while this value was 1–3% with [^{14}C]-2-propanol. Mostly n-alkanes were produced in the former, and isoalkanes in the latter case.[97] With $^{14}CO_2$ added to the starting mixture, roughly 50% of its initial amount was converted to FT products. The molar radioactivity of alkanes increasing with their C number corresponded to a mechanism with CO_2 initiating the chain reaction.[101] This pointed to an adsorbed CO_2-like initiator in the "oxygenate" mechanism on Fe-based catalysts (Scheme 6),[86,101] as opposed to the "carbide" chain initiation *via* M=CH_2 species on Co.[86,87] The importance of CO_2 intermediate was confirmed by ^{14}C studies in methanol synthesis too.[60,61] Chain propagation takes place by insertion of CO or CH_x units.[66]

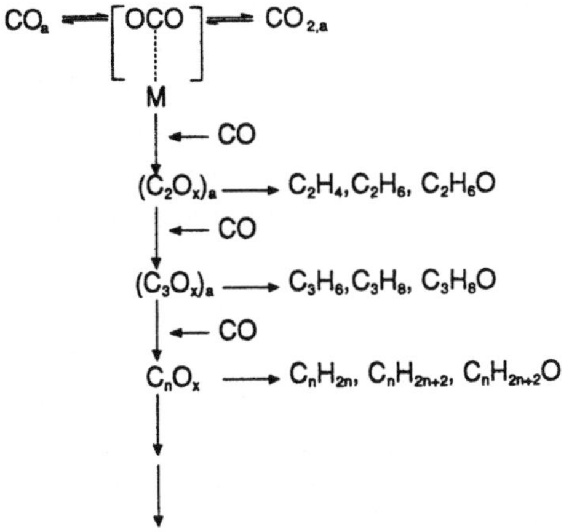

CO_a ⇌ [OCO ⋮ M] ⇌ $CO_{2,a}$
↓ ← CO
$(C_2O_x)_a$ → C_2H_4, C_2H_6, C_2H_6O
↓ ← CO
$(C_3O_x)_a$ → C_3H_6, C_3H_8, C_3H_8O
↓ ← CO
C_nO_x → C_nH_{2n}, C_nH_{2n+2}, $C_nH_{2n+2}O$
↓
↓

Scheme 6. The "oxygenate" mechanism of Fischer-Tropsch reaction on Fe catalyst, with CO_2 as chain initiator and CO as chain propagator. After Ref. 86.

A side reaction of this "oxygenate" initiation is the production of CO_2 from the added tracer alcohols. Its C–O–H group gets chemisorbed in the vicinity of an M=O entity, which obtained its oxygen from the dissociation of CO. CO_2 produced from 1-[^{14}C]-1-pentanol tracer was radioactive but it was inactive if 2-[^{14}C]-1-hexanol was added. The complicated reaction sequence of FT is still not clarified in every detail. The role of "reactor hold-up in determining the distribution of alkanes and alkenes in higher C-number products has been pointed out recently.[102] Possible suggested reaction routes ("current mechanism and futuristic needs") have been summarised by Davis.[103]

5. ^{14}C Tracers — Past or Future?

The development of other more sensitive methods (e.g., GC–MS, microwave or nuclear magnetic resonance spectroscopy, NMR) permitted the use of stable isotopes, avoiding radioactivity which has been meeting more and more resistance by the general public. They are also suited to obtaining more information. "Transient isotopic kinetic studies" involve a sudden switch of labelled and unlabelled feed, the monitoring of which requires continuous sampling which can be carried out by MS, replacing the use of ^{14}C with the stable ^{13}C isotope. Such results, as well as ^{13}C NMR, have added valuable information e.g., to our knowledge on the Fischer-Tropsch reaction.[86] These more recent techniques also permit the determination of the position of the label in the product molecules. Methods like these gradually displaced the use of ^{14}C tracers in heterogeneous catalysis, although this did not lead to its disappearance from chemistry. A recent electronic search in the literature gave hits almost exclusively on the biochemical/biological applications of this isotope. Nevertheless, this development did not make obsolete the *results* obtained by the ^{14}C technique in heterogeneous catalysis. They constitute a permanent mark in the scientific heritage of the 20th Century.

References

[1] Derbentsev Yu I, Isagulyants GV, *Uspekhi Khimii* **38**: 1597, 1969.
[2] Campbell KC, Thomson SJ, *Progr Surface Membrane Sci* **9**: 163, 1975.
[3] Ozaki A, *Isotopic Studies of Heterogeneous Catalysis*, Kodansha, Tokyo–Acad. Press, New York, 1977.
[4] Happel J, *Isotopic Assessment of Heterogeneous Catalysis*, Academic Press, New York, 1986.
[5] Schay Z, in *Radiotracer Studies of Interfaces* (Horanyi G, ed.) Ch. 3, p. 9, Elsevier, Amsterdam, 2004.
[6] Hightower JW, Emmett PH, *J Am Chem Soc* **87**: 939, 1965.
[7] Hightower JW, Gerberich HR, Hall WK, *J Catal* **7**: 57, 1967.
[8] Hightower JW, Hall WK, *J Phys Chem* **71**: 1014, 1967.
[9] Balandin AA, *Z Phys Chem B* **2**: 3124, 1929; Balandin AA, *Adv Catal* **10**: 96, 1958.
[10] Tétényi P, Babernics L, Thomson SJ, *Acta Chim Hung* **34**: 335, 1963.
[11] Tétényi P, Babernics L, Schächter K, *Acta Chim Hung* **58**: 321, 1968.
[12] Tétényi P, Guczi L, Paál Z, Babernics L, *Metal Catalyzed Heterogeneous Reactions of Hydrocarbons*, Akadémiai Kiadó, Budapest, 1974.
[13] Derbentsev Yu I, Paál Z, Tétényi P, *Z Phys Chem Neue Folge* **80**: 51, 1972; Tétényi P, Paál Z, *Z Phys Chem Neue Folge* **80**: 63, 1972.

[14] Derbentsev Yu I, Balandin AA, GV Isagulyants, *Kinet Katal* **2**: 741, 1961.
[15] Tétényi P, Paál Z, Dobrovolszky M, *Z Phys Chem Neue Folge* **102**: 267, 1976.
[16] Satterfield CN, *Heterogeneous Catalysis in Industrial Practice*, McGraw Hill, NewYork, 1991.
[17] Sharan KM, *Catal Rev — Sci Eng* **26**: 141, 1984.
[18] Davis BH, *Catal Today* **53**: 443, 1999.
[19] Paál Z, *Adv Catal* **29**: 273, 1980.
[20] Paál Z, Tétényi P, *Acta Chim Acad Sci Hung* **54**: 175, 1967.
[21] Paál Z, in *Encyclopedia of Catalysis* (Horvath IT, ed.) Wiley, New York, 2003, Vol. 3, p. 9.
[22] Isagulyants GV, Bragin OV, Greish AA, Liberman AL, *Izv Akad Nauk SSSR Ser Khim* p. 820, 1976.
[23] Isagulyants GV, Rozengart MI, Kazanskii BA, Derbentsev Yu I, Dubinskii Yu G, *Dokl Akad Nauk SSSR* **191**: 600, 1970.
[24] Kazanskii BA, Rozengart MI, *Dokl Akad Nauk SSSR* **197**: 1085, 1971.
[25] Shi B, Davis BH, *J Catal* **147**: 38, 1994.
[26] Rozengart MI, Mortikov ES, Kazanskii BA, *Dokl Akad Nauk SSSR* **158**: 911, 1964; Rozengart MI, Krimond T Ya, Kazanskii Ba, *Dokl Akad Nauk SSSR* **199**: 365, 1971.
[27] Paál Z, Rozengart MI, *Acta Chim Acad Sci Hung* **49**: 395, 1966.
[28] Paál Z, Tétényi P, *Acta Chim Acad Sci Hung* **53**: 193, 1967.
[29] Paál Z, Tétényi P, *Acta Chim Acad Sci Hung* **58**: 105, 1968.
[30] Paál Z, Tétényi P, *J Catal* **30**: 350, 1973.
[31] Ponec V, Bond GC, *Catalysis by Metals and Alloys*, Elsevier, Amsterdam, 1995.
[32] Paál Z, in *Catalytic Naphtha Reforming: Science and Technology* (Antos GJ, Aitani AM, eds.) 2nd, revised edition, Marcel Dekker, New York, 2004.
[33] Gál D, Danóczy E, Nemes I, Vidóczy T, Hajdu P, *Ann NY Acad Sci* **213**: 51, 1973.
[34] Zimmer H, Rozanov VV, Sklyarov AV, Paál Z, *Appl Catal* **2**: 51, 1982.
[35] Margitfalvi JL, Szedlacsek P, Hegedüs M, Tálas E, Nagy F, *Proceedings of the 9th International Congress on Catalysis, Calgary*, Vol. 3, p. 1283, 1988.
[36] Manninger I, Xu XL, Tétényi P, Paál Z, *Appl Catal* **51**: L7, 1989.
[37] Wootsch A, Pirault-Roy L, Leverd J, Guérin M, Paál Z, *J Catal* **208**: 490, 2002.
[38] Barron Y, Maire G, Muller JM, Gault FG, *J Catal* **5**: 428, 1966; Gault FG, *Adv Catal* **30**: 1, 1981.
[39] Anderson JR, Baker BG, *Nature* **187**: 937, 1960; Anderson JR, Avery NR, *J Catal* **5**: 446, 1966.
[40] de Jongste HC, Ponec V, *Bull Soc Chim Belg* **88**: 453, 1979.
[41] Zimmer H, Dobrovolszky M, Tétényi P, Paál Z, *J Phys Chem* **90**: 4758, 1986.
[42] Sárkány A, *J Catal* **105**: 65, 1987.

[43] Sárkány A, *J Catal* **89**: 14, 1984.
[44] O'Donohoe C, Clarke JKA, Rooney JJ, *J Chem Soc Chem Commun* 648, 1979; *J Chem Soc Faraday Trans I* **76**: 345, 1980.
[45] Sárkány A, Tétényi P, *J Chem Soc Chem Commun* 535, 1980.
[46] Paál Z, Dobrovolszky M, Tétényi P, *J Chem Soc Faraday Trans 1* **80**: 3037, 1984.
[47] Paál Z, Tétényi P, *Nature* **267**: 234, 1977.
[48] Van Schaik JRH, Dessing RP, Ponec V, *J Catal* **38**: 273, 1975.
[49] Paál Z, Tétényi P, *React Kinet Catal Lett* **12**: 131, 1979.
[50] Paál Z, Tétényi P, *React Kinet Catal Lett* **7**: 39, 1977.
[51] Sárkány A, Pálfi S, Tétényi P, *Acta Chim Hung* **111**: 633, 1982.
[52] Guczi L, Kálmán J, Matusek K, *React Kinet Catal Lett* **1**: 51, 1974; Guczi L, Matusek K, Tétényi P, *React Kinet Catal Lett* **1**: 291, 1974.
[53] Margitfalvi J, Hegedüs M, Göbölös S, Kwaysser E, Koltai L, Nagy F, *Acta Chim Hung* **111**: 573, 1982.
[54] Swift HE, Bozik JE, *J Catal* **12**: 5, 1968.
[55] Tétényi P, Schächter K, *Acta Chim Hung* **65**: 253, 1970.
[56] Paál Z, Péter A, Tétényi P, *React Kinet Catal Lett* **1**: 121, 1974.
[57] Manninger I, Péter A, Paál Z, Tétényi P, *Kémiai Közlemények* **44**: 35, 1975 (in Hungarian).
[58] Manninger I, Paál Z, Tétényi P, *J Catal* **48**: 442, 1977.
[59] Dobrovolszky M, Paál Z, Tétényi P, *J Catal* **74**: 31, 1982.
[60] Kagan Yu B, Rozovskii A Ya, Liberov LG, Slivinskii EV, Lin GI, Loktev SM, Bashkirov AN, *Dokl Akad Nauk SSSR* **221**: 1093, 1975.
[61] Rozovskii A Ya, *Kinet Katal* **21**: 97, 1980.
[62] Tétényi P, Paál Z, *Z Phys Chem Neue Folge* **80**: 63, 1971.
[63] Rupprechter G, Somorjai GA, *J Phys Chem B* **103**: 1623, 1999.
[64] Neiman MB, Gál D, *The Kinetic Isotope Method and Its Applications*, Akadémiai Kiadó, Budapest — Elsevier, Amsterdam, 1971.
[65] Bauer F, Dermietzel J, *Isotopenpraxis* **14**: 300, 1978.
[66] Dermietzel J, Bauer F, Rösseler F, Jockisch W, Franke H, Klempin J, Barz HJ, *Isotopenpraxis* **12**: 57, 1976.
[67] Isagulyants GV, Komarova EN, Balandin AA, *Dokl AN SSSR* **163**: 1607, 1965.
[68] Mitchell JJ, *J Am Chem Soc* **80**: 5848, 1958.
[69] Cannings FR, Fisher A, Ford JF, Holmes PD, Smith RS, *Chem Ind* 228, 1960.
[70] Pines H, Chen CT, *2nd International Congress on Catalysis*, Paris, 1960, Vol. 1, p. 361.
[71] Pines H, Chen CT, *J Org Chem* **26**: 1057, 1961.
[72] Csicsery SM, Pines H, *J Am Chem Soc* **84**: 3939, 1962.
[73] Csicsery SM, Goetschel CT, Pines H, *J Org Chem* **28**: 2713, 1963.
[74] Pines H, Goetschel CT, *J Org Chem* **30**: 3530, 1965.
[75] Nogueira L, Pines H, *J Catal* **70**: 404, 1981.
[76] Feighan JA, Davis BH, *J Catal* **4**: 594, 1965.
[77] Davis BH, Venuto PB, *J Org Chem* **46**: 337, 1971.

[78] Davis BH, *J Catal* **29**: 398, 1973.
[79] Davis BH, *J Catal* **46**: 348, 1977.
[80] Csicsery S, Burnett RL, *J Catal* **8**: 74, 1967.
[81] Huang C-S, Sparks DE, Dabbagh HA, Davis BH, *J Catal* **134**: 269, 1992.
[82] Pines H, Shaw AW, *J Am Chem Soc* **79**: 1474, 1959.
[83] Wetzel K, Dermietzel J, Bauer F, Jockisch W, Rösseler F, Wienhold Ch, Klempin J, Franke H, Thomas W, *Isotopenpraxis* **17**: 50, 1981.
[84] Kopinke F-D, Dermietzel J, Jockisch W, Räuber G, *Isotopenpraxis* **22**: 388, 1986.
[85] Eidus Ya T, *Uspekhi Khimii* **36**: 338, 1967.
[86] Raje A, Davis BH, in *Catalysis Specialists Periodical Reports* (Spivey J, ed.), The Royal Society of Chemistry, London, 1996, Vol. 12, p. 52.
[87] Schulz H, *Topics Catal* **26**: 73, 2003.
[88] Kummer JT, DeWitt TW, Emmett PH, *J Am Chem Soc* **70**: 3632, 1948.
[89] Tau L-M, Dabbagh H, Bao S, Davis BH, *Catal Lett* **7**: 127, 1990.
[90] Tau L-M, Dabbagh H, Chavla B, Davis BH, *Catal Lett* **7**: 141, 1990.
[91] Pichler H, Schulz H, *Chemie-Ing-Techn* **42**: 1162, 1970.
[92] Schulz H, Erdöl, *Kohle–Erdgas–Petrochem* **30**: 123, 1977.
[93] Golovina OA, Dokunina S, Roginskii SZ, Sakhharov MM, Eidus Ya T, *Dokl AN SSSR* **112**: 864, 1957.
[94] Kokes RJ, Hall WK, Emmett PH, *J Am Chem Soc* **79**: 2289, 1957.
[95] Golovina OA, Roginskii SZ, Sakharov MM, Eidus Ya T, *Dokl AN SSSR* **108**: 864, 1956.
[96] Tau L-M, Dabbagh H, Davis BH, *Energy and Fuels* **5**: 174, 1991.
[97] Tau L-M, Dabbagh H, Halász J, Davis BH, *J Mol Catal* **71**: 37, 1992.
[98] Shi B, Davis BH, *Catal Today* **26**: 157, 2003.
[99] Blyholder G, Emmett PH, *J Phys Chem* **63**: 962, 1959; Blyholder G, Emmett PH, *J Phys Chem* **64**: 470, 1960.
[100] Roginskii SZ, Golovina OA, Dokukina ES, Roginskii SZ, Sakharov MM, Eidus Ya T, *Dokl AN SSSR* **112**: 864, 1957.
[101] Xu L, Bao S, Houpt DJ, Davis BH, *Topics Catal* **36**: 347, 1997.
[102] Shi B, Davis BH, *Appl Catal A: General* **277**: 61, 2004.
[103] Davis BH, *Fuel Proc Technol* **71**: 157, 2001.

[78] Dalla [illegible] RH, J Catal 29: 298, 1973.
[79] Dalla [illegible] J Catal 48: 348, 1977.
[80] [illegible] J Catal 8: 74, 1967.
[81] [illegible] J Catal [illegible] 293 [illegible]
[82] [illegible] J Am Chem Soc [illegible] 1954.
[83] [illegible]
[illegible] 17 [illegible]
[84] [illegible] 22 [illegible]
1986.
[85] [illegible] Acta Chem [illegible] 1985.
[86] [illegible]
[illegible] Vol. [illegible]
[87] [illegible] Chem [illegible] 2004.
[88] [illegible] 1995.
[89] [illegible]
[90] [illegible] 1990.
[91] [illegible]
[92] [illegible]
[93] [illegible]
[illegible]
[94] [illegible]
[95] [illegible]
108 [illegible]
[96] [illegible]
[97] [illegible]
[98] [illegible]
[99] [illegible]
[illegible]
[100] [illegible]
[illegible]
[101] [illegible]
[102] [illegible]
[103] [illegible]

Chapter 3

Use of ^{35}S Radiotracer in Catalytic Studies

P Tétényi

Introduction

Sulfur in catalysis is an object of outstanding interest both as atom(s) in compounds being transformed catalytically, and as a part of catalysts applied in chemical transformations, especially in hydrotreatment processes. Catalytic methods — both homogenous and heterogeneous — of sulfuric acid production were known before the concept of catalysis was formulated; the heterogeneous SO_2-SO_3 oxidation was one of the phenomena leading Berzelius to introduce the concept and the name of catalysis.

The role of sulfur in catalysis studies can be compared with that of oxygen, both being converted in the form of compounds and also existing as elements of oxide or sulfide catalysts. The advantage of the sulfur isotope is its existence in radioactive form, with a low energy of β-radiation. This offers the possibility to apply it in very low concentrations and practically excludes any role of the isotopic effect. Observing the basic rules in radioisotope practice, the studies can be carried out in an ordinary chemistry laboratory. Being a low-energy β-emitter, such as ^{14}C and tritium (^{3}H), identical radioactivity measurement techniques can be applied for these radioisotopes. This makes it possible for some parallel studies to be carried out, as in the case of ^{35}S and ^{3}H.[1,2]

Different from ^{14}C-tracer studies, these being connected mostly with reaction mechanism problems, sulfur isotopes are connected with applications to the role of sulfur in solid catalysts to a great extent. This is connected with the wide application of sulfide catalysts. Metal-sulfur bond strengths are substantially ($\sim$100 kJ/mol) larger than those of the metal-oxygen bonds, and as a consequence sulfur compounds often poison metal catalysts. Sulfur is present in metals in different forms — as sulfites, sulfates and oxysulfites. All these specifities explain why studies with radiosulfur are targeted to the role of sulfur in catalysts.

This is well demonstrated in one of the earliest studies[3] contributed to a Conference in Moscow in 1956 entitled *Isotopes in Catalysis*. The study was directed to the behaviour of some metal sulfides under the conditions of CH_3OH conversion. ZnS, CdS and PbS were labelled with ^{35}S. Besides ^{35}S, some $^{35}SO_4^{2-}$ was also added to the catalysts. The authors found substantial differences in the catalyst sulfur behaviour. The radioactivity of ZnS and CdS was reduced to a minimal extent, PbS dissociated to Pb and SO_2. ZnS and CdS were stable in iso-C_3H_7OH conversion at low temperature, whereas at high temperature, they convert to oxides, *viz*

$$ZnS + 3ZnSO_4 = 4ZnO + 4SO_2 \quad (1)$$

This early study indicated the wide spectrum of possibilities to study the chemical behaviour of a sulfided catalysts in reaction conditions with tracers.

Tracers were not limited however to catalyst sulfur; catalysed sulfur exchange reactions were studied too. An early study of sulfur exchange was also preformed in the liquid phase at 273 K, catalysed by chlorides, namely RbCl and $AlCl_3$.[4] A much higher activity of rubidium chloride in comparison with that of $AlCl_3$, has been reported.

Another direction worth mentioning, was the study of the detailed kinetics of oxidation by labelled sulfur. It has been stated[5] that the stoichiometric number[5,6] was equal to 2, for the rate determining step of catalytic SO_2 oxidation to SO_3. Identical values for the stoichiometric number of this reaction[7,8] indicated the reliability of the method.

A short paper, reporting on the results of a basic study, was published in 1962 on the "... isotopic exchange method for measuring the surface area of supported transition metal sulfides".[9] The aim of the study was to determine the fraction of the surface area, contributed by the catalytically active part of the surface. The original aim, to determine the active area was not achieved; comparison with crystallographic data indicated large discrepancies, but "... it routinely has provided information... " on the area available for reaction in supported Ni, Mo, W and mixed sulfide samples. However, the method has been recently widely applied in determining the amount of exchangeable sulfur in or on different catalysts.

Numerous examples indicate that labelled sulfur can be applied both for studying reaction mechanism and changes in the state of the catalyst during the catalytic conversions. Nevertheless, radiosulfur was not applied widely. For example, it is surprising that no original contribution was presented with this tracer on the meeting held in New York in September 1972, bearing the title *The Use of Tracers to Study Heterogeneous Catalysis*. ^{35}S was mentioned only in Horiuti's survey[10] and some references were given by Gaál[11] to the studies mentioned before,[5–8] with respect to the stoichiometric number of SO_2 oxidation.

Strict limitations, introduced in the 1970s, with respect to sulfur emission in exhaust gases, focused attention on catalytic hydrodesulfurization (HDS). In this process, sulfur containing organic compounds are removed from hydrocarbon feedstocks with the aim to produce environmentally acceptable liquid fuels.[12] This hydrotreating process became an integral part of oil refining.[13] The conventional catalysts for the industrial hydrotreatment process are *sulfided* molybdenum and tungsten, promoted by cobalt and nickel. These are sulfided both before (pre-sulfidation) and during the catalytic conversion, where the catalyst sulfur is exchanged with the sulfur of organic origin contained in the hydrocarbon feedstocks. The important role of catalyst sulfur in catalysts' activity and selectivity became apparent. As such, fundamental studies became necessary to clarify the role and behaviour of sulfur on working catalysts.

Different from the former studies directed towards "Sulfur Poisoning of Metals",[14] studies on the "Role of Sulfur in Catalytic Hydrogenation Reactions"[15] and in hydrodesulfurization emerged.

The main problems which arise with respect to the role of sulfur in HDS are:

- Correlation between sulfur uptake and HDS activity, including the role of the reversible/irreversible sulfur uptake ratio;
- Correlation between the amount of exchangeable sulfur (S_{exc}) in and HDS activity of the catalyst;
- The behaviour of catalyst sulfur during the catalytic act.

Hydrogen sulfide and organic compounds, labelled with the sulfur isotope, appeared to be a useful tool providing solutions to these problems.[16] This explains the substantial increase of studies with isotope-labelled sulfur from the 1980s.

1. Catalyst Sulfur Uptake

1.1. *Single crystals*

Radiosulfur is a highly applicable tool for determining the adsorbed amounts on samples with extremely low specific surface areas, such as single crystals, applied in surface chemistry studies. A successful combination of LEED and AES studies with ^{35}S tracer measurements, allowed the differences in adsorbed sulfur amounts to be determined on different platinum and nickel faces[17,18] at extremely low (10^{-3}–10^{-5} Pa) H_2S pressure at different adsorption states. It has been stated before[12] that at coverage, $\theta^- =$ 0.42, sulfur represented a saturated layer on Pt (111) at high (~623 K) temperature. The sulfur coverage was higher with $\theta = 0.59$ at lower (~473 K) temperature, as some weakly bonded sulfur desorbed at high temperature.

The maximal concentrations of adsorbed sulfur on different metals[18] preceding the appearance of solid sulfide were determined. It has been stated that in the case of both platinum and nickel, the saturation states are different for different crystal faces, i.e., the metal: S atomic ratios are ~2 on |100| and slightly lower than 1:2 on the |111| face. On the |110| face of Ni and Pt, the ratios are 1.41 and 1.23 respectively.

These data indicate that sulfur preferentially adsorbs on sites of lowest coordination. Different types of sulfur adsorption were also observed by the Berkeley group on Mo|100| single metal surfaces.[19] ^{35}S was deposited on the Mo-surface by decomposition of $C^{35}S_2$, whereas carbon is dissolved into the bulk of the sample by heating. It was found that sulfur, adsorbed at coverages >0.67 is removed from the surface easily, whereas sulfur adsorbed at lower coverages remains on the surface permanently. This was explained by the differences in the adsorption structure: at $\theta_S \leq 0.67$, sulfur is adsorbed on the "fourfold hollow sites"; at higher coverages, sulfur is adsorbed on a second, less tightly binding, site. Addition of thiophene increases dramatically the sulfur reduction (Fig. 1). This may be caused by a change in the metal-sulfur bonding due to the presence of co-adsorbed thiophene. Another possible explanation may be due to the high amount of surface H_2S produced in thiophene HDS, with 5000 times higher turnover frequency (TOF), in comparison with that of sulfur hydrogenation, displacing the pre-adsorbed sulfur of both adsorption structures.

Sulfur coverage resulted in a decrease in HDS activity of Mo and the catalyst lost its HDS activity after prolonged (~12 hr) thiophene treatment. In the author's opinion, this is a consequence of MoS_2 overlayer formation.

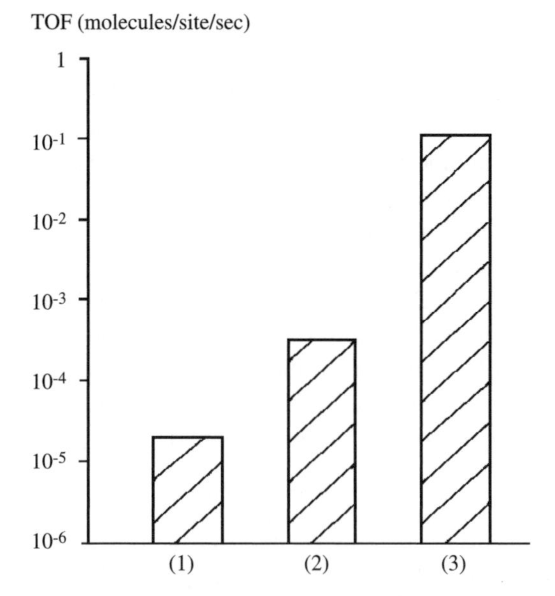

Fig. 1. Relative rates of thiophene HDS. (1) Sulfur hydrogenation in a thiophene/H_2 atmosphere; (2) On the Mo (100) surface; (3) Sulfur hydrogenation in a pure H_2 environment; $P(H_2) = 1.04\,kPa$; $P(Th) = 333\,Pa$; $T = 613\,K$. *Reproduced from* Ref. 19 with permission.

This contradicts, however, the HDS activity of sulfides including MoS_2, observed by a high number of authors (see e.g., Refs. 20–25) long before this study on molybdenum single crystals. The loss of activity upon the long thiophene treatment was possibly a consequence of the formation of carbonaceous deposits, the by-products of thiophene HDS.

Summing up, deposition studies on single crystal with radiosulfur indicated that the sulfur uptake and the metal:sulfur ratio is different on crystal planes with different indices, and a part of the adsorbed sulfur is easily removable, whereas another part is held more strongly by the surface metal atoms.

1.2. *Non-supported catalysts*

One of the first radiotracer studies on sulfur uptake was that on a molybdenum disulfide catalyst.[26] MoS_2 was exposed to $[^{35}S]H_2S$, admitted at 623 K into the evacuated vessel containing the catalyst. ($[^{35}S]H_2S$ in the following will be denoted for simplicity by $H_2{}^{35}S$, indicating ^{35}S labelled, but not carrier free radioactive H_2S). $8.70 \times 10^{16}\,H_2\,S/mg$, i.e., 2.3% of the MoS_2 content of the sample was taken up by the catalyst. 26% of the adsorbed H_2S was dissociated, as calculated from the H_2 amount released into the gas phase, i.e., $2.26 \times 10^{16}\,S/mg$ was taken up in the form of S^{2-}. This equals to the number of M^{4+} ions/mg, assuming that the BET surface area of the sample ($4.7\,m^2/g$) was equal to the sum of the surface S^{2-} ions ($9.5 \times 10^{-2}\,nm^2$). This indicates a $S^{2-}_{upt}/Mo_{surf} = 1$.

Hydrogen pretreatment of the MoS_2 resulted in a H_2 uptake of $\sim 6.5 \times 10^{16}\,H/mg$ i.e., a ~1:4 H:(S_{surf} + Mo_{surf}) ratio. It is not known how far this uptake was irreversible. The H_2S uptake became strongly reduced in that it was only 1.98×10^{16} molecules of H_2S/mg, i.e., about 23% of the value at saturation of the fresh sample. This was explained by the authors with the shift of equilibrium of hydrogen sulfide adsorption

$$H_2S \rightleftharpoons H^{\delta+} + SH^{\delta-} \tag{2}$$

and indirect evidence for heterolytic dissociative H_2S adsorption, indicated by other authors before[27,28] and supported later.[29]

The effect of H_2S-uptake on MoS_2 selectivity in butadiene hydrogenation was also monitored [Fig. 2.] An increase of 2-butenes and a decrease of 1-butene was observed at $\sim 1.2 - 1.3 \times 10^{19}\,mol \cdot g^{-1}$ uptake of H_2S. The character of selectivity changes upon H_2S exposure was similar in the case of samples treated previously with H_2, however the 1-butene → 2 butenes selectivity transition occurred at lower H_2S uptakes. Substantial changes at increasing H_2S uptake are also observed in the trans/cis 2-butene ratio: H_2S treatment of MoS_2 resulting in its increase from 1.2–1.5 up to 2–2.5. A similar effect was observed at treatment by thiophene, due to the effect of H_2S formed in thiophene HDS. These selectivity changes upon H_2S exposure were explained by the reduction of the concentration of coordinatively unsaturated sites (CUS) on the basis of a model introduced earlier.[30–32]

Different types of H_2S uptake were also observed in pulse-system studies on non-supported MoO_x and $CoMoO_x$.[33] Pulses of thiophene and $H_2{}^{35}S$ were injected in a sequence of thiophene-$H_2{}^{35}S$-thiophene. The method is detailed in Sec. 1.3. The total amount of retained hydrogen sulfide (S_T) was found to be ~16% of the number of Mo-atoms, in MoO_x, whereas about 8% was strongly held, not eluted either by H_2 or by thiophene — sulfur. This was substantially higher than the uptake at substantially lower (~1.2 kPa) H_2S pressure reported in Ref. 26. The H_2S uptake of the Co-promoted molybdenum oxide ($CoMoO_x$) was substantially higher: 2×10^{18} molecules/mg. i.e., the degree of sulfiding by strongly held sulfur was

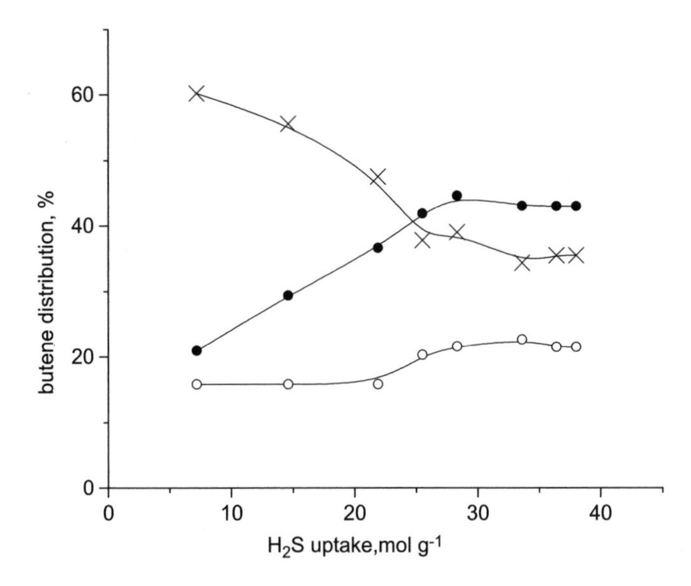

Fig. 2. Butene product distribution of hydrogen sulfide uptake: x, but-1-ene; •, trans-but-2-ene; o, cis-but-2-ene. *Adapted from* Ref. 26.

~32%, if MoS_2 and Co_9S_8 are assumed for complete sulfiding. Formation of strongly bound sulfur was the highest during the first thiophene pulse, and its amount became constant after several pulses. The thiophene HDS activity of $CoMoO_x$ samples decreased strongly upon increasing sulfur uptake. This was explained by the formation of MoS_2-layers indicated by electron microscopy.[34]

Non-supported MoS_2 was also prepared in ^{35}S labelled form by sulfiding molybdenum oxide with labelled elemental sulfur at high pressure.[35] The degree of sulfiding was 52.5%, substantially higher than that in pulse system at sulfidation with H_2S. A small ratio (~2%) of the sulfur was exchangeable, as witnessed by the specific radioactivity of H_2S formed in HDS of thiophene.[36] The uptake is increased by the addition of Co promoter to the MoO_x sample. It is seen that the amount of sulfur uptake and the ratio of the different forms of sulfur in unsupported MoO_x samples, depend strongly on the conditions of the process.

1.3. *Supported catalysts*

Hydrotreatment processes, including hydrodesulfurization, are carried out in practice on supported catalysts, mostly on alumina or silica supports, on noble metals and mainly on molybdenum and tungsten oxides, promoted with different Group 8–10 metals. Studies of interest are also performed with TiO_2, being active in HDS.[37,38] Several groups used radiosulfur to investigate the sulfur uptake.

The sulfidation of an alumina supported $CoMoO_x$ catalyst and labelling with elemental ^{35}S has been studied by the Moscow research group.[36,39] The process was carried out at 6 MPa in an H_2 atmosphere. Hydrogen sulfide formed during the process was deposited in a $(CH_3COO)_2Cd$ solution. Catalyst granules were boiled in benzene for the elution of non-reacted sulfur. Dispersion of sulfur in the catalyst pellets was homogenous, as determined by the radioactivity of the cuts taken from the pellets. The maximal sulfur uptake (6.7 wt%) was near the value given by MoS_2 and Co_9S_8 stoichiometry for granules of 8.5×10^{-2} g. The amounts taken up decrease with increasing mass: the S-uptake was only ~5% by large granules with mass of 0.12 g. The sulfidation kinetics is illustrated in Fig. 3. It is seen that the maximal uptake amounts are reached for 60 minutes. These values are near to equal in the presence of H_2 in temperature interval of 527–633 K, and lower at 453 K and in the presence of N_2. The sulfidation rates increase with increasing temperature.

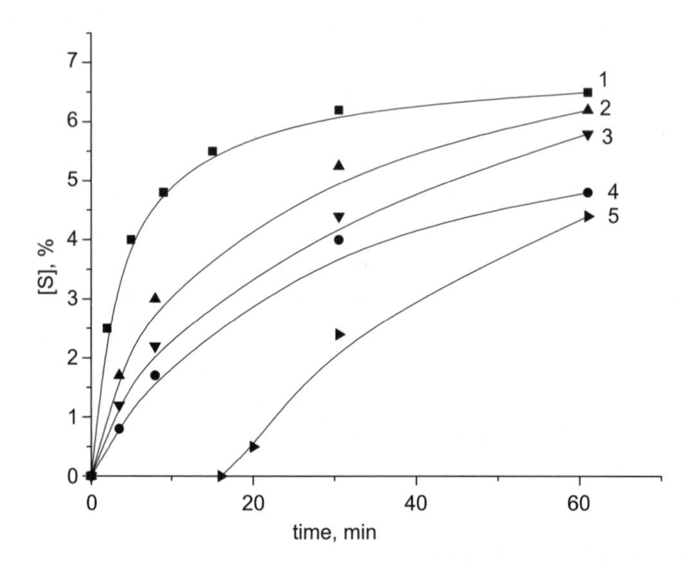

Fig. 3. Sulfidation kinetics of a $CoMoO_x/Al_2O_3$ catalyst in presence of H_2 at (1) 633 K, (2) 588 K, (3) 527 K, (5) at 453 K and (4) in presence of N_2 at 633 K. *Reproduced from* Ref. 36 with permission.

Sulfidation by $H_2{}^{35}S$, represented in Refs. 26, 40–44 is a widely used method for determining the sulfur uptake of catalysts, based on the radioactivity balance in between the introduced $H_2{}^{35}S/H_2$ and the outflow-, or line-out gases[45]:

$$Y_T = m_{so}(1 - I_1/I_o) = S_T \tag{3}$$

where Y_T = the total S uptake S_T; m_{so} = the total amount of H_2S in the initial mixture injected (pulse regime) or in the gas phase at the start of the run (circulation regime); I_o = the radioactivity of the initial mixture; I_1 = the radioactivity in the outflow-gases, (pulse regime), or at equilibrium, i.e., constant final gas phase radioactivity (circulation regime). Mobile or reversibly (S_{mob} or S_{rev}) and strongly or irreversibly held sulfur (S_{cat} or S_{irr}) were distinguished:

$$S_T = S_{irr} + S_{rev} \tag{4}$$

S_{mob} is determined in the pulse-flow system from the radioactivity of fraction containing the reversibly taken up and the desorbed, by flowing gas (e.g., by H_2 or N_2) of I_3 radioactivity collected between injections of two hydrogen sulfide pulses and calculated by the expression:

$$S_{mob} = m_{so}(I_3/I_0) \tag{5}$$

In case of the circulation system, the vessel is evacuated for the removal of S_{rev}. Introduction of the $H_2{}^{35}S/H_2$ mixture of initial radioactivity is repeated, and S_{rev} is calculated from the radioactivity loss (41) by expression (3), applying the radioactivity, I_2, at line out. The S_T value in this run does not contain S_{irr}, as the catalyst was sulfided in Run 1. No S_{exc} is measured either, as the gas phase ^{35}S specific radioactivity is equivalent to that of sulfur in the S_{cat}.

Expression (4) is also applicable for calculation of actual sulfur exchange ($_{a}S_{exc}$) between ^{35}S-labelled catalyst and non-labelled (non-radioactive) H_2S circulated over the catalyst, until constant radioactivity (I_3) is reached.

The reliability of this method for sulfur uptake measurements was confirmed[42] by parallel oxidation experiments with the sulfided samples and the determination of their sulfur content from the radioactivity of oxides (SO_2 and SO_3 in H_2SO_4). The mean arithmetic error of the two series (with 20 experiments each) was ±10%. Comparison of S-uptakes measured by radioactivity with those of XPS-intensities, indicates a similar ratio,[44] or sequence[45] of S-uptakes, confirming the applicability of the method.

The S-uptake by the alumina and silica supports was zero or minimal in comparison with that of the supported MoO_x,[42] $NiMoO_x$, and WO_x.[46] Sulfur in Pt/Al_2O_3 catalysts is reversibly bonded to the support,[41] or as was found[45] with Pd- and $PtMoO_x/Al_2O_3$ separate bulk phases of Pd and Pt, containing S, are present.

The degree of near to total sulfidation (~90–100%), is reached for Co- and Ni-promoted MoO_x/Al_2O_3 both in pulse systems[42,47] at high (~1–5 MPa) H_2S pressure and in circulation systems from substantially lower, (6–7 kPa) H_2S pressure, irrespective of the H_2:H_2S ratio.[48] The sulfidation degree of non-promoted molybdena depends substantially on conditions: $MoS_{2.2}/Al_2O_3$ was observed at high pressure,[42] whereas only $MoS_{0.9}$ was found at 4–27 kPa in the circulation system,[48] and the S/Mo ratio in pulse system at 2 kPa was only 0.1.[44] 50–70% of sulfidation degrees were observed for Pd- and $PtMoO_x/Al_2O_3$, and for $NiWO_x$ at 4–27 kPa H_2S.

Using $H_2{}^{35}S/H_2$ pulses, the sulfidation of TiO_2 catalyst samples of different (18, 70 and 120 m^2/g) specific surface area has been studied.[37] A direct relationship was observed between the surface area of the samples and the amounts of sulfur uptake by them. The amounts of sulfur accumulated on the catalyst increased with increasing temperature. The maximal sulfidation degree at 773 K was 4% for the sample of the lowest, and 16.6% for that of the highest surface area. The surface was covered with S^{2-} ions, and approximately half of the sulfur was incorporated into the bulk. Sulfidation of MoO_x/TiO_2 samples with $H_2{}^{35}S/H_2$ pulses[38] resulted in a lower maximal S-uptake, in comparison with that measured on TiO_2, as a substantial part of the support was covered by molybdena.

In all studies, increase of S_{cat} was observed with increasing temperature up to 673 K. It was stated by the Tokyo Group (T Kabe and colleagues) that the maximal S_{cat} was approached at 473 K on the non-promoted molybdena, whereas the maximal uptake by Co- and Ni-promoted MoO_x/Al_2O was reached only at 573 K and 673 K.[42,47] It was concluded that molybdenum oxide was preferably sulfided in the lower temperature interval, whereas higher temperatures were necessary for sulfiding nickel and cobalt species.

Results of some studies indicate[48–50] that the maximal degree of sulfidation depends strongly on the preparation and pretreatment method, i.e., 52% and 86% for two Co-promoted, alumina supported catalysts of near to equal chemical composition, and 80% and 100% for two nickel promoted samples.[49] A further example of the effect of the preparation method is seen with chromia-alumina.[51] Sulfidation by $H_2{}^{35}S$ pulses resulted in an increasing S-uptake up to 673 K by forming CrS_x-species. In the 573–673 K region, the S-uptakes decreased ~25% due to the formation of Cr_2S_3 entities in amounts depending on the Cr-content of the sample. It is noteworthy that both the S-uptake and the Cr_2S_3 ratio was substantially higher on the sample sulfided directly at 673 K, rather than on the stepwise sulfided one.

Besides sulfidation, reversible S-uptake was also observed in all $H_2{}^{35}S$ uptake experiments. Its amount, in contrast to the irreversible one, decreased with increasing temperature,[42,44] and reached a constant value at higher (25–45 kPa) H_2S pressure, than the irreversible one.[48]

Besides H_2S, sulfur uptake studies were carried out with ^{35}S labelled organic compounds, mostly with thiophene and dibenzothiophene (DBT). In these methods, the sulfur uptake is determined from the balance of radioactivity-loss of the S-labelled compound and the radioactivity of $H_2{}^{35}S$ produced in HDS. It was established[36] that $CoMo/Al_2O_3$ samples, sulfided with ^{35}S-labelled thiophene pulses, contained sulfur about 2.5 times less than the same catalyst, presulfided with elemental sulfur at high pressure. S-uptake, determined by this method, represents uptake by active in HDS sites; those sites that are non-participating in the reaction are involved due to the surface migration of sulfur.

Experiments with ^{35}S labelled thiophene indicated[35] that part of sulfur in cobalt promoted, previously sulfided molybdena-alumina is replaced by thiophene sulfur.

Comparison of irreversible S-uptake from $H_2{}^{35}S$ pulses by samples exposed to thiophene with the non-exposed ones supports this conclusion: the irreversible uptakes by thiophene treated Pd-, Ir-, Pt-promoted and non-promoted MoO_x samples were about 50–70% of those by the non-exposed samples.[40,44]

The S-uptake by a non-presulfided MoO_3 catalyst was also determined[52,53] in a pressurized fixed bed flow reactor, by injecting pulses of a 1 wt% decalin solution of ^{35}S-labelled dibenzothiophene (^{35}S-DBT), up to the steady state radioactivity of $H_2{}^{35}S$ formed due to HDS of DBT.[53] This took a substantial period of time, as sulfur (SH) formed initially on catalytic sites, or part of it, accommodated on other S-vacancies. The steady state gas phase $H_2{}^{35}S$ radioactivity is reached after the equilibrium between the formation of S-vacancies (due to H_2S release), and their occupation (by S-attachment) is approached. The S/Mo ratio from the amount of sulfur,

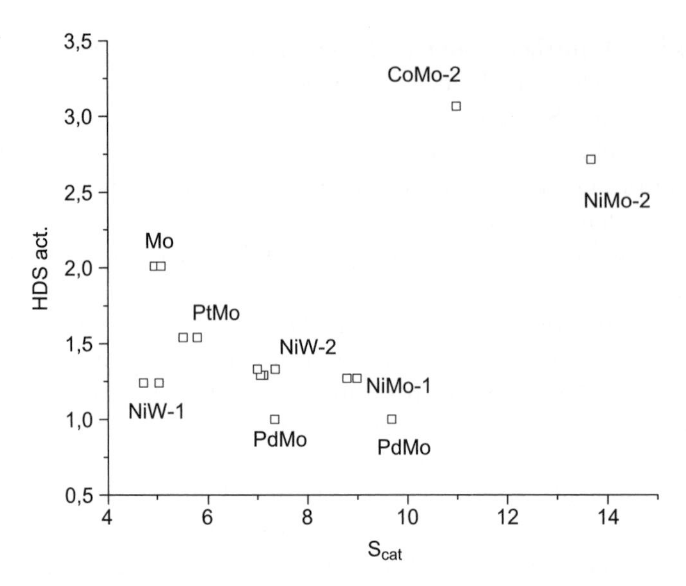

Fig. 4. HDS rates (10^{17} mol./mg.s) versus S-uptake (10^{17} S_{at}/mg). *Compiled from data in* Refs. 43, 48 and 50.

accommodated on the catalyst, calculated by Eq. (3) equals to 1.92 at 633 K, near to the value determined from $H_2{}^{35}S$ data for MoO_x/Al_2O_3.[42] The S-uptake was irreversible as no $H_2{}^{35}S$ was produced at treatment with H_2. The method was applied for alumina supported 2 wt% Pd and Pt, where the S/metal ratio was ~1.88 and 3.04 respectively. Taking into consideration the S-uptake by alumina, 1.6 S/Pd(Pt) is calculated at 533 K and 5 MPa.[54]

Comparison of the extent of sulfur uptake by different catalysts with their HDS activity indicates the absence of a definite correlation between them. Sulfidation of some noble metal promoted molybdenum oxide catalysts results even in a decreased HDS activity.[40] The absence of such correlation is demonstrated in Fig. 4. It follows from this that the character and the properties of sulfur in sulfided catalysts should play a decisive role in the behaviour of the HDS catalysts.

2. Catalysts' Sulfur Mobility

Results of sulfur uptake studies, detailed in the preceding section, indicate the heterogeneous character of sulfur bonding to different solids, such as single crystals, non-supported and supported samples. There exists a great variety of sulfur bonding to solids such as surface and bulk sulfur ions, reversibly and irreversibly adsorbed sulfur species, adsorbed as H_2S, $SH^{\delta+}$ and $SH^{\delta-}$. The strength of bonds between the reacting atoms and the catalyst surface atoms, calculated from the activation energies of some catalytic reactions,[55,56] indicated near to equal bonding energies for Mo- and W-oxides and sulfides. This implies a direct bonding of the reacting atoms to Mo- and W-ions. Type and strength of S-bonding should be a central question in the explanation of catalyst activity in HDS. Pecoraro and Chianelli indicated in a fundamental study[24] that the general correlation between catalysts' activities and reactant-catalyst interaction strength, represented by a volcano shaped curve,[55] is valid for the function of a metal's HDS activity on the heat of formation of its sulfide.

Labelled sulfur offers an excellent possibility for conclusions with respect to the mobility of catalyst sulfur, to determine its extent and to distinguish its different kinds (as reversibly adsorbed, or eluted with H_2, or displaced by different other molecules, including S-containing ones, and that of sulfur exchange). There exist different non-isotopic methods for sulfur uptake determination; these are different from sulfur mobility studies which is difficult to perform without applying isotope tracer.

2.1. *Catalyst sulfur exchange with hydrogen sulfide*

The most simple way to determine the amount of mobile sulfur is to expose the sulfided catalyst to $H_2{}^{35}S$

$$S_{cat} + H_2{}^{35}S \rightarrow {}^{35}S_{cat} + H_2S \qquad (6a)$$

or the reverse procedure of treating the ^{35}S labelled catalyst with non-labelled H_2S:

$$^{35}S_{cat} + H_2S \rightarrow S_{cat} + H_2{}^{35}S \qquad (6b)$$

followed by the loss [Eq. 5(a)] or increase [Eq. 5(b)] of gas phase radioactivity at catalyst interaction with gas phase hydrogen sulfide. The actual exchange ($^{a}S_{exc}$) of sulfur is determined from sulfur uptakes in several runs calculated by Eqs. (4) and (3) respectively. The total uptakes contain both reversibly and irreversibly adsorbed H_2S. The amount of the exchangeable sulfur atoms should be calculated[48] by expressions

$$S_{exc} = (S_T - S_{rev})/(1 - S_T/m_{so}) \quad \text{and} \qquad (7a)$$

$$S_{exc} = m_{so}aS_{exc}/(m_{so} - S_{rev} - {}^{a}S_{exc}) \qquad (7b)$$

for Eqs. 5(a) and 5(b) respectively. The S_{rev} values are determined from the two-step experiment[41] referred to in Sec. 2.1. Experiments with alumina supported MoO_x, $CoMoO_x$ and $PtMoO_x$ indicated[48] that the differences between the amounts of exchangeable strongly bound sulfur <10%, is determined by these two procedures.

Sulfur exchange (and uptake) experiments are performed usually applying H_2/H_2S gas mixtures of different $H_2{:}H_2S$ ratio. Comparison of $H_2/H_2{}^{34}S$ (with 90% content of the stable sulfur isotope (^{34}S)) experiments with those of $Ar/H_2{}^{34}S$, indicated[57] that the presence of H_2 had no essential influence on the extent of S exchange, whereas the introduction of $H_2{}^{34}S$ resulted in a sharp evolution of H_2, indicating that H_2S adsorption displaced H_2 quickly. Consequently, different amounts of H_2 at different $H_2{:}H_2S$ ratio should not affect the S-exchange and uptake. The differences in results can be assigned to the differences in H_2S pressure.

The first systematic sulfur exchange study was performed with $H_2{}^{35}S$ on a Mo/Al_2O_3 of 4–30% wt% Mo and on a $NiMoO_x/Al_2O_3$ catalyst.[58] The amount of exchangeable sulfur was only 8–10% of the irreversibly bound sulfur at 373 K for catalysts of 1.8 S/Mo ratio, and no discrete molybdenum phase was observed by electron microscopy. The low ratio of exchange at 373 K indicated that the $MoS_{1.8}$ surface is different from that of the non-supported molybdenum disulfide, with all its surface sulfide ions exchangeable at room temperature.[9] This is in agreement with the monolayer coverage of alumina by the molybdenum phase,[45,59] and with the absence of a crystal phase in MoO_x/Al_2O_3,[40] that the local environment of Mo in the sulfided catalyst, contains oxy-sulfide species, and Mo-sulfides are bound to alumina by Mo-O-Al bonds.[27] The exchange of the irreversibly bound sulfur increased from 10 to 30% of the catalyst S-content in the 373–673 K temperature interval, whereas the S-amounts, desorbed by evacuation (named "displace exchange") decreased from 8.3 to 3% of the irreversibly bound sulfur. Nickel (1.5 wt%) did not affect the exchange at 373 K, whereas at 588 K, the exchange on $NiMoO_x$ was nearly two times higher than on the non-promoted sample, similar to the observation made[42,47] with respect to the effect of Co and Ni on the sulfur uptake of MoO_x. This is explained by the difficulty in sulfiding Co- and Ni-oxides at lower (~573 K) temperatures, and indicates the heterogeneity of surface S-bonding. From experiments with non-supported tungsten disulfide[60] it was concluded that surface sulfide was exchangeable at 373 K when no bulk exchange was observed. Bulk S-exchange was extremely slow even at high (673 K) temperature.

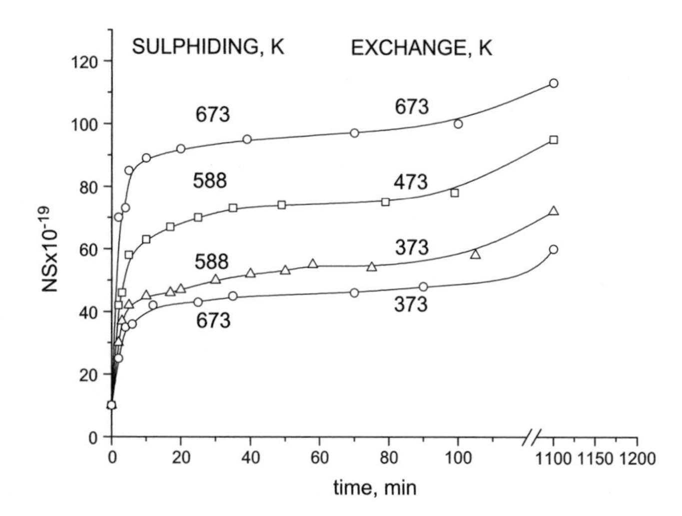

Fig. 5. Exchange of sulfur with time for 12 wt% molybdena-alumina as a function of presulfiding and exchange temperature. Sample size 2.79 g. *Adapted from* Ref. 58.

The exchange rates at different temperatures were near to equal, as demonstrated in Fig. 5. Three regions are distinguishable on the S_{exc} *vs.* time graph: a rapid increase of uptake was related to displace exchange, whereas two different exchange rates of irreversibly bound sulfur were separated. This indicates the heterogeneity of Mo-S bonding, resulting from different solid phases. Comparison of the exchangeable sulfur amounts with desulfurisation activities of catalysts, with different Mo content, showed a strong relationship. The percentage desulfurisation and exchangeable sulfur increased up to 5×10^{17} Mo/mg, but remained constant in the region up to 1.9×10^{18} Mo/mg.

$H_2{}^{35}S$ hetero-exchange studies were carried out by the Novosibirsk Group too.[61] Silica supported Mo, NiMo and NiW were sulfided with non-radioactive H_2S, determining the amounts and rates of exchange. The differences in exchange rates on different catalysts were mainly the function of the preparation method, and the catalyst chemical composition affected the exchange rates and amounts, to a much lower extent. High initial rates were observed, and were followed by substantially lower exchange rates, indicating the heterogeneity of S-bonding in/on the catalysts. The higher exchange rates in this case were coupled with lower thiophene HDS rates, observed with the alumina supported MoO_x catalysts.[58] This difference was possibly caused by the differences in the catalyst structures caused by the different supports and their different acidities.

Sulfur exchange between sulfided $CoMoO_x/Al_2O_3$ catalyst and gas phase radioactivity of H_2S has been studied by Qian *et al.* (1997) at high (1–5 MPa) pressure in a pressurised fixed bed flow reactor. The H_2S sulfided catalyst was treated with $H_2/H_2{}^{35}S$ pulses (0.1% of $H_2{}^{35}S$). The differences in the flow rates did not affect practically the measured S_{exc}/S_{cat} ratios. These increased with temperature from 15 to 44% of the total sulfur in the 503–673 K region.

Sulfur exchange with $H_2{}^{35}S$ has been studied on alumina supported molybdena in non-promoted and in Co-, Ni-, Pd-, Pt-promoted form, on $NiWO_x$ supported by alumina or by amorphous silica-alumina.[43,48] Sulfur uptake was determined with $H_2{}^{35}S/H_2$ mixtures of ~53 kPa total pressure (10 and 50 vol% H_2S). Both the exchange of ^{35}S-catalyst with S in non-labelled H_2S, and that of the non-labelled catalyst sulfur with ^{35}S labelled hydrogen-sulfide (i.e., Proc. 5b and 5a respectively) were determined, whereas reversibly adsorbed sulfur was removed by evacuation. The S_{exc}/S_T fraction was different for catalysts of different chemical content and of different preparation, i.e., 30–100% for MoO_x, $NiMoO_x$, $CoMoO_x$, and 10–20% for $PdMoO_x$, $PtMoO_x$ and $NiWO_x$. The S_{exc} and the S_{exc}/S_T values were substantially higher for catalysts with zero valent Co or Ni on the surface in the reduced state of the samples, than those with no surface Co^0 and Ni^0.[48,50,51] It is noteworthy that the ratio of exchangeable sulfur was ~45% (taking into consideration that the S-uptake by this catalyst was 86% of the stoichiometric S-content), i.e., equal with that determined at a much higher H_2S pressure.[62]

Different from the observation on the absence of a correlation between sulfur uptake and HDS activity, a definite correlation was found (Fig. 6) between the amounts of exchangeable sulfur atoms (S_{exc}) and the thiophene HDS activity of the different catalysts, as mentioned before in the case of some other catalysts. This observation suggests some parallelism of exchangeable sulfur with the catalyst sulfur, and with the ratio of catalytic sites among sites of sulfur uptake.

A sulfur exchange study has also been carried out with stable sulfur isotope (^{34}S) on a sulfided (with ^{32}S) Mo/Al_2O_3 catalyst in a wide (434–803 K) temperature region.[57] The extent of exchange was determined by temperature programmed oxidation of the $H_2{}^{34}S$ treated catalyst, whereas the amounts of evolved ^{32}S- and ^{34}S-oxides have been measured. The exchange ratio increased with increasing temperature from 19% at 434 K up to 46% at 712 K, and no further increase was observed at higher temperature. Two temperature regions were distinguishable with respect to the oxidation rate

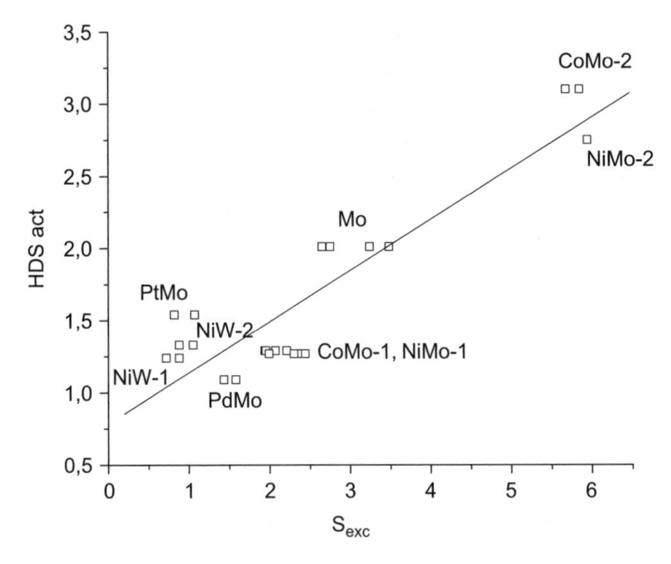

Fig. 6. HDS rates (10^{17} mol./mg) versus S_{exc} (10^{17} Sat/mg). *Compiled from data in* Refs. 43, 48 and 50.

of catalyst sulfur. ^{34}S oxidation was substantially faster in the low temperature region in comparison with that of ^{32}S, i.e., the exchanged sulfur was located mostly in more accessible surface positions. (It is of interest to note that the advantage of this method is that the difference in the strength of S-bonding to the surface cannot be stated with the temperature programmed oxidation of the regularly sulfided molybdenum oxide.) No difference was found, however, between the two S-isotopes in the high temperature region. Apparently the oxidation at high temperature is rapid enough to oxidise both strongly and weakly-bound sulfur. These observations supported the conclusions previously referred to the different strengths in S-bonding to the catalyst.

2.2. *Sulfur exchange via thiophene HDS*

Exchange of radioactive sulfur and the hydrodesulfurisation mechanism has been studied intensively for the last two decades by the Moscow Zelinsky Institute Group. The first studies appeared in the 80s,[35,36] and an increasing research activity has been apparent in the last few years.[2,63] The method of their exchange studies is based on the preparation of ^{35}S-labelled catalyst samples,[39] followed by testing in the thiophene HDS reaction and monitoring the radioactivity of the $H_2{}^{35}S$ produced. An important general conclusion was the absence of any sulfur exchange between catalyst sulfur and thiophene, since radioactivity was only detected in H_2S. It was also stated that a part (~60% in these studies) of the initial S-content of the sulfided $CoMo/Al_2O_3$ remained in the catalyst. The specific radioactivity of $H_2{}^{35}S$, formed via thiophene HDS, decreased as H_2S production developed, as the ^{35}S catalyst sulfur was substituted increasingly by non-radioactive sulfur from thiophene.

Experimental data indicated that the isotope exchange process is described by first order equations,[36,64] and in most cases, e.g., on supported $CoMo/Al_2O_3$ catalysts by a superposition of two curves representing two types of sulfur mobility, the more and the less mobile i.e., rapidly and slowly exchangeable sulfur. In general, the H_2S molar radioactivity α (in percent of the initial molar radioactivity of catalyst sulfide sulfur), as a function of the produced H_2S-[X(cm^3)], is given (Fig. 7) as a superposition of curves[65]

$$\alpha = \Sigma A_i \exp(-\lambda_i X) \tag{8}$$

where λ_i = intensity, characteristic of the given type active sites;

A_i is the proportion in thiophene, converted on the sites of the given type.

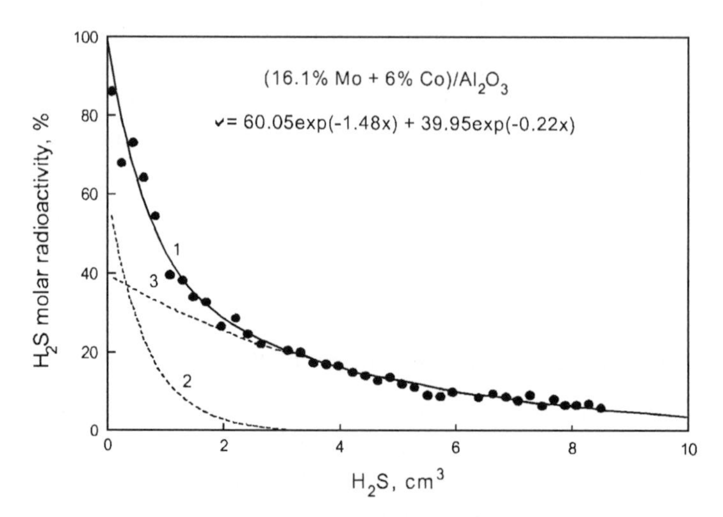

Fig. 7. Dependence of molar radioactivity on the amount of H_2S formed in the course of thiophene HDS Reproduced from 1: experimental curve; 2: "rapid" component; 3: "slow" component.

Note that a near to similar ratio of rapidly and slowly exchangeable sulfur was determined when the catalyst was sulfided originally with thiophene instead of elemental ^{35}S, irrespective of the 2.5 times lower S content of the sample, sulfided by thiophene. This indicates the similarity of active site distribution for catalysts of equal chemical composition.

Experiments indicated[36] that the molar radioactivity function of H_2S production on MoS_2 can be described by one equation ($A = 100, i = 1$); biexponential equations were required for promoted catalysts[64,65] with different A_i and λ_i. This indicates that the catalyst sulfur was not bonded homogenously, like that of the observations with $H_2{}^{34}S$ or $H_2{}^{35}S$ exchange.[57,62] An important difference from those observations is that, in this case, we deal with $H_2{}^{35}S$, replaced by sulfur, formed in HDS, so that the H_2S molar radioactivity represents sulfur formed on the (or via) catalytically active sites. Consequently, the fact that function (7) is described by a biexponential equation with two A-s and λ-s, indicates that there exist two types of catalytic sites, active in HDS on promoted Mo-based catalysts.

The number of coordinatively unsaturated sites (CUS) was evaluated[62] by the expression

$$\mathrm{CUS} = 0.4\mathrm{S}_{\mathrm{stoi}} - \mathrm{S}_{\mathrm{mob}} \tag{9}$$

i.e., the difference between 40% of the maximal number of catalyst SH groups, possible (S_{stoi} = the stoichiometric number of sulfur) and the number of exchangeable sulfur, determined experimentally (S_{mob}, or S_{exc} is denoted here via S_{mob} if the S exchange is performed via sulfur formed in HDS reaction). The sum of participating in the reaction (reacting) sites V and empty sites ES, gives the number of CUS

$$\Sigma\mathrm{CUS} = \Sigma\mathrm{V} + \Sigma\mathrm{ES} \tag{10}$$

Assumptions made in these calculations are

i) MoS_2 forms a monomolecular layer on the alumina surface as previously stated[45,66];
ii) Mobile sulfur atoms form surface SH groups.

The relative number of CUS and surface SH groups was evaluated from the radioisotope testing of the different catalysts. As a result, the number and the ratio of the two active site types (if any) and their productivity during one thiophene pulse of γ conversion ratio was determined by expression

$$\mathrm{P}_{\mathrm{i}} = 0.284\gamma\lambda \tag{11}$$

Experiments with increasing Mo-content of non-promoted samples indicated that increased S_{mob} was paired with increasing conversion. This was the consequence of the increased number of active sites when their productivity practically did not change. To some extent, the S_{mob}/S_{cat} decreased resulting in an increase of the CUS/SH ratio, indicating the comparatively lower density of SH groups, and a decreasing V/ES ratio. These parameter changes were caused by the MoS_2 slabs' growth, as explained by the authors.

Addition of Co resulted in the appearance of a second type of site, characterised in general with higher S uptake, but of a much lower SH mobility and density of functioning vacancies. In agreement with this, the productivity of sites characterised with higher S_{exc} was of 3–10 times higher than that of sites with a lower one.

It is of interest to compare in absolute values, the HDS activity and ΣV of two catalysts. (Data for calculation are taken from Ref. 65) For the 2 wt% Mo ΣV was $3-4 \times 10^{15}$ functioning site $\cdot$ mg^{-1}, the converted thiophene $3,5 \times 10^{15}$ mol/pulse. These values were 4×10^{16} f $\cdot$ site $\cdot$ mg^{-1} and $4,37 \times 10^{16}$ mol/pulse respectively for the Co(4 wt%) Mo (8 wt%) catalyst, i.e., the ratio of V values was 11.76, and that of the thiophene conversions $-12,5$. This seems to indicate the reliability of the applied method. It is seen from the data that the maximal density of sites with the higher $H_2{}^{35}S$ release rate within the 0.3–0.4 Co/(Co + Mo) ratio is typical for the CoMoS phase. At higher Co content, the Co_9S_8 phase formation hinders the formation of vacancies with the high sulfur release rate required for this.

Experiments indicated[36] that the fraction of exchangeable sulfur is only 4% of the sulfur content in non-supported MoS_2. The ratio S_{mob}/S_{cat} in alumina supported catalysts was substantially higher: ~15% in the case of non-promoted molybdena, and increased with increasing Co:Mo ratio up to 35% at 633 K.[64] The ratio of S_{mob} increased with ~20–40% at increasing temperature in the 573–673 K temperature region, in the case of both non-promoted and Co promoted molybdena-alumina. The amount of the rapidly exchangeable sulfur was higher on non-promoted MoO_x.[64] The density of "rapid" sites increased with increasing temperature in the case of $CoMoO_x$ in comparison with that in the non-promoted one.[67]

It is of interest to compare the S_{mob}/S_{cat} data with S_{exc}/S_{cat} obtained in studies on the S-exchange of catalyst sulfide with H_2S.[48] The degree of MoO_x sulfidation was 63% if sulfided with thiophene,[64] that by the H_2S-sulfided sample — 48%.[48] The fraction of exchangeable sulfur in the thiophene sulfided sample was 0.39,[64] that in the H_2S sulfided sample — 0.68 at 4.04 kPa and 0.53 at 27.2 kPa.[48] The sulfidation degree of thiophene sulfided $CoMoO_x$ samples was 0.67[67] and 86% by the H_2S sulfided ones.[48] The S_{mob}/S_{cat} and S_{exc}/S_{cat} values were 0.92[67] and 0.86[48] for thiophene and H_2S sulfided samples respectively. These differences are insignificant, considering the substantial differences in sulfur uptake and exchange data obtained by identical methods, for two $CoMoO_x$ catalysts on different alumina supports and different preparation procedures.[43,48]

Comparison of thiophene HDS conversion with the ratio of mobile (exchangeable) sulfur indicated[65,67] that the higher ratio of mobile sulfur was paired with higher HDS activity of the $CoMo_x/Al_2O_3$ catalysts.

A radioisotope testing of alumina and silica supported reduced Co, Ni and Ni + Co, indicating[2] that the $H_2{}^{35}S$ molar radioactivity versus H_2S formed correlation, were described by a mono-exponential equation of type 7. Sulfidation with H_2/H_2S mixture resulted in a much higher degree in comparison with that performed with thiophene. The fraction of mobile sulfur was in the range of 10–45%, if the sample was sulfided with H_2S/H_2. This is higher in Ni containing samples than in the Co-catalysts for both supports, possibly due to the higher dispersion of NiS. The S_{mob}/S_{cat} fraction in the case of alumina supported samples is higher than that of the silica supported ones, due to the lower dispersion of sulfided particles on silica.

2.3. *Sulfur exchange via HDS of thioaromatic compounds*

The exchange of H_2S formed in hydrodesulfurization of DBT with ^{35}S-labelled surface sulfur of alumina supported $CoMoO_x$ has been studied by Gachet *et al.*[68] They found that catalyst sulfur was replaced by non-radioactive sulfur from DBT and released in $H_2{}^{35}S$.

The Tokyo Group (Kabe and colleagues) has performed a high number of studies with ^{35}S labelled thioaromatic compounds, mostly DBT, on sulfided alumina supported MoO_x, Co, $CoMoO_x$, Ni and $NiMoO_x$,[52,53,69–71] and noble metals.[54,72,73]

Referred to in Sec. 2.3, the sulfur uptakes of the non-presulfided samples were determined from $H_2{}^{35}S$-^{35}S-DBT radioactivity balance at HDS of ^{35}S-labelled DBT,[52,53] by addition of a 1% decalin solution of this compound. However, it was observed that a substantial difference in the behaviour of presulfided and non-presulfided samples, the introduction of [^{35}S] DBT to presulfided samples, resulted in a rapidly approached steady state of DBT radioactivity; whereas in the case of the non-presulfided sample, approximately 140 min was needed to reach the steady state, and unlike the observation with the presulfided sample, $H_2{}^{35}S$ was not detected until ^{35}S DBT was reacted for 45 min.[53] This offered a possibility to study the S-exchange.

The following method is applied. The catalyst is sulfided by pumping a solution of 1wt% non-radioactive DBT solution in decalin (Fig. 7), until its conversion becomes constant (for ~3 h), as determined by GC analysis. ^{35}S-DBT is then pumped into the reactor until the formation of $H_2{}^{35}S$ becomes constant (~3.5 hr). For this period of time, the exchangeable part of ^{32}S catalyst sulfur is replaced by the ^{35}S-labelled one. Different from $H_2{}^{35}S$, the radioactivity of ^{35}S-DBT reaches steady state immediately; its conversion calculated from the loss of its radioactivity is in agreement with that which is estimated from the GC analysis. After the $H_2{}^{35}S$ formation becomes constant, non-radioactive DBT is added again. This results in an immediate and total decrease of the radioactivity of unreacted DBT, but more than 2 hr is required until the total loss of $H_2{}^{35}S$ radioactivity is achieved. This duration is actually required for replacing the catalyst ^{35}S with ^{32}S. Calculations indicated that the amount of ^{35}S incorporated was near to equal with that of ^{35}S released in the second part of the procedure (see areas A and B on Fig. 8). From the amount of radioactivities of the incorporated/released $H_2{}^{35}S$, the amount of mobile (named labile by the authors) sulfur was calculated. This was about 38.4% of total sulfur in MoS_2/Al_2O_3 at 633 K.[53]

The fraction of S_{mob}/S_{cat} and the amounts of mobile sulfur for $CoMoS_2$ were 8–43% and $2.8–7.5 \times 10^{17}$ S atoms/mg respectively in the 533–673 K interval,[42,48] similar to the values obtained with $H_2{}^{35}S$ experiments.[62]

On alumina supported Pd and Pt, the S_{mob}/S_{cat} was 0.2 and 0.12; the amount of mobile sulfur was $3–3.9 \times 10^{16}$ S atoms/mg at 533 K.[54] According to the authors' opinion, the difference between the molybdena based and the noble metal catalysts, is a consequence of the attachment of a part of sulfur to the alumina support of the noble metal catalysts. The data indicate that almost all sulfur accommodated on the noble metals was mobile, i.e., exchangeable and participated in the HDS reaction. This was different from the case with molybdena-based catalysts. This observation was confirmed recently that S_{mob}/Pd(Pd+Pt) ratio was 0.2, both for alumina supported Pd and Pd-Pt, irrespective of the H_2S partial pressure and the amount of the total sulfur uptake.[73]

The mobile sulfur amounts were determined for alumina supported Cr- and W- oxide[74]: $4.8–5.65 \times 10^{16}$ and $4.6–8.3 \times 10^{16}$ mol/mg respectively in the 613–653 K temperature interval. Like the S-uptake, the S_{mob}/S_{cat} ratio would be different in the case of the stepwise- and for the directly sulfided chromium-oxide/Al_2O_3 sample.[51]

An observation is made recently,[75] comparing the mobile sulfur data of a presulfided with $H_2{}^{35}S$ Ru-Cs/Al_2O_3 catalyst, with those determined for a non-presulfided sample described before by HDS of ^{35}S-DBT. S_{mob} was found to be much lower for the presulfided samples, indicating a lower density of active sites (vacancies) in samples sulfided by hydrogen sulfide.

A minimal amount of mobile sulfur was observed[37] on TiO_2 but ~10% of that was determined[53] on Mo/Al_2O_3 at similar HDS activities. This indicates a different reaction mechanism in this case. It has been stated from the results of a study of S-exchange on Mo/TiO_2 samples[38] that the S_{mob}/S_{cat} and the amount S_{mob} was on this catalyst at 653 K at ~0.15 and 1.22×10^{17} at/mg respectively, i.e., lower than the respective values determined for molybdena-alumina. In contrast to this, the HDS conversion was 83.2%, whereas on MoO_x/Al_2O_3, it was only 36.9% for the same Mo content.[53] This indicates that TiO_2 of lower propensity to exchange cooperates with MoO_x.

The effect of methyl substituents is a widely discussed issue in HDS studies.[13] Kabe *et al.* found[76] different S-exchange values for 4-methyl-DBT, 4,6-dimethyl-DBT and DBT. The amount of exchangeable sulfur was higher in the case of DBT than that in the case of the methyl-substituted DBT-s. In agreement with this, the HDS rate of DBT was higher than those of the methyl substituted ones. The lower amount of mobile sulfur is presumably the consequence of steric hindrance caused by methyl groups, as it follows from the sequence of the HDS rates

$$\mathrm{DBT} > \mathrm{4M - DBT} > \mathrm{4,6DM - DBT}$$

This indicates again that increased S-exchange is paired with increasing HDS activity. This seems to indicate that the concentration of HDS active sites equals to the HDS conversion of these compounds, and equals with that of the sites for S-exchange. S_{exc} and S_{mob} data indicate, in general, the amount and the ratio of mobile or exchangeable sulfur, determined by S-isotope as tracer of high reliability. The data are near to equal, as determined by the different methods. Both the amount of mobile, or exchangeable sulfur and its fraction in catalyst sulfur is different for different catalysts and they increase with temperature. Two types of S-exchange were observed in the studies.

Definite correlation was found between the amount of exchangeable or mobile sulfur and the HDS activity. A deeper insight can be gained into the mechanism of the sulfur-catalyst interaction and of the catalytic conversion, on the basis of these data.

3. On the Role of Catalyst Sulfur in Catalytic Hydrodesulfurisation: Some Conclusions from Tracer Studies

This section aims to provide an outlook of some conclusions with respect to the role of sulfur on catalyst HDS activity, gained from tracer experiments with sulfur isotope. This survey is not, as it cannot be, exhaustive.

- The conditions applied in different studies are substantially different. The strong and different pretreatment methods on sulfur uptake and thiophene HDS activity, for different catalysts effect, is demonstrated by a comparison of the effect of six different procedures applied for preparation and sulfidation of supported $NiMoO_x$ and two $NiWO_x$ samples.[77] Note that even a 50 K difference by similar pretreatment results in substantial differences in HDS activity, and the effect is different for different catalysts;
- A general explanation cannot be given, which is valid for different catalysts and reactions, even if these are limited to HDS. One of the characteristic examples is the essential difference in the mechanism of catalytic effect on sulfided noble metals in comparison with that of molybdenum and tungsten-oxide based catalyst, due to large differences in the energy of their sulfide formation.[45,54] This difference in HDS mechanism is also well seen from the rather different effect of sulfidation on precious metal promoted MoO_x, in comparison with $CoMoO_x$, in the initial period of sulfidation.[40]
- An exhausting survey should first of all include the results gained by the most advanced physical methods (STM, EXAFS), and by theoretical (DFT) calculations. This would be an object for a special monograph.

In this respect, attention should be drawn to an outstanding study of Kogan[16] on the catalyst structure and mechanism of C-S bond breaking, studied by isotopes, applying the results obtained by other developed methods.

3.1. *On the mechanism of sulfur exchange*

Sulfidation of Mo- and W-oxides with H_2/H_2S[26–29] results in the formation of sulfides and surface SH groups.

Different examples, presented in Secs. 1 and 2,[19,26,36,52,57,64,69] indicated that sulfur is bonded heterogeneously to the catalyst surface, and the strength of bonding, including that of mobile and exchangeable sulfur, is varying, depending on its surface position.

The most general observation is to be made on the basis of the tracer studies that only a part of surface sulfur is mobile or exchangeable on catalysts with sulfides and oxides. Supported Co and Ni contain only mobile sulfur, if sulfided by thiophene and contain both mobile and immobile sulfur, if they are sulfided by H_2S/H_2.[78] This is interpreted with the differences in the metal-sulfur bond strengths, as the edge S-atoms are held more weakly than the top S-atoms of the slab.[61] This follows from the results of some radiosulfur tracer studies of metal single crystals that different surface S-species are formed at different surface sulfur concentrations, and on different crystal faces of Pt, Ni[17,18] and Mo.[19] Note that the surface monolayers of different structures PtS_2 or PtS are formed in different experimental conditions, as witnessed by studies with sulfur-35.[79]

Exchange experiments with $H_2{}^{35}S$ indicated[58] the effect of the diversity of the local environment of molybdenum on the strength of Mo-S bond in sulfided molybdena-alumina. More and less strongly bonded surface sulfur regions on MoS_2 were found.[57] From the comparatively low 0.46 ratio of ^{34}S-exchange, a conclusion was made that the total sulfur exchange represents the fraction of S on edges, and the partial exchange at the lower temperature, indicates that some S-atoms at the edges are less strongly bonded than others. Calculations based on geometric data and on the stoichiometric approach resulted in hexagonal slab lengths in the range $\sim 1.2-3.5$ nm, assuming no top S-atoms exchanged. Several explanations were given to the partial exchange of the less strongly bonded sulfur, such as the weakness of terminal S-bonding to one Mo in comparison with the strongly bonded S, bridged to two Mo atoms;[80] another explanation, given by the authors of Ref. 57 is based on a scheme of the exchange of some top S atoms, with an adjacent top SH group, removed during exchange, as H_2S, leaving a vacancy in the top site. This approach is a likely explanation of the results, indicating the existence of different types of active sites for mobile sulfur formation and the HDS of sulfur-organic compounds.

The slab sizes in different Mo (and W) based catalysts determined the S-edge:S-total ratio in them. This ratio depends on the slab sizes, and its value is the highest for slabs of the lowest number of Mo atoms. Geometric calculations of the S-edge:S-total ratio, made on the basis of EXAFS and TEM data, indicate a reasonable correlation of this indicator with the S_{exc}/S_{cat} fraction, determined for different catalysts by radioisotope tracer.[48] The linear relationship, HDS conversion versus S_{exc}, represented in Fig. 6, indicates that HDS conversion is associated with S-vacancies, in agreement with the conclusion made in Ref. 57 on the correlation between vacancies and S-exchange. The lower HDS activities and the lower S_{exc} values are possibly caused by the absence of Co^o and Ni^o on the surface.[49,50] Accommodated under MoS, Co and Ni-oxides probably cause some increase of the slab sizes, and the decrease of the S-edge:S-total ratio. This explains the lower S_{exc}/S_{cat} values of these catalysts in comparison with the non-promoted molybdena alumina.

Comparison of the behaviour of $NiMoO_x$ catalysts on different supports also indicates that increasing S_{mob} is paired with increasing thiophene conversion.[81] The HDS conversion increases with increasing Ni content on the alumina supported samples. A substantially higher sulfur mobility on Mo/C, in comparison with that on Mo/Al_2O_3, is paired with higher HDS conversion. On carbon supported samples, however, the HDS conversion remains constant at increasing Ni content, and S_{mob} increases only in a narrow range. Analysis of the radioactivity versus H_2S correlation gives 0.4 and 1 for the value of the S_{mob}/Mo ratio for alumina and carbon supported samples respectively. This is explained in the assumption that the reactive sulfur on alumina supported samples is attached to the edges of the Mo slabs only, whereas on carbon supported samples, mobile sulfur is accommodated on the top sites too. The density of the more active ("rapid") sites decreases with increasing amounts of Ni, as NiS particles block them. This decrease is stronger on carbon supported samples, as NiS blocks both edge and basal plane active sulfur.

3.2. *On the reaction mechanism of desulfurisation*

Data discussed in Sec. 3.1 seem to indicate that HDS activity is associated with sulfur vacancies, as exchange requires sulfur attachment to the surface. Exchange mechanism via surface SH groups was also assumed by Massoth and Zeuthen.[57]

The role of surface SH groups in exchange and HDS was studied with tritium (^3H) labelled hydrogen.[36] Thiophene HDS, performed on $CoMoO_x/Al_2O_3$, containing irreversibly bonded ^3H resulted radioactivity solely in H_2S, as tritium was practically absent in the non-converted thiophene and in C_4-hydrocarbons — products of HDS. The observed dependence of ^3H_2S specific radioactivity, on the amount of hydrogen sulfide formed, is approximated by equation of type (7) with constants (A_i and λ_i), identical to those calculated for $H_2{}^{35}S$ radioactivity versus hydrogen sulfide formation in experiments with ^{35}S labelled thiophene. The total amount of H_2S formed in HDS with the participation of irreversibly adsorbed (labelled) hydrogen, was equivalent to the amount of $H_2{}^{35}S$ formed in the parallel experiment with ^{35}S labelled thiophene. It was also stated[35–39] that the SH groups formed at sulfidation participate in the formation of the H_2S molecules. Consequently, the surface density of SH groups should also be an important factor in catalyst HDS activity besides that of the S-vacancies.

The study was later supplemented[1,16] with thiophene HDS experiments in a ^3H labelled hydrogen atmosphere on $CoMo/Al_2O_3$ sulfide. Comparison of thiophene and C_4-hydrocarbon radioactivity data indicated a rapid $^3H \rightarrow {}^1H$ exchange between gas phase and thiophene hydrogen, preceding the thiophene HDS. Analysis of radioactivity distribution between thiophene, C_4-hydrocarbons and hydrogen sulfide indicated that the formation of hydrogen sulfide is accompanied by an isotope effect; presumably this process should be the rate limiting step in the HDS reaction.

This conclusion is in agreement with a near to general result of the sulfur exchange study with ^{35}S-DBT, where first order $H_2{}^{35}S$ radioactivity versus time plots were obtained from both the decreasing and increasing periods of radioactivity, and the rate constants of H_2S release and activation energy values were calculated for alumina supported Co, Ni, $CoMoO_x$, $NiMoO_x$ and MoO_x.[70,71,76,82] The calculated release rates were equal to the DBT desulfurization rates. This also indicates H_2S-release being the rate determining step in HDS. The activation energies were near to equal for the supported promoter metals and the molybdena-alumina ($\sim$40 kJ/mol) promoted by them.

It has been found in HDS of ^3H-labelled thiophene[16] that the radioactivity of hydrogen sulfide formed from the first pulses, was much lower than that of the C-4 hydrocarbons produced. A measurable radioactivity appeared only upon treating the catalyst with ^3H-labelled thiophene for a longer time. The time delay in ^3H_2S release contradicts to a direct ^3H transfer between ^3H-thiophene and H_2S, formed in HDS. Instead, the surface SH groups take up ^3H, and form gas phase ^3H_2S.

Consequently, the rate determining step of the whole process should be the interaction of the surface SH group (equal with S_{mob} with one of the H atoms originated from the gas phase hydrogen. According to this concept, S in thiophene attaches to a surface anionic vacancy; its interaction with hydrogen results in the formation of a surface SH-group. Consequently, the number of vacancies (V) and the V:SH ratio decreases. This would stop the catalytic reaction; however, the V:SH ratio becomes re-established due to surface SH interaction with gas phase hydrogen (possibly via H_2 dissociation) and H_2S release. Following this, the ratio of the acting vacancy, the mobile SH-group number, is the determining factor in the HDS activity of the catalyst. The proposed "forcing out" or "displace" mechanism is represented in Fig. 9. It should be added to this scheme that the C-S bond cleavage is preceded presumably with hydrogenative thiophene transformation into 2,3-dihydrothiophene, as the C-S bond strength in the latter is substantially (55 kJ mol^{-1}) lower than that in thiophene. The high (~3.5 h′) delay in approaching steady state $H_2{}^{35}S$ formation, in comparison with that reached immediately by ^{35}S-DBT (Fig. 8), indicates a near

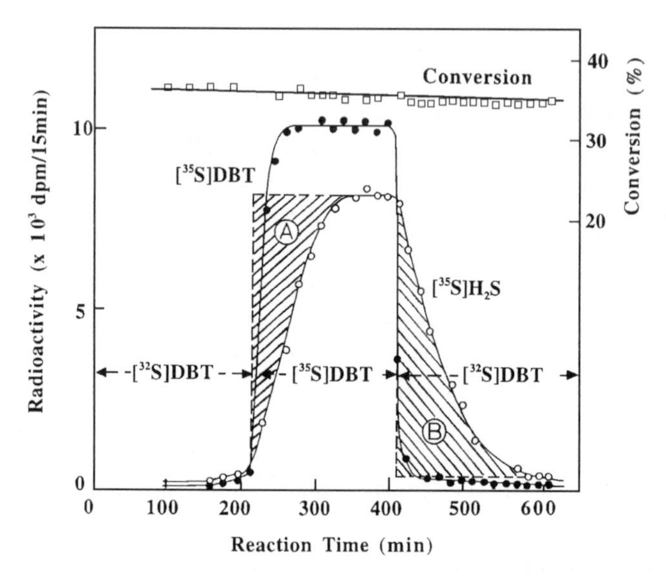

Fig. 8. Operation procedure in hydrodesulfurization of ^{35}S-DBT on Mo/Al_2O_3, temperature 633 K, pressur 2.5 MPa. *Reproduced from* Ref. 52 with permission.

to similar mechanism in HDS of DBT. In the proposal by the Tokyo Group mechanism for HDS of DBT, the sulfur compound is adsorbed on a vacancy of the $CoMoO_x/Al_2O_3$ catalyst, the C-S bond is cleaved by H_2, the sulfur remains on the catalyst, another mobile sulfur releases in the form of H_2S and a new vacancy is formed. This mechanism supposes vacancy migration and seems to be similar to that given in the scheme in Fig. 9.

Comparison of CoMo and NiMo sulfide with monometallic Co or Ni indicates[62] that the functioning vacancies/empty sites (V/ES) ratio is of a 10^{-2}–10^{-3} order of magnitude for the monometallic catalysts, whereas for the Co and Ni promoted MoO_x/Al_2O_3 samples, the V value reaches 10–20% of the ES. However, in the case of some catalysts, it reaches even 50–100%.[66] Comparison of these data with HDS conversion activities indicates that the number of functioning vacancies plays no role in the activity of the monometallic Co and Ni catalysts. Their coverage with SH groups is the decisive element in their HDS activity. This indicates the difference in HDS mechanism of sulfided metal catalysts in comparison with that of the MoO_x and WO_x based ones.

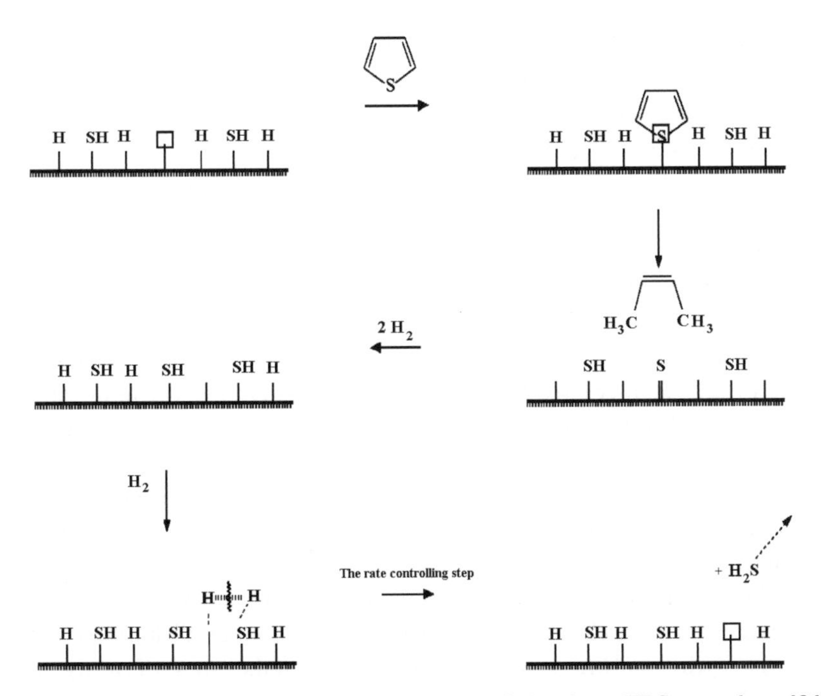

Fig. 9. Scheme of the "displacement" mechanism of thiophene HDS over the sulfide catalyst. *Reproduced from* Ref. 16 with permission.

Sulfur uptake and exchange data[54,73] referred in Sec. 2.3 indicated substantial differences of alumina supported Pd and Pt, in comparison with the molybdena based catalysts. The mobile sulfur amounts were lower [0.25 S_{mob}/Pd (or /(Pt)] than those experienced with the Mo and W based catalysts. Near to all sulfur, accommodated on the noble metal was mobile and — different from the Mo- and W-based catalysts — participated in the reaction; the sulfur mobility was substantially higher[54] due to the much lower S-Pd and S-Pt bond strengths. This indicates that the HDS mechanism, detailed above for Mo based catalysts cannot be valid for the noble metals. Again, the number of vacancies for bonding sulfur-organic compounds is not limiting, as in the case with monometallic Co and Ni.

Data for sulfur uptake and HDS activity of precious metals promoted molybdena-alumina,[40,44,45] also indicate the differences in the behaviour of these metals and molybdena. Sulfidation results in decreasing HDS activity of PtMo and PdMo, but this trend is changing upon increasing intensity of sulfidation. The metals act presumably as C-S bond breaking sites, being active in hydrogenative dissociation. With increasing sulfidation of MoO_x it presumably takes over the classical HDS mechanism.

The high activity of monometallic catalysts in HDS supports the afore given explanation that the absence of surface zero valent Co and Ni results in a decreased S_{cat}, S_{exc} and HDS activity.

Reviewing the tracer studies on the mechanism of catalytic hydrodesulfurisation, it can be concluded that at this time, a general mechanism for these reaction cannot be given. Possibly, it does not exist at all. It is well seen, however, that sulfur-catalyst interaction influences the catalysts behaviour; substantially the activity, to a large extent, depends on the S_{mob}/S_{irr} correlation, S-uptake affects the number of vacancies and consequently the value of the vacancy:empty sites ratio. It was seen that H_2S treatment affects the catalyst selectivity.[26] There are a number of examples on this, mostly for HDS/hydrogenation selectivity ratio, not referred here.

The effect of H_2S uptake on the catalyst behaviour is an outstanding example of the concept of the catalysts system based on the principle that, "the original catalyst, the reacting compounds and the products of conversion form the system responsible for the catalytic reaction",[83] and expressed by Delmon more thoroughly, "Probably all kinds of active sites in hydrotreating catalysts result from some sort of a dynamic process".[84]

References

[1] Kogan VM, Greish AA, Isagulyants GV, *Proceedings of the 2nd European Congress on Catalysis*, Maastricht, 3–8, September, 1995.
[2] Kogan VM, Parfenova NM, Gaziev RG, Rozhdestvenskaya NN, Korshevets IK, *Kinet Catal* **44**: 583, 2003.
[3] Krylov OV, Fokina EA, Problems of kinetics and catalysis (eds. Roginszky SZ, Krilov OV) Vol 10. p. 294 Moscow, Acad. of Sci. USSR 1957.
[4] Köster R, *Angew Chem* **73**(2): 66, 1961.
[5] Kaneko Y, Odanaka H, *J Res Inst Catal* **13**: 29, 1965.
[6] Kaneko Y, Odanaka H, *J Res Inst Catal* **14**: 213, 1967.
[7] Happel J, Odanaka H, Roche P, *Chem Eng Progr Symposium* **67**: 60, 1971.
[8] Roitr VA, Korneichuk GP, Volikovskaya NA, Golodets GI, *Kinet Katal* **1**: 408, 1960.
[9] Lukens HR, Jr, Meisenheimer RG, Wilson JN, *J Phys Chem* **66**: 469, 1962.
[10] Horiuti J, Proc. the use of tracer to study heterogeneous catalysis, *Annals N Y Acad Sci* **213**: 1, 1973.
[11] Gaál D, Danóczy E, Nemes I, Vidóczy T, Hajdú P, The use of tracers to study heterogenous catalysis, *Annals N Y Acad Sci* **213**: 51, 1973.
[12] Gates BC, Katzer JR, Schuit GCA, *Chemistry of Catalytic Processes*, McGraw-Hill, New York, 1979.
[13] Topsoe H, Clausen BS, Massoth FE, *Hydrotreating Catalysis*, Springer, Berlin, 1996.
[14] Bartolomew CH, Agrawal PK, *Adv Catal* **31**: 135, 1978.
[15] Barbier J, Larmy-Pitara E, Marecot P, Boitiaux JP, Cosyns J, Verna F, *Adv Catal* **37**: 279, 1990.
[16] Kogan VM, Transition metal sulfides, *Chemistry and Catalysis 3*, High Technology NATO ASI Series, Kluwer Academie Publishers, Dordrecht p. 235, 1998.
[17] Berthier Y, Perdereau M, Oudar J, *Surf Sci* **36**: 225, 1973.
[18] Oudar J, *Catal Rev Sci Eng* **22**: 171, 1980.
[19] Gellman AJ, Bossel ME, Somorjai GA, *J Catal* **107**: 103, 1987.
[20] Varga J, *Brennstoff-Chem* **9**: 277, 1928.
[21] Weisser O, Landa S, *Sulphide Catalysts. Their Properties and Applications*, Academia, Praha, 1972.
[22] Massoth FE, *Adv Catal* **27**: 266, 1978.
[23] Grange P, *Catal Rev Sci Eng* **21**: 135, 1980.
[24] Pecoraro TA, Chianelli RR, *J Catal* **67**: 430, 1981.
[25] Zdrazil M, *Appl Catal* **4**: 107, 1982.
[26] Campbell KC, Mirza ML, Thomson SJ, Webb G, H *J Chem Soc Farad Trans* **80**: 1689, 1984.
[27] Massoth FE, *J Catal* **36**: 164, 1975.
[28] Wright CJ, Sampson, Fraser D, Moyes RB, Wells PB, Riekel C, *J Chem Soc Farad Trans* **76**: 1585, 1980.
[29] Leglise J, Finot L, van Gastel JNM, Duchet JC, *Stud Surf Sci Catal* **127**: 51, 1999.

[30] Siegel S, *J Catal* **30**: 139, 1973.
[31] Tanaka K, Okuhara T, *Catal Rev Sci Eng* **15**: 249, 1977.
[32] Michell PCH, in Catalysis (Spec. Per. Rep). The Royal Soc. Chem. London 1981. Vol. 4. P. 175.
[33] Dobrovolszky M, Tétényi P, Paál Z, *Chem Eng Comm* **83**: 287, 1989.
[34] Günter JR, Marks O, Korányi TI, Paál Z, *Appl Catal* **39**: 285, 1988.
[35] Isagulyants GV, Greish AA, Kogan VM, Vunova GM, Antoshin GV, *Kinet Catal* **28**: 550, 1987.
[36] Isagulyants GV, Greish A, Kogan VM, in *Proceedings of the 9th International Congress on Catalysis*, Calgary, Ottawa Ontario, Canada, Vol. 1, p. 35, 1888.
[37] Wang D, Qian W, Ishihara A, Kabe T, *J Catal* **202**: 322, 2001.
[38] Wang D, Qian W, Ishihara A, Kabe T, *Appl Catal A* **224**: 191, 2002.
[39] Isagulyants GV, Greish AA, Kogan VM, *Kinet Catal* **28**: 555, 1987.
[40] Dobrovolszky M, Matusek K, Paál Z, Tétényi P, *J Chem Soc Farad Trans* **86**: 3137, 1993.
[41] Jackson SD, Casey NJ, *J Radioanal Nucl Chem Lett* **200**: 465, 1995.
[42] Kabe T, Qian W, Tanihata K, Ishihara A, Goda M, *J Chem Soc Farad Trans* **93**: 3709, 1997.
[43] Koltai T, Massoth FE, Tétényi P, *React Kinet Catal Lett* **71**: 85, 2000.
[44] Dobrovolszky M, Paál Z, Tétényi P, *Appl Catal A* **142**: 159, 1996.
[45] Paál Z, Koltai T, Matusek K, Manoli JM, Potvin C, Muhler M, Wilde V, Tétényi P, *Phys Chem Chem Phys* **3**: 1535, 2001.
[46] Korányi TI, Dobrovolszky M, Koltai T, Matusek K, Paál Z, Tétényi P, Fuel, *Proc Technol* **61**: 55, 1999.
[47] Quian W, Ishihara A, Aoyama V, Kabe T, *Appl Catal A* **196**: 103, 2000.
[48] Massoth FE, Koltai T, Tétényi P, *J Catal* **203**: 33, 2001.
[49] Koltai T, Dobrovolszky, Tétényi P, in Proc. 2nd Eur. Symp.Hydrotreating a. Hydrotreating of Oil Fractions (Forment GF, Delmon B, Grange P, eds.) Elsevier, 1999, p. 137.
[50] Tétényi P, Galsán V, *React Kinet Catal Lett* **74**: 251, 2001.
[51] Dumeignil F, Amano H, Wang D, Qian EW, Ishihara A, Kabe T, *Appl Catal A* **249**: 255, 2003.
[52] Qian W, Ishihara A, Ogawa S, Kabe T, *J Phys Chem* **98**: 907, 1994.
[53] Qian W, Zhang Q, Okoshi Y, Itshihara A, Kabe T, *J Chem Soc Farad Trans* **93**: 1821, 1997.
[54] Kabe T, Qian W, Hirai, Li L, Ishihara A, *J Catal* **190**: 191, 2000.
[55] Balandin A, *Adv Catal* **19**: 1, 1969.
[56] Tolstopyatova AA, Balandin AA, Naumov VA, *Zh Fiz Him* **38**: 1619, 1964 (Russian).
[57] Massoth FE, Zeuthen P, *J Catal* **145**: 216, 1994.
[58] Scarpiello DA, Montagna AA, Freel J, *J Catal* **96**: 276, 1985.
[59] Russel AS, Stokes JJ, *Ind Eng Chem* **38**: 1071, 1946.
[60] Freel J, Larsson JG, Adams JF, *J Catal* **96**: 544, 1985.
[61] Startsev AN, Artomonov EV, Yermakov Yu I, *Appl Catal* **45**: 183, 1988.
[62] Qian W, Ishihara A, Wang G, Tsuzuki T, Godo M, Kabe T, *J Catal* **170**: 286, 1997.

[63] Kogan VM, Gaziev RG, Lee SW, Rozhsdestvenskaya NN, *Appl Catal A* **251**: 187, 2003.
[64] Kogan VM, Greish AA, Isagulyants GV, *Catal Lett* **6**: 157, 1990.
[65] Kogan VM, Roshdestvenskaya NN, Korsevets IK, *Appl Catal A* **234**: 207, 2002.
[66] Millman WS, Koii-Ichi Segawa, Smrz D, Hall WK, *Polyhedron* **5**: 169, 1986.
[67] Kogan VM, *Appl Catal Gen* **237**: 161, 2002.
[68] Gachet CG, Dhainaut E, de Mourgues L, Candy JP, Fouilloux P, *Bull Soc Chim Belg* **90**: 1279, 1981.
[69] Kabe T, Qian W, Ogawa S, Ishihara A, *J Catal* **143**: 239, 1993.
[70] Kabe T, Qia W, Ishihara A, *J Phys Chem* **98**: 912, 1994.
[71] Kabe T, Qian W, Ishihara A, *J Catal* **149**: 171, 1994.
[72] Qian W, Yoda Y, Hirai H, Ishihara A, Kabe T, *Appl Catal A* **184**: 81, 1999.
[73] Qian EW, Otani K, Li L, Ishihara A, Kabe T, *J Catal* **221**: 294, 2004.
[74] Kabe T, Ishihara A, Qian W, Funato A, Kawano T, *React Kinet Catal Lett* **68**: 69, 1999.
[75] Ishihara A, Li J, Dumeignil F, Higashi R, Wang A, Qian EI, Kabe T, *Appl Catal A* **217**: 59, 2003.
[76] Kabe T, Qian W, Wang W, Ishihara A, *Catal Today* **29**: 197, 1996.
[77] Koltai T, Galsán V, Tétényi P, *React Kinet Catal Lett* **67**: 391, 1999.
[78] Kogan VM, Parfenova NM, in Proc. 1st Eur.Symp. Hydrotreating a. Hydrocracking of Oil fractions (Froment GF, Delmon B, Grange P, eds.) Elsevier B.V. 1997, P. 449.
[79] Horányi C, Ritzmayer EM, *J Electroanal Chem* **206**: 297, 1986.
[80] Kasztelan S, Toulhoat H, Grimblot J, Bonnelle JP, *Appl Catal* **13**: 127, 1984.
[81] Kogan VM, Nguen Thi Dung, Yakerson VI, *Bull Soc Chim Belg* **104**: 303, 1995.
[82] Startsev AN, Burmistrov VA, Yermakov Yu I, *Appl Catal* **45**: 191, 1988.
[83] Tétényi P, Guczi L, Paál Z, *Acta Chim Acad Sci Hung* **83**: 37, 1974.
[84] Delmon B, *Bull Soc Chim Belg* **104**: 173, 1995.

Chapter 4

Tritium in Catalysis

John R Jones and Shui-Yu Lu

1. Introduction

The catalysis of organic reactions frequently involves the breakage and formation of one or more carbon-hydrogen bonds.[1] Such reactions can be performed under homogeneous or heterogeneous conditions. In order to improve our understanding of the mechanistic aspects of such reactions, it is advisable to simplify the problem and one of the most powerful tools at the disposal of the chemist is that of isotopic substitution. As hydrogen possesses two isotopes, deuterium (^2H or D, stable)[2] and tritium (^3H or T, radioactive),[3] various approaches can be adopted. Thus, in the study of hydrogen isotope exchange one can use a source of deuterium (usually D_2O) or tritium (HTO) to follow the rates of deuteriation or tritiation. These reactions are easier to understand when the label is introduced at one particular site. An alternative approach, which is frequently used, is to label the compound first with deuterium or tritium, and then follow the rates of dedeuteriation or detritiation. In this way, it is possible to measure more easily primary kinetic hydrogen isotope effects, the magnitude of which can provide information concerning the rate-determining step and the nature of the transition state.

$$R_3C{-}H \longrightarrow R_3C{-}D/R_3C{-}T \quad (A)$$

Hydrogen isotope exchange (A) is therefore a very versatile reaction and as it can be acid-, base- or metal-catalysed it has been extensively used.[4,5] It is also ideal for studying solvent effects in reaction kinetics. Recent attention has focused on energy-enhanced reactions, particularly those involving microwaves.[6–8] For those interested in the preparation of deuterium- and tritium-labelled compounds, the results of such investigations can be very useful prior to adopting the optimum labelling conditions.

Another "simple" reaction that has been extensively studied in catalysis is that of hydrogenation, either under homogeneous or heterogeneous conditions.[9] Rather than measure the uptake of gas, it is customary, at least in the case of deuterium, to measure changes in the ^{1}H NMR spectrum of the substrate. Recently, with higher and more stable magnetic fields, greater use is being made of ^{2}H NMR spectroscopy[10] despite the less favourable properties of this nuclide (Table 1). With a nuclear spin of unity the signals are broader than those for ^{1}H and resolution problems may arise. The "dynamic range" (100% — natural abundance) is a further problem. On the other hand, with modern day instrumentation spectra can be obtained in a matter of minutes and the kind of reactions that can be investigated can be widened to include hydrogenation, epoxidation, hydroxylation, methylation and oxidation.

$$RCH{=}CH_2 + {}^*H_2 \longrightarrow R\overset{{}^*H}{\overset{|}{C}}H{-}\underset{{}^*H}{\underset{|}{C}}H_2 \qquad ({}^*H = D, T)$$

Because tritium is radioactive, the number of laboratories that are set up to use this radionuclide in the study of catalysis is very small. This is unfortunate because the radioactivity can be put to good advantage, *e.g.*, in following the rates of very slow reactions in a relatively short time. Nevertheless, proper radiochemical facilities and appropriate experience is required prior to embarking on research using tritiated compounds.

2. Radiochemical Facilities

Before embarking on any tritium work the personnel must become designated radiation workers and be familiar with the appropriate rules and regulations. Ideally there should be both a "high level" radiochemistry laboratory, as well as a "low level" laboratory with a separate room for the scintillation counter. One should only use the minimum amount of radioactivity, consistent with the requirements of the research project. Furthermore, it is often useful to carry out initial training studies using deuteriated compounds, even though the subsequent tritium work will be carried out on a much smaller scale and rely on one or more radiochromatographic methods[11,12] for the purification of the labelled compounds. With appropriate rules and regulations in place, a radiochemical laboratory need not be any more hazardous than an ordinary chemistry laboratory.

There are two separate units of radioactivity in use, the first being the Curie (Ci) which is defined as an activity of 3.7×10^{10} disintegrations per second. This is a large unit — bear in mind that with a modern day liquid scintillation counter radioactivity levels down to a few hundred counts per minute can be easily measured — so it is very common to use smaller subunits such as the millicurie (10^{-3} Ci) and the microcurie (10^{-6} Ci). The second, and more recently introduced unit, is the Becquerel (Bq). At one disintegration per second this is an extremely small amount of radioactivity. The conversions are

$$1 \text{ Bq} = 2.703 \times 10^{-11} \text{ Ci (or 27.03 pCi)}$$
$$1 \text{ Ci} = 3.7 \times 10^{10} \text{ Bq (or 37 GBq)}$$

In a "high level" radioactivity laboratory one would be expected to handle many millicuries, even curies, whilst in a "low level" laboratory one would try and limit the radioactivity to less than a millicurie, and often work at the microcurie level.

Good laboratory practice should ensure that the radiochemical work be carried out over spill trays in a fumecupboard with excellent ventilation. Plastic gloves, sometimes one pair worn over the other, should be used at all times and any radioactive waste be placed in a container filled with vermiculite. Regular urine samples should provide the necessary reassurance that the work is being performed with due diligence. Satisfactory arrangements for the disposal of radioactive waste should be put in place during early consultation with the radiochemical inspectorate.

For the purification of tritiated compounds, radio-thin layer chromatography or radio-high performance liquid chromatography are frequently used, whilst radio-gas chromatography has its uses when the compounds are highly volatile.[11] In addition, radio-size exclusion chromatography[12] has been developed for tritiated macromolecules.

Tritium is the only radionuclide with a nuclear spin of 1/2 (Table 1) and the technique of ^{3}H NMR spectroscopy has been in use for more than 30 years.[13] Although its sensitivity to detection is the highest of all nuclides — some 21% higher than for ^{1}H — the fact remains that (a) because NMR spectroscopy is not a particularly sensitive technique

Table 1. Important properties of ^{1}H, ^{2}H (D) and ^{3}H (T).

Nucleus	Natural Abundance (%)	Nuclear Spin	NMR Sensitivity	Radiation Properties
^{1}H	99.985	1/2	1.0	Not radioactive
^{2}H	0.015	1	9.65×10^{-3}	Not radioactive
^{3}H	$<10^{-16}$	1/2	1.21	weak β^- emitter (E_{max} = 18 KeV), $t_{1/2}$ = 12.3 years, Maximum specific activity = 29 Ci mmol^{-1}

(at least up to now), and (b) that tritium is radioactive, ^{3}H NMR spectroscopy is not widely used beyond the pharmaceutical industry (where it is invaluable) and a small number of academic/research centres specializing in tritium chemistry.

In the early studies using a 60 MHz NMR spectrometer, it was a normal practice to use ~20 mCi of tritiated compound in order to obtain a satisfactory ^{3}H NMR spectrum; this would usually require an overnight accumulation of scans. Now, with a 500 MHz NMR spectrometer a 1 mCi sample is usually sufficient.

Recently, the sensitivity of the technique has received a major boost through the development of a tritium cryoprobe. Cooling the RF coils of a probe to cryogenic temperature improves the RF efficiency and reduces the noise generated by the coils. Further improvements can be achieved if the preamplifier is also cooled to cryogenic temperatures, as in this way the noise generated in the circuit is reduced. These objectives, despite the formidable challenges presented by the need to keep the sample temperature stable at close to room temperature, whilst the RF coils nearby are cooled to below 35 K, led to a signal (S) to noise (N) ratio improvement of around 4.3, equating to approximately a $(4.3)^2 \approx 20$-fold saving in time or alternatively, a significant reduction in the required radioactivity. Under favourable conditions (all the tritium located at one site) a satisfactory ^{3}H NMR spectrum can now be obtained from a sample containing only 11 μCi.[14]

Until recently all signal processing relied on, and was restricted by, conventional passive analysis where a signal must be some measurable parameter greater than the background in order to be discriminated from the background. A new revolutionary process — quantum resonance interferometry (QRI)[15] — is an example of active signal processing in which new data is derived by interference with the original data. This makes it possible to improve detection sensitivity. Thus far, it has only been applied to gene expression microarrays although its extension to mass spectrometry and NMR spectroscopy has been anticipated.

The main features of ^{3}H NMR spectroscopy are well known (Table 2). To illustrate several of the points made above, Fig. 1(A) shows the ^{3}H NMR (^{1}H decoupled) of a sample of acetophenone, tritiated by a base-catalysed hydrogen isotope exchange procedure. The sample contains 11 μCi of

Table 2. Main features of ^{3}H NMR spectroscopy.

	Feature	Comments
1	Chemical Shift (δ_T)	The ^{3}H chemical shifts are virtually the same as those for ^{1}H, thereby simplifying spectral interpretation.
2	^{1}H Decoupling	The signals are very sharp and at low (<1%) isotopic incorporation, it is customary to decouple the ^{1}H interactions so that each labelled site gives a sharp singlet.
3	Internal Reference	There is no need to synthesise tritiated TMS — a "ghost referencing" procedure is adopted.
4	Background Signal	There are no tritium background signals at the levels of radioactivity used for NMR analysis.
5	T-T Coupling	For work involving high specific activity compounds, the T-T coupling constants are related to the corresponding H-H coupling.
6	Differential NOE	Differential nuclear Overhauser effects are either small or absent, so that integrals of peak areas can be used to calculate the relative % incorporation.
7	Sample Handling	Apart from an NMR spectrometer there is also a need for a tritium probe before one can obtain a ^{3}H NMR spectrum. For safety reasons, it is best to insert the radioactive solution in a narrow cylindrical tube which is then sealed at the top prior to being inserted in a conventional glass NMR tube. This "double containment" procedure has now been in operation for many years without any recorded breakages.
8	Isotope Effect	Small but significant primary and secondary isotope effects on chemical shifts can be seen and these can be put to good effect in analysing for isotopomers, *e.g.*, compounds containing a tritiated methyl group —CT_3, —CT_2H and —CTH_2.

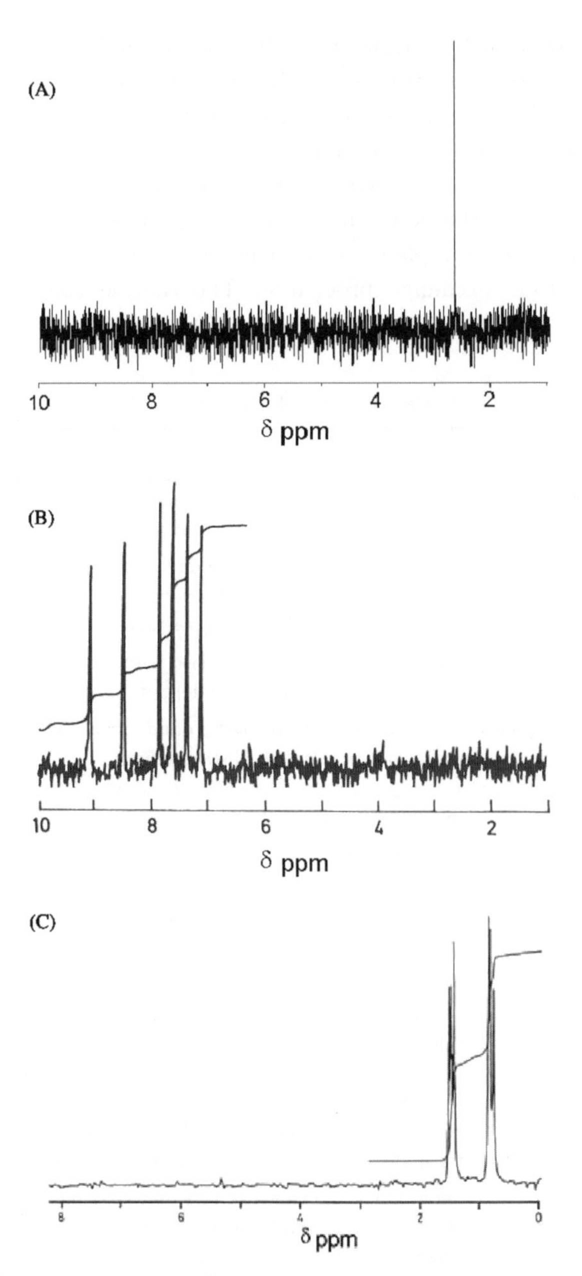

Fig. 1. (A) ^3H NMR spectrum (^1H decoupled) of tritiated *o*-methoxyacetophenone (*o*-CH_3O-$C_6H_4COCH_2T$, 11 μCi) with a proton/tritium cryoprobe at 533.5 MHz, (B) ^3H NMR spectrum (^1H decoupled) of [G-^3H]quinoline (∼10 mCi) at 96 MHz and (C) ^3H NMR (^1H decoupled) of [^3H]-dihydroalprenolol showing T-T coupling and multiplicity of isotopomers. Reproduced with permission.

radioactivity, all located in the side-chain. This example is one of the first to illustrate the benefits of using a tritium cryoprobe.[14]

The second example shows a ^{3}H NMR spectrum (^{1}H decoupled) of a sample of generally labelled quinoline, *i.e.*, [G-^{3}H]-quinoline [Fig. 1(B)].[16] This is prepared by heating the compound in the presence of a small amount (5 μl) of tritiated water (5 Ci ml^{-1}) and pre-reduced PtO_2 catalysts at 120°C for 16 hrs. There are 5 signals of approximately the same intensity and one signal (where the chemical shifts of the two sites are virtually the same) of twice the intensity. These results clearly signify that equilibrium has been reached and that very even incorporation of tritium has been achieved.

The third example shows the very even addition of tritium that has taken place, when the reduction of alprenolol with tritium gas (T_2) is studied in the presence of a heterogeneous catalyst (Pd/C). The signals now are of multiplets, with evidence of T-T couplings and a multiplicity of isotopomers [Fig. 1(C)].[17]

^{3}H NMR spectroscopy can also be used to follow rates of detritiation but at the present time more attractive alternatives are available. This is because the time required to produce the spectrum, despite the inherently high sensitivity of tritium, limits the approach to fairly slow reactions. Secondly, there is the problem of having to restrict the instrument to fairly long times on a single project when others, less time consuming, but equally important, require attention. With the possible improvements in sensitivity, the above-mentioned difficulties will be greatly reduced and the attractions of the technique for studies in the catalysis area greatly enhanced.

Studying the kinetics of heterogeneous reactions is more difficult and less accurate than the kinetics of homogeneous reactions; problems of catalyst reproducibility, stirring speed and diffusion come into play. On the other hand studies on the rates of detritiation of organic compounds have been widely investigated.[5] By being able to prepare a compound at very high specific activity, one can then use a very small quantity, *e.g.*, 1 μl in a 100 ml of reaction medium, so that even the most difficult of compounds to solubilise can be studied under homogeneous conditions. The second advantage relates to the rates of reaction. Invariably, 1st order kinetics are observed with the first order rate constant k^T being given by Eq. (1):

$$k^T = \frac{1}{t} \ln \left(\frac{a}{a - x} \right) \tag{1}$$

If the rate of detritiation is followed by measuring the increase in the radioactivity of the solvent *e.g.*, water, Eq. (1) now takes the form of

$$k^T = \frac{1}{t}\ln\left(\frac{C_\infty - C_0}{C_\infty - C_t}\right) \tag{2}$$

where C_0, C_t and C_∞ are the radioactivity of water at t = 0, t and also at the end of the reaction, respectively. C_0 is virtually zero, so that Eq. (2) becomes Eq. (4)

$$k^T = \frac{1}{t}\ln\left(\frac{C_\infty}{C_\infty - C_t}\right) \tag{3}$$

$$= \frac{1}{t}\ln\left(\frac{1}{1-x}\right) \tag{4}$$

For the initial, *i.e.*, less than 3% completion, stages of the reaction when $\ln(1-x) \approx -x$, Eq. (4) simplifies further to Eq. (5)

$$k^T = \frac{x}{t} \tag{5}$$

so that the radioactivity of the solvent increase linearly with time, *i.e.*, zero-order kinetics. C_t values can be of the order of a few thousand counts min^{-1} whilst C_∞ can be a few million counts min^{-1}. Therefore, it is possible to measure very slow rates of reaction in relatively short times and this is an extremely useful ability.[18]

The versatility of the technique can be illustrated by other examples, *e.g.*, when exchange occurs from twin sites[19] or when exchange is accompanied by other processes:[20]

A $\xrightarrow[k]{\text{hydrolysis}}$ B

A: detritiation $\downarrow k_1$ B: $k_2 \downarrow$ detritiation

Tritium labelled compounds can also be used as true tracers for following the rates of homogeneous reactions. Thus, in the reaction of amines with epoxides, one can label the amine and by taking samples at fixed time intervals, quenching and subjecting them to radio-HPLC one can follow the disappearance of amine, the increase in the formation of a 1:1 complex and in some cases, the appearance of further complexes until reaction is complete (Fig. 2).[21] With so much information becoming available it is then possible to model many of these reactions.[22]

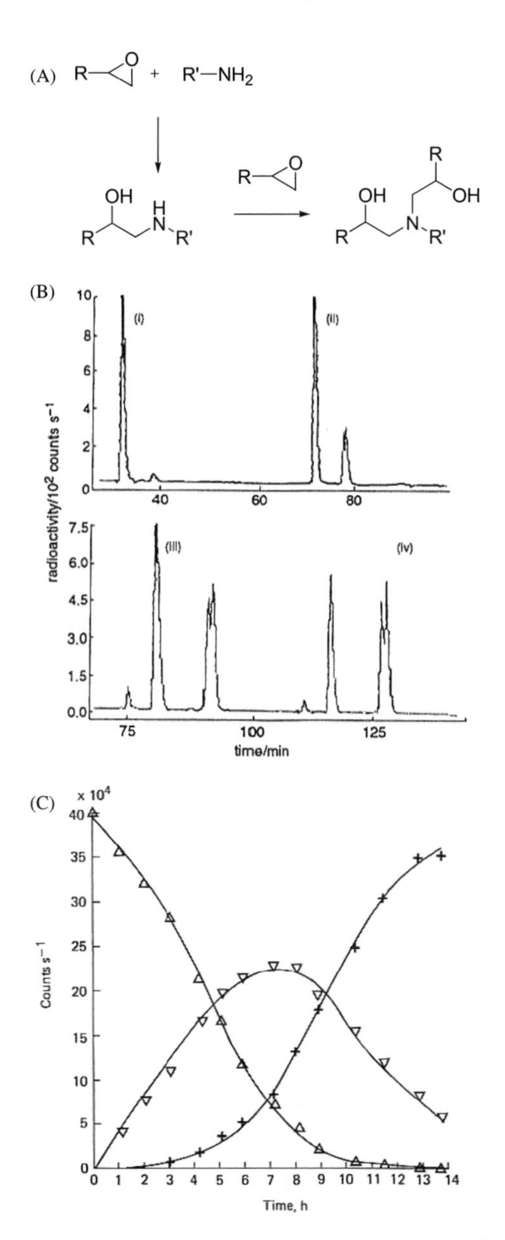

Fig. 2. (A) Scheme for the reaction of phenylglycidyl ether (PGE) and [G-^{3}H]aniline. (B) Radio-HPLC separation of primary amine and adducts for the reaction between PGE and aniline at various stages and (C) Variation in radioactivity (counts s^{-1}) for the reactions: (Δ) decrease in [G-^{3}H]aniline concentration, (∇) appearance of the 1:1 complex and (+) formation of the 2:1 complex. Reproduced with permission.

3. Tritiation Procedures

The most widely used methods for tritiating organic compounds include the following. Some methods are able to introduce the tritium at specific sites whilst others are used to prepare more generally labelled compounds.

(a) *Hydrogen isotope exchange* (which can be catalysed by acids, bases and metals):

catalyst
T_2 or HTO

R — T

catalyst
T_2 or HTO

R′ R″ — T T T R′ R″

(b) *Catalytic hydrogenation* (with palladium on carbon being the preferred catalyst):

T_2 Pd/C — $CHT{-}CH_2T$

(c) *Catalytic aromatic dehalogenation* (usually debromination using Pd/C)

H_3C Br — T_2 Pd/C — H_3C T

(d) *Methylation* (using ^{3}H-methyl iodide)

N H — CT_3I NaH — N CT_3

(e) *Sodium [^{3}H]borohydride reduction*

O H — $NaBT_4$ — HO H T

Whilst tritiation technology has progressed greatly in recent years there is still room for improvement. Many of the reactions are slow and require many hours to come to completion. Sometimes the regiospecificity is low and in other cases the degree of radioactive waste produced is high. Thus, in the catalytic aromatic debromination only 50% (maximum) of the tritium is incorporated, decreasing to 25% for borohydride reduction. Some of these problems have been addressed through the development of microwave-enhanced procedures, where the reactions are much faster and where the tritiated donor can be varied.

4. Discussion

In the case of homogeneous acid catalysed hydrogen isotope exchange at low $[H^+]$ concentrations the first-order tritiation rate constant, k^T, is linearly related to $[H^+]$. As the acid concentration is increased, it is the plot of $\log k^T$ and the acidity function H_0 that is linear. Many organic compounds are unable to withstand these hostile conditions and a combination of low $[H^+]$ concentration and high temperature has been used[23] to compensate for this difficulty. Acidic conditions tend to be used more for aromatic compounds and where separation of the product can be difficult, ion exchange resins can be usefully employed. Clear evidence that the extent of labelling was influenced by the nature of the polymer support matrix emerged. Dowex-50W-X8, Amberlyst 15 and Amberlyst 101 C resins were all effective with the much stronger acid, Nafion, being particularly so.[24] Zeolites have also been studied in the same way.[25]

$$[(CF_2CF_2)_nCFCF_2]_x$$
$$|$$
$$(OCF_2CF)_mOCF_2CF_2SO_3$$
$$|$$
$$CF_3 \qquad \text{(Nafion)}$$

Base catalysed hydrogen isotope exchange in highly basic media is a very popular and reasonably fast way of tritiating many organic compounds. As far as bases are concerned the hydroxide ion, either as NaOH or $(CH_3)_4NOH$, is the most frequently used. Unlike highly acidic media there is no need to increase the base concentration — a much more attractive alternative is to use a dipolar aprotic solvent such as DMSO. The hydroxide ion becomes gradually desolvated — $OH^-(H_2O)_3 \rightarrow OH^-(H_2O)_2 \rightarrow OH^-(H_2O) \rightarrow OH^-$ — as the DMSO concentration is increased. The basicity of the medium is now expressed by the H_- acidity function which increases from 12 for a 0.01 M $(CH_3)_4NOH$ solution in water to ~26 for a 99.6% DMSO solution, *i.e.*, a 10^{14} increase in basicity.[26] This means that many rates of detritiation are dramatically increased by using highly basic media. The reverse of this is that many weakly acidic carbon acids (pK_a's as high as 30) can now be readily tritiated. Anion exchange resins (Amberlite 402 or A26) were also found to be good base catalysts but could not be used because of thermal instability at high temperatures.[27] A polystyrene supported pyridinium hydroxide resin turned out to have fewer disadvantages.

High temperature solid-state hydrogen isotope exchange has been pioneered by Myasoedov *et al.* to tritiate organic compounds to high specific activities.[28] A large number of variables are able to influence the degree of isotopic incorporation and as with several of the catalysed hydrogen isotope exchange procedures we are some way from being able to predict the likely success of a particular reaction.

Metal catalysed hydrogen isotope exchange reactions have been extensively investigated by catalysis groups worldwide and their popularity for tritiating organic compounds has increased considerably with the development of ^3H NMR spectroscopy. Initial satisfaction with the production of a generally-labelled compound has since been replaced by an appreciation of the need to introduce tritium at metabolically stable positions. With metal catalysts the task is to improve the selectivity so that time consuming synthetic routes can be avoided. The early work of Garnett, Long and co-workers[29] served to draw attention to the potential of metal catalysts in tritium chemistry and homogeneous $RhCl_3$ enabled the highly regiospecific *ortho*-tritiation of arylcarboxylic acids and arylcarboxamides to be achieved.[30] The work was soon extended to the tritiation of a number of drugs containing different functional groups.[31] Ruthenium acetonylacetate was also identified as an excellent promoter of *ortho*-exchange in benzoic acids.[30] More recently studies of the organometallic chemistry of other transition metals, particularly iridium, have led to even better catalysts[32,33] that are able to operate at room temperature and where D_2 can be used as a donor. In the near future, it is likely that tritium gas, rather than tritiated water, will be used for these reactions, enabling products of high specific activity to be obtained. The advantages of these one-step metal-catalysed tritiation procedures are neatly illustrated by the example of pentamidine.[34] For ^{14}C synthesis, a 7-stage procedure taking approximately 1-man month is necessary and the product is obtained in low yield (19%) and specific activity ($24\,mCi\,mmol^{-1}$). The $RhCl_3$ procedure gave specific *ortho*-tritiation merely by heating at 110°C for 24 hrs; the specific activity was $90\,mCi\ mmol^{-1}$ (Fig. 3). Now, with

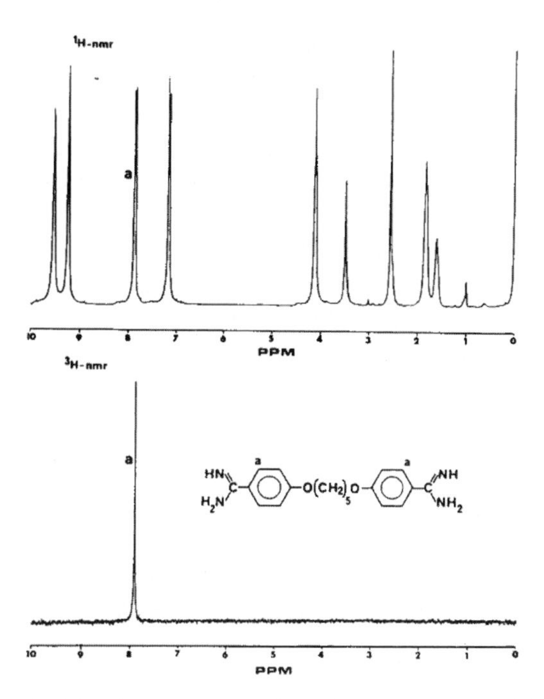

Fig. 3. Regiospecific tritiation of pentamidine using homogeneous $RhCl_3$ as catalyst. Reproduced with permission.

the use of microwaves, such compounds can be tritiated in even shorter time.

Many pharmaceutical agents are unable to withstand these high temperatures and are also required at higher specific activities. As mentioned above the need is for catalysts that can operate at ambient temperatures and where T_2 gas rather than tritiated water is used as the donor; parallel and combinatorial-type studies will accelerate this process.[35]

The above mentioned catalysts have been used for labelling aromatics. A successful example of the rapid regiospecific tritiation of aliphatics concerns the homogeneous tris-triphenylphosphine ruthenium(II) catalyst.[36] Specific exchange at the α-methylene group of a number of primary alcohols takes place. In this case, it is important to keep the reaction time short (<5 min), otherwise the β-methylene hydrogens also become labelled. The successful tritiation of ethanol, heptanol, benzyl alcohol and 3-phenylpropanol, together with the failure to tritiate tertiary alcohols, such as adamantol, strongly implies an oxidation-reduction mechanism.

Heterogeneous catalysts are also capable of achieving high regiospecificity. Thus Raney Nickel, a catalyst that has not been widely used in tritiation studies, labels only the side chains of toluene and butylbenzene under mild conditions.[36] On the other hand the Lewis acids, $EtAlCl_2$ and BBr_3, tritiate the equivalent 2,4,6-positions of 1,3,5-trimethylbenzene despite the obvious steric hindrance.

At an early stage in the use of ^{3}H NMR spectroscopy for studying hydrogenation reactions, using both homogeneous and heterogeneous catalysts, it was noticed that the tritium had not added evenly across the double bond and that some had migrated to adjacent sites. A typical example concerned *N*-acetylvaline (Fig. 4):[37]

$$(H_3C)_2C{=}C(NHCOCH_3)(COOH) + T_2 \xrightarrow{Pd/C} (H_3C)_2CT{-}CT(COOH){-}NHCOCH_3$$

In compounds such as styrene[38] the unsymmetrical pattern of labelling is thought to arise from a combination of exchange and hydrogenation reactions:

$$RCH_a{=}CH_2 \xrightarrow{T_2} RCH_a{=}CHT \xrightarrow{T_2} RCH_aT{-}CHT_2$$

No exchange of the H_a hydrogen takes place. For compounds of the kind $RCH_2CH{=}CH_2$, all three side chain positions are labelled and

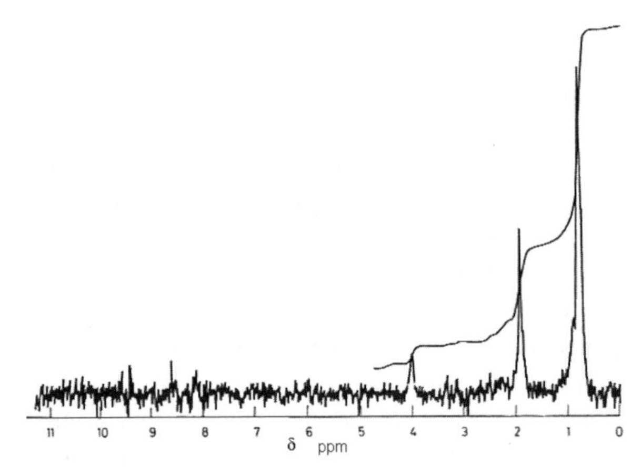

Fig. 4. ^{3}H NMR (^{1}H decoupled) spectrum of tritiated *N*-acetylvaline. Reproduced with permission.

there is evidence of vinylic, allylic and benzylic exchange as well as double bond migration and double bond hydrogenation. With homogeneous Wilkinson's catalyst similar patterns to those obtained for heterogeneous catalysts are observed. Only when the catalyst concentration is high is the labelling entirely specific and the distribution even. Some double labelling also occurs, implying that the $H_2 + T_2 \rightarrow 2HT$ reaction is not as fast as under heterogeneous conditions. By the time the heterogeneous catalytic (Pd/C) hydrogenation of β-methylstyrene to n-propylbenzene was studied a number of multiple pulse NMR techniques had been developed, and these, when applied to tritium,[39] demonstrate the power of ^{3}H NMR spectroscopy in the analysis of complex isotopic mixtures. A double quantum filter was applied to the one-dimensional spectra in order to directly observe the labelled molecules containing spin systems with greater than one tritium atom. J-resolved $^3H-^1H$ correlation and phase sensitive $^3H-^3H$ COSY spectra were also obtained whilst selective ^{3}H homodecoupling and $^3H-^1H$ DEPT spectra were reported for the first time. Once again the expected simple doublet at each methylene chemical shift is absent. The most abundant tritiated methyl species has 3 tritium atoms in the methyl carbon and there are clearly species present with at least 5 tritium atoms in the alkyl sidechain. All in all, evidence of in excess of 30 isotopic species, admittedly at widely different concentrations, was obtained. So called, well-understood reactions, can clearly benefit from further analysis.

Finally, as hydrogen isotope exchange is a reversible process, it can be used to detritiate a large number of organic compounds. Raney Nickel, for example, has been used[40] under microwave enhanced conditions to detritiate contaminated oils. In this way the tritiated water formed can then be used for tritiation purposes and the process repeated, thereby greatly reducing the problem associated with the storage of tritiated waste and contributing to Green technology.

5. Conclusion

The future is challenging and full of promise. The central issue is the need to be able to do more radiochemistry in a shorter time interval using less radioactivity whilst at the same time minimizing the amount of radioactive waste produced. The recent development of microwave-enhanced tritiation technology means that many labelled compounds can now be prepared in a matter of minutes. As a result of the improvement in NMR spectrometer performance ^{3}H NMR spectra can be obtained using less radioactivity than ever before. Liquid scintillation counting is extremely sensitive — detection at the Becquerel level (1 dps) is routine — and with ^{3}H accelerator mass spectrometry (AMS) this is a factor of 10–100 lower, taking us down to the attomole range. Will it be possible one day to obtain ^{3}H NMR spectra at this level of radioactivity? If so, it will be a wonderful achievement, opening the way to a large number of new applications.

As far as catalysis is concerned it is already benefiting from new developments such as supercritical fluids, ionic liquids, sol-gel technology, solid phase reactions, parallel and combinatorial chemistry and some of these can be coupled to microwave irradiation. Already a number of very efficient organometallic (mainly iridium based) catalysts have been produced although we are still some way from the acid, base situation where tritiation (or detritiation) rates can be estimated from a knowledge of acid-base strengths.

Finally the above improvements mirror those that are taking place in chemistry at large — the need to design more efficient catalysts so that reactions can proceed rapidly, more efficiently and with little waste formation. For these reasons alone the initial cost of setting up a tritium radiochemical facility are outweighed by the benefits.

6. Acknowledgements

The research carried out at the University of Surrey has been funded over many years by various research councils (SRC, SERC, EPSRC), NATO, The Royal Society and a large number of industrial concerns, too numerous to name individually. Shui-Yu Lu is supported by the Intramural Research Program of the NIH (National Institute of Mental Health).

References

[1] Gates BC, *Catalytic Chemistry*, John Wiley & Sons, New York, 1992.
[2] Thomas AF, *Deuterium Labelling in Organic Chemistry*, Appleton-Century-Croft Publishing, New York, 1971.
[3] Evans EA, *Tritium and Its Compounds*, 2nd ed., Butterworth, London, 1974.
[4] Jones JR, *The Ionisation of Carbon Acids*, Academic Press, London, 1973.
[5] Jones JR, in *Isotopes: Essential Chemistry and Applications*, Elvidge JA and Jones JR (eds.), The Chemical Society, London, 1980.
[6] Jones JR, Lu SY, in *Microwaves in Organic Synthesis*, Loupy A (ed.), Wiley-VCH, Weinheim, pp. 435, 2002.
[7] Elander N, Stone-Elander S, *J Label Compd Radiopharm,* **45**: 715, 2002.

[8] Elander N, Jones JR, Lu SY and Stone-Elander S, *Chem Soc Rev*, **29**: 239, 2000.
[9] Freifelder M, *Catalytic Hydrogenation in Organic Synthesis*, John Wiley & Sons, New York, 1978.
[10] Mantsch HH, Saito H and Smith ICP, *Prog NMR Spectroscopy*, **11**: 211, 1977.
[11] Lockley WJS, in *Isotopes: Essential Chemistry and Applications II*, Jones JR (ed.), Chapter 3, The Royal Society of Chemistry, London, 1988.
[12] Aspin IP, Hamerton I, Howlin BJ, Jones JR, Parker MJ, Russell JC and Vick TA, *J Chromatogr A*, **727**: 61, 1996.
[13] Evans EA, Warrell DC, Elvidge JA and Jones JR, *Handbook of Tritium NMR Spectroscopy and Applications*, John Wiley & Sons, Chichester, 1985.
[14] Bloxsidge JP, Garman RN, Gillies DG, Jones JR and Lu SY, in *Synthesis and Applications of Isotopically Labelled Compounds*, Vol. 8, Dean DC, Filer CN and McCarthy KE (eds.), John Wiley & Sons, Chichester, pp. 381, 2004.
[15] Gulati S, Method and system for signal detection in arrayed instrumentation based on quantum resonance interferometry, *US Patent 6671625*, 2003.
[16] Elvidge JA, Jones JR, Mane RB and Al-Rawi JMA, *J Chem Soc, Perkin II*, 386, 1979.
[17] Bloxsidge JP, Elvidge JA, Gower M, Jones JR, Evans EA, Kitcher JP and Warrell DC, *J Label Compd Radiopharm*, **18**: 1141, 1981.
[18] Jones JR and Taylor SE, *Chem Soc Rev*, **10**: 329, 1981.
[19] Buncel E, Davey JP, Buist GJ, Jones JR and Perring KD, *J Chem Soc, Perkin II*, 169, 1990.
[20] Jones JR and Taylor SE, *Intl J Chem Kinetics*, **12**: 141, 1980.
[21] Buist GJ, Hagger AJ, Jones JR, Barton JM and Wright WW, *Polym Commun*, **29**: 5, 1988.
[22] Buist GJ, Barton JM, Howlin BJ, Jones JR and Parker MJ, *J Mater Chem*, **5**: 213, 1995.
[23] Werstiuk NH, in *Isotopes in the Physical and Biomedical Sciences*, Vol. 1A, Buncel E and Jones JR (eds.), Chapter 5, Elsevier, Amsterdam, 1987.
[24] Brewer JR, Jones JR, Lawrie KWM, Saunders D and Simmonds A, *J Label Compd Radiopharm*, **34**: 391, 1994.
[25] Garnett JL and Long MA, in *Isotopes in the Physical and Biomedical Sciences*, Vol. 1A, Buncel E and Jones JR (eds.), Chapter 4, Elsevier, Amsterdam, 1987.
[26] Dolman D and Stewart R, *Can J Chem*, **45**: 911, 1967.
[27] Brewer JR, Jones JR, Lawrie KWM, Saunders D and Simmonds A, *J Label Compd Radiopharm*, **34**: 787, 1994.
[28] Shevchenko VP, Nagaev IY and Myasoedov NF, *Usp Khim*, **72**: 471, 2003.
[29] Blake MR, Garnett JL, Gregor IK, Hannan W, Hoa K and Long MA, *J Chem Soc, Chem Commun*, 930, 1975.
[30] Hesk D, Jones JR and Lockley WJS, *J Label Compd Radiopharm*, **28**: 1427, 1990.
[31] Hesk D, Jones JR and Lockley WJS, *J Pharm Sci*, **80**: 887, 1991.

[32] Hickey MJ, Jones JR, Kingston LP, Lockley WJS, Mather AN, McAuley BM and Wilkinson DJ, *Tetrahedron Lett*, **44**: 3959, 2003.
[33] Heys R, in *Synthesis and Applications of Isotopically Labelled Compounds*, Vol. 8, Dean DC, Filer CN and McCarthy KE (eds.), John Wiley & Sons, Chichester, pp. 37, 2004.
[34] Hesk D, Jones JR, Lockley WJS and Wilkinson DJ, *J Label Compd Radiopharm*, **28**: 1309, 1990.
[35] Crabtree RH, *J Chem Soc, Chem Commun*, 1611, 1999.
[36] Al-Rawi JMA, Elvidge JA, Jones JR, Mane RB and Saieed M, *J Chem Res-S*, 298, 1980.
[37] Evans EA, Warrell DC, Elvidge JA and Jones JR, *J Radioanal Chem*, **64**: 41, 1981.
[38] Elvidge JA, Jones JR, Lenk RM, Tang YS, Evans EA, Guilford GL and Warrell DC, *J Chem Res-S*, 82, 1982.
[39] Williams PG, Morimoto H and Wemmer DE, *J Am Chem Soc*, **110**: 8038, 1988.
[40] Jones JR, Langham PR and Lu SY, *Green Chem*, **4**: 464, 2002.

Chapter 5

Isotopic Exchange of Oxygen Over Oxide Surfaces

J S J Hargreaves and I M Mellor

1. Introduction

The isotopic oxygen exchange reaction is a useful tool for investigating the reactivity of various oxygen species on oxide surfaces. Most studies in this area have used $^{18}O_2$ as a source of isotopic oxygen, although other oxygen carriers such as $H_2{}^{18}O$,[1,2] $C^{18}O_2$[3] and $C^{18}O$[4] have been employed as well. There have also been studies employing $^{17}O_2$,[5] which is a useful isotope for ESR studies. The isotopic composition of oxygen is reported in Table 1.[6] Within this chapter, some of the literature employing $^{18}O_2$ will be reviewed. The early work on this reaction has also been the subject of a number of reviews elsewhere.[7–9] The application of isotopic oxygen in steady state transient kinetic studies and the measurement of surface diffusion will be reviewed in other chapters in this book. The intention of this chapter is to demonstrate the salient features of the reaction rather than being an extensive bibliography of the literature on this subject.

Two general approaches to the study of the exchange reaction have been made namely the homomolecular reaction (e.g., [10]) and the heterolytic reaction (e.g., [11]). In the former, a mixture of $^{18}O_2/^{16}O_2$ is scrambled over the oxide surface and in the latter, gas-phase $^{18}O_2$ is reacted directly with the oxide itself. The former reaction leads solely to the production of $^{34}O_2$:

$$^{16}O_{2(g)} + {}^{18}O_{2(g)} \rightleftharpoons 2{}^{16}O^{18}O_{(g)}$$

Table 1. Isotopic composition of oxygen.

Nuclide	Atomic mass	Natural abundance (%)	Nuclear spin I	Nuclear magnetic moment μ
^{16}O	15.99491	99.762	0	—
^{17}O	16.9991	0.038	5/2	−1.8937
^{18}O	17.9992	0.200	0	—

No long-lived radioactive isotopes.
^{15}O has $t_{1/2} = 124s$.

This reaction is sometimes termed as the R_0 mechanism, whereas in the case of heterolytic exchange, a number of different possibilities have been reported:

(i) the R_1 mechanism in which a single atom is exchanged in each reaction event:

$$^{18}O_{2(g)} + {}^{16}O_{(s)} \rightleftharpoons {}^{18}O^{16}O_{(g)} + {}^{18}O_{(s)}$$

(ii) the R_2 mechanism in which a double exchange event occurs leading to the production of $^{32}O_2$:

$$^{18}O_{2(g)} + 2^{16}O_{(s)} \rightleftharpoons {}^{16}O_{2(g)} + 2^{18}O_{(s)}.$$

An alternative route to R_2 exchange is via "place exchange", in which chemisorbed O_2 is displaced from the surface, e.g., [12, 13].

Once there is a mixture of isotopes in the gas-phase, a number of other possibilities can occur with the R_1 and R_2 mechanisms, e.g.,

$$^{16}O^{18}O_{(g)} + {}^{16}O_{(s)} \rightleftharpoons {}^{16}O_{2(g)} + {}^{18}O_{(s)}(R_1)$$
$$^{18}O^{16}O_{(g)} + 2^{16}O_{(s)} \rightleftharpoons {}^{16}O_{2(g)} + {}^{18}O_{(s)} + {}^{16}O_{(s)}(R_2)$$
$$^{18}O^{16}O_{(g)} + 2^{18}O_{(s)} \rightleftharpoons {}^{18}O_{2(g)} + {}^{18}O_{(s)} + 16O_{(s)}(R_2) \textit{ et cetera.}$$

At first sight, R_1 and R_2 mechanisms appear to be better probes of Mars–van Krevelen type processes[14,15] in which substrates are oxidised via lattice oxygen, generating vacancies which are replenished from the gas-phase. However, despite the different experimental approaches employed to study the homomolecular and heterolytic processes, consideration must be given to the possibility of the simultaneous operation of the R_0, R_1 and R_2 mechanisms. The following kinetic model, for the determination of individual rate constants for these processes, has been developed by Klier and co-workers:[16]

$$-\frac{dx}{dt} = \frac{R_0}{a}\left[x - \frac{1}{a}\left(\frac{w}{2}\right)^2\right] + \frac{R_1}{a}\left[x - \frac{w(c-w)}{2m}\right] + \frac{R_2}{a}\left[x - \frac{a(c-w)^2}{m^2}\right],$$

where x = the number of $^{18}O_2$ molecules in the gas-phase, a = the total number of oxygen molecules in the gas-phase, m = the total number of exchangeable sites in the solid, c = the total number of exchangeable atoms in the system and w = number of ^{18}O atoms in the gas-phase. This was then expanded to obtain an expression describing the dependence of the number of $^{18}O_2$ molecules in the gas-phase with time:

$$\begin{aligned} x = {} & \frac{ac^2}{(2a+m)^2} + (w_0 - w_\infty)\left(\frac{c}{2a+m}\right)\exp\left[-(2R_2+R_1)\left(\frac{2a+m}{2am}\right)t\right] \\ & - \left\{\frac{1}{2m}(w_0 - w_\infty)^2 \frac{\left[1+\left(\frac{m^2R_0}{4a^2R_2}\right)-\left(\frac{mR_1}{2aR_2}\right)\right]}{\left[2+\left(\frac{m}{2a}\right)-\left(\frac{mR_0}{2aR_2}\right)+\left(\frac{R_1}{R_2}\right)\right]}\right. \\ & \left. \times \exp\left[-(2R_2+R_1)\left(\frac{2a+m}{am}\right)t\right]\right\} \\ & + \left\{x_0 - \frac{ac^2}{(2a+m)^2} - (w_0 - w_\infty)\left(\frac{c}{2a+m}\right) + \frac{1}{2m}(w_0 - w_\infty)^2\right. \\ & \left. \times \frac{\left[1+\left(\frac{m^2R_0}{4aR_2}\right)-\left(\frac{mR_1}{2aR_2}\right)\right]}{\left[2+\left(\frac{m}{2a}\right)-\left(\frac{mR_0}{2aR_2}\right)+\left(\frac{R_1}{R_2}\right)\right]}\right\} \times \exp\left[-\frac{1}{a}(R_0+R_1+R_2)t\right], \end{aligned}$$

where x_0 = the value of x at time zero.

This expression has subsequently been evaluated and modified by Ponec and co-workers,[17] illustrating that the rate of diffusion of labelled oxygen into the bulk can be significant in exchange studies. This modified expression allows the determination of the apparent diffusion parameter in addition to the three individual R_0, R_1 and R_2 rate constants.

Cunningham and co-workers[18,19] have monitored the *atomic fraction 18* $\left({}^{18}f_a\right)$ to distinguish the homomolecular and heterolytic exchange processes:

$$^{18}f_a = \frac{P_{34} + 2P_{36}}{2(P_{32} + P_{34} + P_{36})},$$

where P = the partial pressures of the oxygen molecules with the specified masses. When employing this method, R_0 will not be accompanied by a change in $^{18}f_a$, unlike R_1 and R_2.

Studies of the homomolecular and heterolytic exchange processes are generally in the form of the measurement of rates under isothermal conditions. However, studies have also been made of temperature programmed isotopic exchange, in which the oxide is subjected to a temperature ramp under the reaction atmosphere, and the partial pressures of various isotopic oxygen species is determined as a function of temperature (e.g., Refs. 20–21). The photoactivation of oxygen exchange has also been reported in a number of studies which have been performed under UV irradiation (e.g., Refs. 18, 22, 23).

2. Activity Patterns

Historically, the approach taken in isotopic exchange studies was to screen large numbers of oxides and to make activity comparisons related to the patterns of oxidation activity for various reactions. In general, very good agreement has been found between the rates of the scrambling of $^{18}O_2/^{16}O_2$ gas-phase mixtures,[10,17] as well as the direct reaction of oxides with $^{18}O_2$,[17] for a variety of different reactions involving oxygen transfer. Figure 1 reproduces some of the pioneering work of Boreskov,[10] in which the activity of homomolecular exchange over some first row transition metal oxides is shown to correlate well with the activity patterns of hydrogen oxidation, methane oxidation and nitrogen oxide decomposition. In Fig. 2, a weak but significant correlation is shown between the temperature required for 30% methanol conversion in methanol oxidation, catalysed by a range of oxides and their published rates of isotopic exchange at 350°C.[24] Both figures demonstrate the applicability of isotopic oxygen exchange to the study of oxidation catalysis.

Winter, another early pioneer of the study of oxygen exchange, has made a thorough investigation of the kinetics of R_1 and R_2 exchange over 38 inorganic oxides, and the results are reproduced in Tables 2 and 3.[11] The

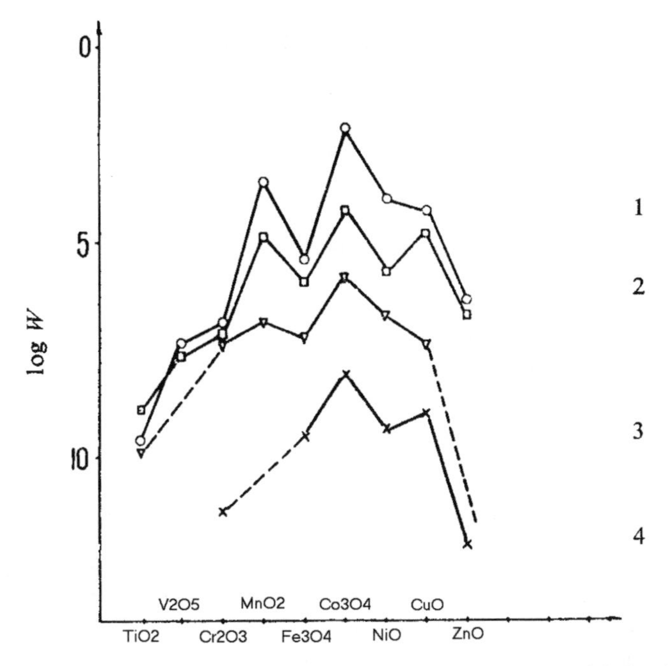

Fig. 1. Activity patterns of first row transition metal oxides at 300°C for (1) homomolecular exchange of oxygen, (2) oxidation of hydrogen, (3) oxidation of methane and (4) nitrogen oxide decomposition. Adapted from reference [10] with permission.

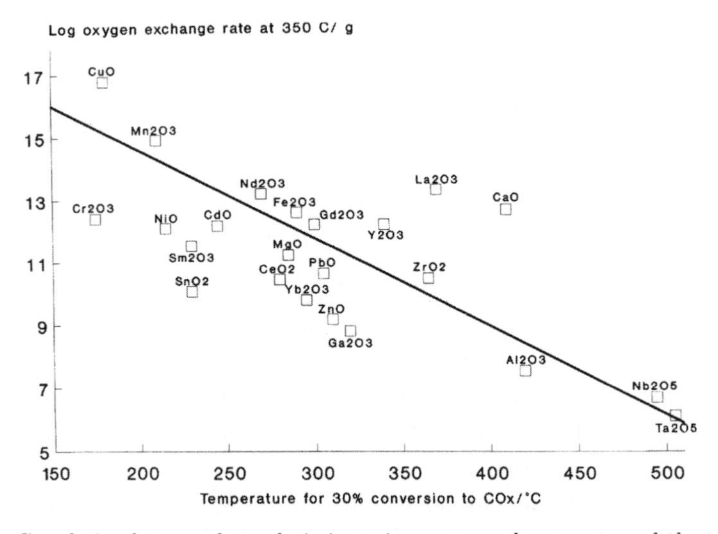

Fig. 2. Correlation between heterolytic isotopic oxygen exchange rate and the temperature for 30% methanol conversion to carbon oxides. Adapted from reference [24] with permission.

Table 2. Oxygen exchange reactions: oxide outgassed.

Oxide	B.E.T. (m^2g^{-1})	T_0 (°C)	Reaction Temp. (°C)	Main Reaction	log_{10} (A_0)	E (kcal mol^{-1})	log_{10} $(rate)_{350}$	log_{10} $(rate)_{300}$	m
MgO	61.5	510	350–440	$R_1(R_0)$	22.90	38	9.66	8.50	0
CaO	29.8	730	260–370	R_1	23.18	36	10.63	9.46	0.5
NiO	3.12	520	200–370	$R_1(R_2)$	23.85	35	11.66	10.59	0
ZnO	5.25	550	360–420	$R_1(R_0)$	25.35	40	11.42	10.19	1.0
$Al_2O_3(\gamma)$	95	600	480–580	R_1	21.33	39	7.75	6.55	0
$Al_2O_3(\delta)$	41	730	620–720	R_1	19.98	39	6.39	5.20	0
Cr_2O_3	14.2	530	270–400	$R_1 > 6R_2$	22.30	32	11.15	10.18	0
$Fe_2O_3(\alpha)$	33.5	450	290–400	$R_1 \approx 4R_2(R_0)$	19.50	19	12.88	12.30	1.0
Sc_2O_3	14.25	620	450–550	$R_1(R_0)$	21.66	36.5	8.94	7.80	0
Y_2O_3	3.64	630	250–400	R_1	20.64	27	11.23	10.40	0
La_2O_3	2.41	620	350–450	R_1	19.34	23	11.32	10.62	0
ZrO_2	10.45	630	550–650	R_1	15.45	15	10.22	9.87	0.9
HfO_2	7.6	600	550–650	R_1	16.25	20	9.28	8.62	0.9
ThO_2	5.9	600	450–550	R_1	19.62	26	10.56	9.77	0.5
TiO_2(R)	5.4	750	630–750	R_1	20.32	35	8.12	7.06	1.0
TiO_2(A)*	25.4	750	630–730	R_1	18.31	32	7.16	6.19	1.0
SiO_2	296	750	650	R_1	—	—	8.0†	—	—
GeO_2	0.83	750	700				Inactive		
Nb_2O_5	4.25	740	600–730	R_1	19.16	39	5.57	4.33	1.0
Ta_2O_5	3.02	740	600–730	R_1	19.21	39	5.62	4.38	1.0

*During reduction of B.E.T. to ~1/10 of original: material stabilised at 20–30 $m^2 g^{-1}$ had $n_s = 3\text{–}6 \times 10^{20}$.
†Rate at 650°C: undetectable at 600°C.
T_0 refers to outgassing temperature.
Rate is defined as molecules cm^{-2} s^{-1}. Rates reported at 300 and 350°C.

data has been determined by applying the following equation:

$$\text{rate} = P_{O_2}{}^{m} A_o \exp(-E/RT),$$

where P_{O_2} is the partial pressure of oxygen, m is the reaction order, A_o the pre-exponential factor and E is the activation energy.

Although the activities have been determined under differing reaction temperatures, they have been extrapolated to 300°C and 350°C as a means of comparison. The data in the tables illustrates a number of features of the reaction. Comparison has been made between the conditions of pre-treatment, such as vacuum versus oxygen atmospheres. The temperature of the vacuum pre-treatment was carefully controlled, as previous studies had shown that when the sample was pre-heated too high in vacuum, a

Table 3. Oxygen exchange reactions: pretreated with oxygen.

Oxide	B.E.T. (m^2g^{-1})	Reaction Temp. (°C)	Main Reaction	$\log_{10}$ (A_0)	E (kcal mol^{-1})	$\log_{10}$ $(rate)_{350}$	$\log_{10}$ $(rate)_{300}$	m
BeO	15.3	660–720	R_1	20.35	44.5	4.84	3.50	0
MgO	61.5	370–450	R_1	22.69	38	9.44	8.29	0
CaO	35	300–400	R_1	21.59	30	11.14	10.22	0.7
NiO	6.04	330–400	$R_1 \approx 4R_2$[d]	26.49[a]	45[a]	10.81[a]	9.44[a]	0.2[a]
ZnO	5.25	350–410	R_1	21.00	36	8.45	7.36	1.0
SrO	0.26	250–320	R_1	16.74	15	11.51	11.05	0.5
CdO	5.5	250–350	R_1	18.41	20	11.44	10.83	0.9
CuO	10.3	230–330	R_2	23.44	22	15.77	15.09	0.5
PdO	20.8	200–300	$R_1(R_0)$	18.82[a]	20[a]	11.85[a]	11.25[a]	0[a]
AgO	2.0	90–130	$R_2 > 10R_1$	29.75[b]	34[b]	17.88[b]	16.87[b]	0.2[b]
PbO	1.56	420–500	$R_2 \approx 10R_1$	14.80[b]	12.5[b]	10.44[b]	10.06[b]	0.35[b]
$Al_2O_3(\alpha)$	15	500–600	R_1	22.04	45	6.36	4.99	0
Cr_2O_3	14.2	270–370	$R_1(R_2)$[d]	22.62	34	10.77	9.74	0
Fe_2O_3	33.5	260–340	$R_1 \approx 2R_2$[d]	20.48[a]	27[a]	11.07[a]	10.25[a]	0.8[a]
Ga_2O_3	14.2	520–620	R_1	21.87	40.5	7.65	6.52	0
Rh_2O_3	1.49	290–370	$R_1 \approx 1.5R_2$	15.23[c]	10[c]	11.75[c]	11.44[c]	0[c]
Sc_2O_3	14.25	450–520	$R_1(R_0)$	23.70	43	8.72	7.41	0
$Mn_2O_3(\alpha)$	4.18	200–280	$R_1(R_2)$	22.29	23	14.28	13.58	0.3
Y_2O_3	3.64	250–400	R_1	18.11	18.5	11.66	11.10	0
In_2O_3	2.65	550–650	$R_1 > 10R_2$	24.70[a]	57[a]	4.84[a]	3.10[a]	0.2
La_2O_3	2.41	220–300	R_1	16.77	11	12.94	12.60	0
Gd_2O_3	11.7	240–320	$R_1(R_0)$	15.33	12	11.15	10.79	0
ZrO_2	10.45	350–450	R_1	19.91	30	9.46	8.54	1.0
CeO_2	1.68	450–520	$R_1 \approx 8R_2$	19.28[a]	26[a]	10.21[a]	9.43[a]	0.79[a]
HfO_2	7.6	350–450	R_1	19.68	30	9.23	8.31	1.0
ThO_2	5.9	320–420	R_1	18.44	22	10.42	9.72	0.8
$TiO_2(R)$	17	500–580	R_1	20.44	35	8.24	7.03	1.0
MnO_2	57	250–320	$R_1 \approx 4R_2$	16.34[a]	15[a]	11.11[a]	10.65[a]	0.05[a]
RuO_2	1.06	320–400	$R_1(R_0)$	18.14	18	11.97	11.32	0
RuO_2	1.06	200–280	R_0	23.45	23	15.43	14.74	0
SnO_2	6.18	450–520	R_1	18.80	27	9.38	8.57	1.0
IrO_2	4.60	150–200	$R_1 \approx 3R_2$	24.60[a]	24[a]	16.24[a]	15.51[a]	0[a]
SiO_2	296	650	R_1	—	—	0*	—	—
GeO_2	0.83	605–700	—	—	—	0*	—	—
Nb_2O_5	4.25	600–700	R_1	20.34	41	6.05	4.80	1.0
Ta_2O_5	3.02	600–700	R_1	19.56	40	5.62	4.40	1.0

[a]Refers to R_1. [b]Refers to R_2. [c]Refers to $R_1 + R_2$. *Inactive. [d]Ratio of R_1 to R_2 variable: results not very reproducible. Rate is defined as molecules cm^{-2} s^{-1}. Rates quoted at 300 and 350°C.

rapid initial decrease of $^{18}O_2$ was observed,[25] resulting from the formation of a defective surface rather than exchange proper. In Table 2, the pre-treatment temperature has been carefully controlled, and a comparison with Table 3 shows that while there may be small influences of the pre-treatment atmosphere in some cases, there are however no significant trends. On inspection of the tables, it is evident that within the pressure range studied, the reaction is frequently zero order in oxygen. This has also been observed in other studies (e.g., Refs. 26, 27) and has been taken as an indication that the desorption of oxygen is a rate determining step in this regime, despite the correlations of this data with oxidation activity, as illustrated in Fig. 2. In addition, the values of activation energy indicate the absence of diffusion limitation in the temperature ranges studied. This was previously a concern in earlier studies, where there was disagreement as to the extent of its influence.[8,11]

Taking MgO as an illustrative example, Table 4 has been constructed to display the range of data which has been reported for a single oxide. It is apparent that there is a marked variation, with reported activation energies ranging from $13\,kJ\,mol^{-1}$ (for a diffusion limited regime) to $256\,kJ\,mol^{-1}$. There is also little agreement as to the pattern of heterolytic exchange i.e., the extent of the R_1 versus R_2 mechanism. Part of this discrepancy lies in the choice of preparation route, with potential contributions from the role of impurities, which may be present in significant concentrations for preparations from some precursors[35] as well as from structure sensitivity. In relation to the former possibility, the presence of chloride, for instance, has been shown to strongly poison exchange in M^{3+}/CeO_2 systems.[36] Preparing MgO from the basic carbonate can lead to MgO containing Na^+ and Ca^{2+}, which can segregate to the surface and may be expected to promote the rate of exchange on the basis of Table 3. It is noted that despite this important consideration, the preparative methods and impurity content of the various oxides tested, are often not adequately described, especially in the earlier studies where activity comparisons between different oxides have been made. The possibility of structure-sensitivity affecting exchange rate and activity patterns has not been much studied in relation to oxygen exchange, which is surprising in view of the well known effect of structure in oxidation catalysis (e.g., Ref. 37). However, it has been shown that the surface area normalised reaction rate is independent of particle size for 100, 500, 1,000 and 2,000 Å Ube MgO samples.[34] The relative proportions of 5, 4 and 3 co-ordinate lattice oxide ions has little effect on the reaction, while other surface sites may be the locus of exchange, as described later in this chapter. Furthermore, the occurence of the R_2

Table 4. A diffusion limitation still apparent at higher temperatures but magnitude not specified.

Author and reference	Reactor type	MgO Precursor	Pre-treatment	Reaction	Ea/kJ mol^{-1}
Winter[28–30]	closed	$MgCO_3$	673–873 K in vacuum, 16 h	heterolytic	$142 < 693$ K $13 > 693$ K
Winter[7,29]	closed	$MgCO_3$	813 K in vacuum, 16 h	homomolecular	113
Winter[7]	closed	$MgCO_3$	813 K in vacuum, 16 h	heterolytic	$150 < 693$ K $32 > 693$ K
Boreskov[8]	closed	no reference	773 K in vacuum, 4 h	homomolecular heterolytic	167 167
Muzykantov *et al.*[30,31]	closed	no reference	in vacuum then in O_2 at exch. Temp.	$2R_2 = R_1$	—
Winter[11]	closed	$MgCO_3$	783 K in vacuum 753 K in O_2	R_1 (little R_0) R_1	159^a 159^a
Martin and Duprez[20]	closed	commercial 100 Å Ube	in vacuum then O_2/H_2 at 723 K	R_1	166
Mellor *et al.*[34]	closed	commercial 100, 500, 1000 and 2000 Å Ube	733 K in vacuum, or in O_2, 16 h	R_1	120 ± 25
Mellor *et al.*[26,27]	closed	$Mg(OH)_2$ and basic magnesium carbonate	733 K in O_2	R_1 and R_2 — exact relative proportions dependent upon preparation temperature	—
Karasuda and Aika[33]	closed	commercial	1173 K in vacuum, 1 h	heterolytic	—
Peil *et al.*[32]	continuous flow	no reference	O_2 at exchange temperature, 30 min.	heterolytic	256

mechanism may be associated with MgO samples prepared from hydroxide containing precursors, such as brucite and basic carbonate.[27]

Despite the lack of study of structure-sensitivity, many workers have sought to relate the activity patterns for exchange with different oxides to various physical parameters, including a relationship between the activation energy for exchange and the heat of formation of the oxide,[8] illustrated in Fig. 3. On the basis of this figure, it was concluded that Co_3O_4 had an optimum strength of binding, whilst CuO was sub-optimum. This was subsequently disputed by Winter,[11] whose data was reproduced in Fig. 4. A wider range of oxides was plotted and it was concluded that such a relationship was generally non-existent. Co_3O_4 was omitted from this study due to its instability. In the latter study, it was concluded that the variation in metal ion valency and the structure-type was so great that no relationship between them was likely. However, subdivisions of the data by valency and/or structure-type were made and relationships between activation energy and [molecular volume]$^{1/3}$ and metal-oxygen nearest distance were evident. In the former, oxides of general formula MO, of the rock salt and wurtzite structure-types, fell on a single line, with the MO_2 oxides falling on another line; the M_2O_3 oxides could be sub-divided between those with the corundum structure and all the rest. The activation energy was observed to decrease with increasing unit cell size. However, within Winter's data, the best correlation was found to exist between the nearest oxygen-oxygen distance as shown in Fig. 5. Since there is a compensation

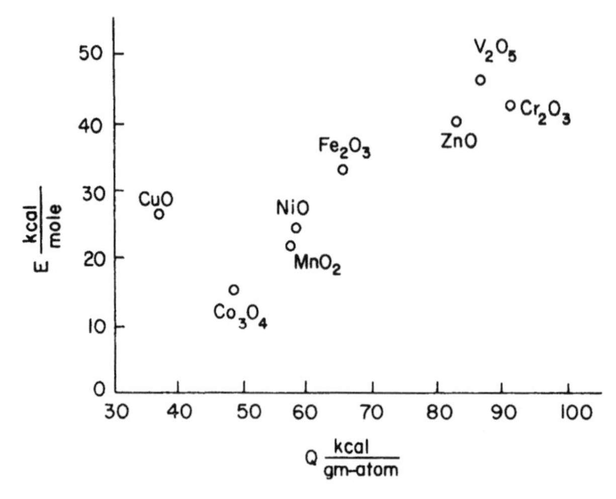

Fig. 3. Relationship between activation energy for oxygen exchange and heat of formation of the corresponding oxide. Reproduced from reference [8] with permission.

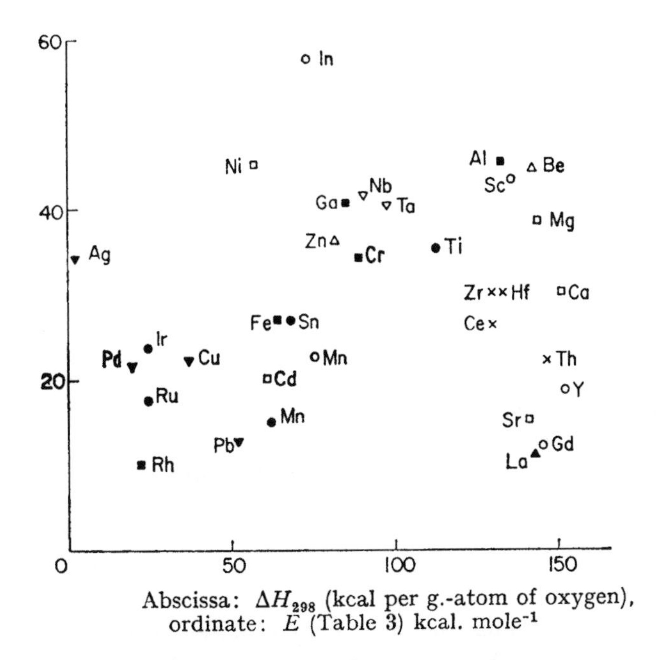

Fig. 4. Relationship between activation energy for exchange and heat of formation of the corresponding oxide. Reproduced from reference [11] with permission.

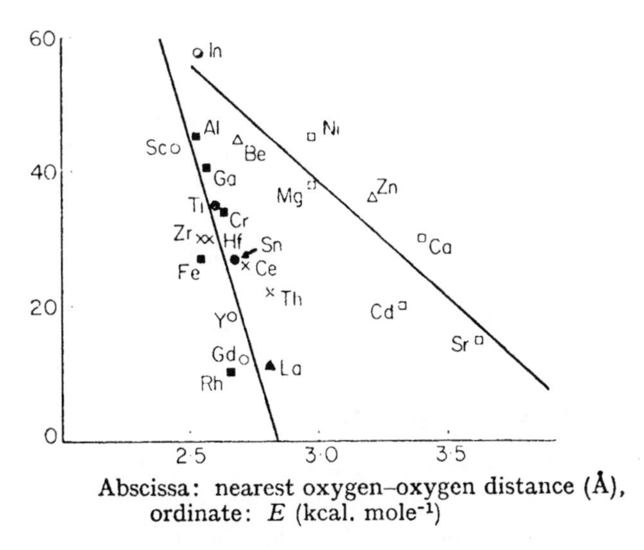

Fig. 5. Correlation between activation energy for exchange and nearest oxygen-oxygen distance. Reproduced from reference [11] with permission.

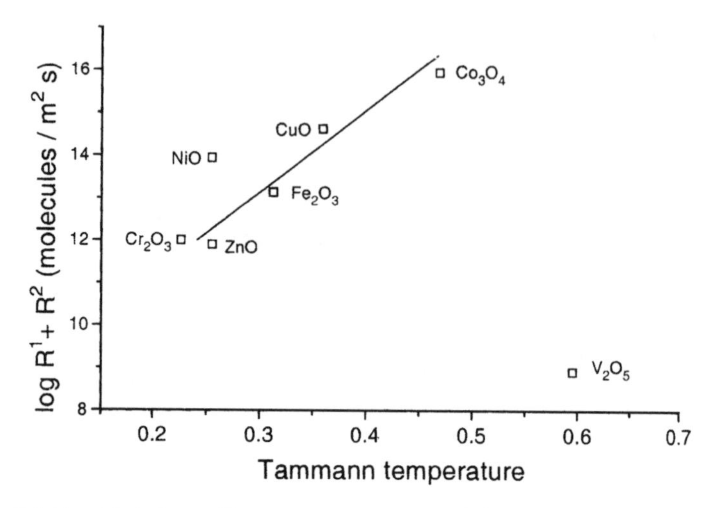

Fig. 6. Relationship between Tamman temperature and the rate of R1 + R2 exchange for some first row transition metal oxides. Reproduced from reference [17] with permission.

effect operative in Winter's data, similar correlations can be made with the log of A (where A is the pre-exponential factor). As n- and p-type semiconducting oxides lose and gain oxygen respectively on heating in air, a relationship may be expected to exist between semi-conductivity type and exchange activity. However, in both Boreskov's and Winter's work, this relationship was not apparent. Inspite of this, semiconducting metal oxides can be activated for exchange using UV light irradiation.[18,22,23]

Ponec and co-workers[17] have recently demonstrated a relationship between the Tamman temperature and the R_1 and R_2 exchange rate for some first row transition metal oxides as shown in Fig. 6. In this study, the Tamman temperature plotted is the ratio of the exchange reaction temperature to the Tamman temperature of the oxide. This was taken to be indicative of the importance of diffusion, with surface diffusion occurring in the 0.2–0.5 temperature range and a bulk diffusion beyond 0.5.

3. Nature of the Exchange Site

In addition to O^{2-} ions of varying degrees of co-ordination, the presence of a number of different oxygen species such as O^-, O_2^-, O_2^{2-}, O_3^-, is well documented on oxide surfaces.[38–40] Within the literature, relatively little attention is devoted to the exact nature of the oxygen species undergoing exchange. The identity of the active species is likely to be highly dependent on the oxide under consideration, the temperature range employed, the particular exchange pattern, i.e., R_0 v R_1 v R_2, and the pre-treatment history. It is very tempting to ascribe the R_1 process to the decomposition of a three-centred intermediate and that of the R_2 and R_0 pathways to the decomposition of four centred intermediates. In the following few pages, we briefly describe a number of proposals which have been made for alkaline earth oxides, which is illustrative of some of the discussion on this topic.

For the alkaline earth oxides, Winter originally suggested that R_2 defect centres, i.e., pairs of associated F centres, were of considerable importance.[11] It was proposed that the activation of oxygen by such centres was rapid and non-activated with the reverse reaction (i.e., the elimination of exchanged oxygen) being responsible for the observed kinetics:

$$O_2(\text{gas}) + (e^-|\square_s^-)(e^-|\square_s^-) \rightleftharpoons 2(O^-|\square_s^-),$$

where e^- = a trapped electron and $\square_s^-$ = a surface vacancy, $(e^-|\square_s^-)$ centres were proposed to be highly mobile on the surface with exchange occurring via processes such as

$$({}^{18}O^-|\square_s^-)_1 + ({}^{16}O^{2-}|\square_s^-)_2 \rightleftharpoons ({}^{18}O^{2-}|\square_s^-)_1 + ({}^{16}O^-|\square_s^-)_2,$$

and rate-limiting desorption of exchanged oxygen occurring after the collision of two $(O^-|\square_s^-)$ centres. Electron donor impurity centres were also proposed to assist in the generation of $(O^-|\square_s^-)$ centres, via the decomposition of intermediate O_2^- formed via the adsorption of gas-phase O_2. On the basis of the energetics and low probability of a four centred collision, the possibility of exchange occurring via the activation of oxygen from $2O^{2-}(s) + O_2 \rightleftharpoons 4O^-(s)$ was ruled out.

In a comparison of the activities of hydrogen-deuterium exchange, heterolytic oxygen isotopic exchange and homomolecular oxygen isotopic exchange over Specpure CaO, Cunningham and Healy remarked the similarity of the optimum pre-treatment conditions required to give the optimal H_2/D_2 and heterolytic exchange rates, principally the out-gassing temperature.[19] Surface hydroxyls, proposed to be of significance for the H_2/D_2 exchange, were also suggested to be involved in heterolytic exchange at 350°C, with the pathway being:

$${}^{18}O_{2(g)} + {}^{16}OH^- \rightleftharpoons {}^{16}O^{18}O_{(g)} + {}^{18}OH^-.$$

At higher temperatures, additional pathways were suggested to be operative and it was observed that homomolecular exchange required the removal of most surface hydroxyls at high temperatures in order to proceed. In addition, Hargreaves *et al.*[34] proposed the involvement of surface hydroxyls in their study of the structure-sensitivity of heterolytic exchange over MgO at 415–460°C, since no direct correlation of exchange rate could be made with the concentration of various low-co-ordinate O^{2-} ions, and also that only a fraction of the oxygen monolayer could be exchanged under these conditions. However, subsequent *in situ* FTIR measurements have demonstrated that $^{16}OH/^{18}O_2$ exchange is extremely rapid on the experimental timescale. This process is almost instantaneous and as such, it is unable to account for the majority of the exchange process, occuring over a longer time frame. Therefore, additional proposals have been made, including the possible involvement of ensembles of low co-ordination sites and O^- sites generated via the decomposition of $OH_{(s)}$.[27]

In terms of the intermediacy of O^- generated via the dehydrogenation of $OH_{(s)}$, Karasuda and Aika[33] have observed that there are at least two exchange processes of differing rates occurring over MgO at 700°C as shown in Fig. 7. The fast and slow rates were a result of the exchange of surface and lattice oxygen, as detailed in the scheme below. On the basis of the quantification of the various numbers of exchangeable oxygen species with altervalent metal ion doping studies, along with correlations with H_2 TPD, O^- and the oxygen ions surrounding it was proposed as the site undergoing very rapid exchange:

$$^{18}O_{2(g)} + {}^{16}O^-_{(s)} \rightarrow {}^{16}O^{18}O_{(g)} + {}^{18}O^-_{(s)}.$$

The active $O^-_{(s)}$ involved was proposed to be generated via the decomposition of some of the hydroxyl groups (hence the relationship with H_2 TPD),

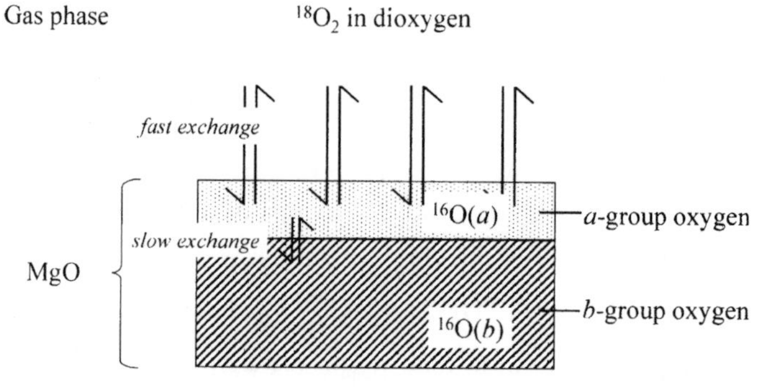

Fig. 7. The two exchange processes reported to occur over MgO at 700°C. Reproduced from reference [33] with permission.

according to the following scheme:

$$Mg(OH)_2 \rightarrow Mg^{2+} + O_2^{2-} + H_2 \rightarrow Mg^{2+} + 2O^- + H_2.$$

The evolution of hydrogen during the thermal decomposition of brucite has also been reported elsewhere.[41]

The involvement of low co-ordination O^- ions in the exchange process over CaO and irradiated MgO has been suggested by Yanagisawa *et al.*,[42,43] who have postulated the involvement of an O_3^- intermediate in thermal desorption studies of pre-adsorbed $^{18}O_2$. This intermediate was proposed to be symmetrical and the consideration of its decomposition pathway was stated to be compatible with the isotopic composition of one of the thermal desorption states. However, it was inconsistent with other thermal desorption pathways and alternative proposals for intermediates in these were made and are mentioned below. Acke and Panas[44] have compared the effects of various pre-treatments on the exchange properties of CaO. Based on activation energy considerations, they have concluded that in order for exchange to occur, a triplet to singlet interconversion (i.e., $^3O_2 \rightarrow {}^1O_2$) is necessary. In the case of pre-reduction, they argued that an $^2O_2$ adsorption channel is accessible, leading to an R_1 process with lower apparent activation energy.

Using ^{17}O and ESR spectroscopy, Tench reported the following process involving O^- and intermediate ozonide, occuring over MgO:[5]

$$O_{2(g)} + O^-_{(s)} \rightarrow O^-_{3(s)}$$

followed by:

$$2O^-_{3(s)} \rightarrow 2O^-_{2(s)} + O_{2(g)}.$$

This mechanism could be expected to lead to the occurrence of some R_2 products.

As described, the potential involvement of O_2^{2-} species as intermediates in the hydroxyl group decomposition suggested by Karasuda and Aika[33] is interesting, as an R_2 exchange pathway could result from exchange with this diatomic active species. However, Acke and Panas[44] have reported in their study that R_2 exchange between $^{18}O_2$ and pre-oxidised CaO occurs via two successive single exchange events, on the basis of pre-exponential factor considerations. In other studies over MgO where a large excess of $^{18}O_2$ has been employed, this was unlikely to happen.[26,27] As mentioned previously, a significant amount of R_2 exchange in conjunction with R_1 has occured over MgO samples prepared from hydroxylated precursors,[26,27] and can be interpreted as taking place with diatomic moieties such as

O_2^{2-}. In this context, it is interesting to note that Yanagisawa *et al.*[43] have only observed the presence of R_2 exchange activity over pre-irradiated MgO in their thermal desorption studies. They have reported that pre-irradiation was not a requirement for the occurrence of the R_2 process over CaO and SrO. On the basis of the isotopic distribution of the thermal desorption state where the R_2 process is observed, they have invoked a non-square O_4^- intermediate. The formation of the low symmetry four-centred intermediate is suggested to occur at kink sites on single atom stepped (111) surfaces. This suggestion raises the interesting possibility that the presence of the R_2 process may be morphological in origin. It is notable that the effect of variation of MgO morphology was not investigated in this study. Since the morphology of the alkaline earth oxides, e.g., MgO,[35] is known to be strongly dependent upon the preparation method/conditions, it may be possible to interpret the observations for non-irradiated MgO in Table 4 on this basis. The same group have extended their observation of the R_2 process on alkaline earth oxides by employing *ab-initio* molecular orbital calculations.[45] They propose the involvement of a charged O_5 intermediate which decomposes to release the exchanged O_2. Modelling was performed on defective (111) surface clusters and the apparent lack of the R_2 process on non-irradiated MgO was explained as a result of its rigid co-ordination not favouring the formation of the key O_5 intermediate. The effect of UV irradiation on oxygen exchange has also been investigated in the TiO_2 system[23,46–48] and room temperature exchange with lattice oxygen has been reported in some cases[23,46] while not in others, although photodesorption was reported.[47] In this system, it has been proposed that UV light enhances the exchange between gas-phase O_2 and adsorbed water.[48]

In view of the above, it is clear that a range of different proposals have been made to account for the exchange activity and patterns observed over the alkaline earth oxides. The extent to which the involvement of different oxygen species may be operative under different conditions still remains uncertain. It is also apparent that parameters such as oxide preparation route, pre-treatment and impurity concentration, could have a pronounced effect on reactivity, even though studies in this area have been rather limited to date.

4. Conclusion

In this short review, we have sought to provide an introduction to some of the extensive literature on the isotopic oxygen exchange reaction over metal oxides, to illustrate some of the salient points. The relationship between the activity patterns and those with reactions involving oxygen transfer, has been clear from an early stage. Despite this, even for the simplest metal oxide systems, the process is still relatively little understood. This is perhaps not surprising, in view of the fact that different active species may potentially be reactive under the various sets of conditions studied. The application of oxygen exchange for the measurement of diffusion parameters and its application in the determination of reaction mechanisms by the steady state transient isotope kinetic technique, are reviewed in other chapters within this book.

References

[1] Whalley E, Winter ERS, *J Chem Soc* 1175, 1950.
[2] Cameron WC, Parkas A, Litz LM, *J Phys Chem* **57**: 229, 1953.
[3] Tsuji H, Shishido T, Okamura A, Gao Y, Hattori H, Kita H, *J Chem Soc, Faraday Trans* **90**: 803, 1994.
[4] Guerro-Ruiz A, Rodrigez-Ramos I, Ferreira-Aparicio P, Volta J-C, *Catal Lett* **45**: 113, 1995.
[5] Tench AJ, *J Chem Soc, Faraday Trans 1* **68**: 1181, 1972.
[6] Emsley J, *The Elements*, Oxford University Press, Oxford, 1989.
[7] Winter ERS, *Adv Catal* **10**: 196, 1958.
[8] Boreskov GK, *Adv Catal* **15**: 285, 1964.
[9] Novakova J, *Catal Rev* **4**: 77, 1970.
[10] Boreskov GK, *Disc Faraday Soc* **41**: 263, 1966.
[11] Winter ERS, *J Chem Soc A* 2889, 1968.
[12] Ben Tarrit Y, Naccache C, Che M, Tench AJ, *Chem Phys Lett* **24**: 41, 1974.
[13] Che M, Shelimov BN, Kibblewhite JFJ, Tench AJ, *Chem Phys Lett* **28**: 387.
[14] Mars P, van Krevelen DW, *Chem Eng Sci* **3** (special supplement): 41, 1954.
[15] Doornkamp C, Ponec V, *J Mol Catal A: Chemical* **162**: 19, 2000.
[16] Klier K, Novakova J, Jiru P, *J Catal* **2**: 479, 1963.
[17] Doornkamp C, Clement M, Ponec V, *J Catal* **182**: 390, 1999.
[18] Cunningham J, Goold EL, Leahy EM, *J Chem Soc, Faraday Trans 1* **75**: 305, 1979.
[19] Cunningham J, Healy CP, *J Chem Soc, Faraday Trans 1* **83**: 2973, 1987.
[20] Martin D, Duprez D, *J Phys Chem* **100**: 9429, 1996.
[21] Ciuparu D, Bozon-Verduraz F, Pfefferle L, *J Phys Chem B* **106**: 3434, 2002.
[22] Cunningham J, Goold EL, Fierro JLG, *J Chem Soc, Faraday Trans 1* **78**: 785, 1982.
[23] Courbon H, Formenti M, Pichat P, *J Phys Chem* **81**: 550, 1977.
[24] Taylor SH, Hargreaves JSJ, Hutchings GJ, Joyner RW, *Appl Catal A:Gen* **126**: 287, 1995.
[25] Winter ERS, *J Chem Soc* 3824, 1955.
[26] Mellor IM, "Isotopic exchange reactions on magnesium oxide", PhD thesis, The Nottingham Trent University, UK, 1999.

[27] Mellor IM, Burrows A, Coluccia S, Hargreaves JSJ, Joyner RW, Kiely CJ, Martra G, Stockenhuber M, Tang WM, *J Catal* **234**: 14, 2005.
[28] Winter ERS, Houghton G, Mass Spectrometry — Inst. Petroleum, 127, 1952.
[29] Winter ERS, *Disc Faraday Soc* **8**: 231, 1950.
[30] Houghton G, Winter ERS, *J Chem Soc* 1509, 1954.
[29] Winter ERS, *J Chem Soc* 1522, 1954.
[30] Muzykantov VS, Jiru P, Klier K, Novakova J, *J Collection Czech Chem Commun* **33**: 829, 1968.
[31] Muzykantov VS, Popovskii VV, Boreskov GK, *Kinet Catal* **5**: 551, 1964.
[32] Peil KP, Goodwin Jr JG, Marcelin GJ, *J Catal* **131**: 143, 1991.
[33] Karasuda T, Aika K-I, *J Catal* **171**: 439, 1997.
[34] Hargreaves JSJ, Joyner RW, Mellor IM, *J Mol Catal A: Chem* **141**: 171, 1999.
[35] Hargreaves JSJ, Hutchings GJ, Joyner RW, Kiely CJ, *J Catal* **135**: 576, 1992.
[36] Cunningham J, Cullinane D, Farell F, O'Driscoll JP, Morris MA, *J Mater Chem* **5**: 1027, 1995.
[37] Smith MR, Ozkan US, *J Catal* **141**: 124, 1993.
[38] Che M, Tench AJ, *Adv Catal* **31**: 38, 1982.
[39] Che M, Tench AJ, *Adv Catal* **32**: 1, 1983.
[40] Gellings PJ, Bouwmeester HJM, *Catal Today* **58**: 1, 2000.
[41] Martens R, Gentsch H, Freund F, *J Catal* **44**: 366, 1976.
[42] Yanagisawa Y, Yamabe S, Matsumura K, Huzimura R, *Phys Rev B* **48**: 4925, 1993.
[43] Yanagisawa Y, Huzimura R, Matsumura K, Yamabe S, *Surf Sci* **242**: 513, 1991.
[44] Acke F, Panas I, *J Phys Chem B* **102**: 5158, 1998.
[45] Huzimura R, Matsumura K, Yamabe S, Yanagisawa Y, *Phys Rev B* **54**: 13480, 1996.
[46] Tanaka K, *J Phys Chem* **78**: 555, 1974.
[47] Yanagisawa Y, Ota Y, *Surf Sci Lett* **254**: L433, 1991.
[48] Muggli DS, Falconer JL, *J Catal* **181**: 155, 1999.

Chapter 6

Oxygen and Hydrogen Surface Mobility in Supported Metal Catalysts. Study by $^{18}O/^{16}O$ and $^{2}H/^{1}H$ Exchange

D Duprez

Glossary of Symbols Used

α^*	:	value of α_g and α_s at equilibrium ($t = t^\infty$)
α_g	:	atomic fraction of *X in gas phase at time t
α_m	:	atomic fraction of *X in the metal at time t
α_s	:	atomic fraction of *X in the support at time t
α_s^o	:	natural isotopic abundance (0.015% for ^2H and 0.2% for ^{18}O)
λ	:	N_g to N_s ratio Eq. (6)
ρ	:	volumic weight of catalyst ($g\,m^{-3}$)
ξ	:	concentration ratio in Eq. (28)
A	:	B.E.T. area of catalyst ($m^2\,g^{-1}$)
C_m^*	:	surface concentration of *X atoms on the metal particles (atoms m^{-2})
C_s^*	:	surface concentration of *X atoms on the support (atoms m^{-2})
C_m	:	surface concentration of X atoms on the metal particles (atoms m^{-2})
C_{mo}	:	surface concentration of metal atoms chemisorbing X and *X species
C_s	:	surface concentration of X atoms on the support (atoms m^{-2})
D	:	metal dispersion (%)
d	:	particle size of metal (m)
D_r	:	coefficient of bulk diffusion of X in the support ($m^2\,s^{-1}$)
D_s	:	coefficient of surface diffusion of X on the support ($m^2\,s^{-1}$)
I_0	:	specific perimeter of the metal particles (m per g of catalyst)

K	:	exchange rate constant
m	:	weight of catalyst (g)
N	:	number of metal particles on the support (particles per gramme)
N_A	:	Avogadro's number
N_e	:	number of exchanged atoms of support at time t
N_g	:	number of $^*X + X$ atoms in gas phase
N_m	:	number of $^*X + X$ atoms in the metal particles
N_s	:	number of exchangeable atoms of support
P_0	:	total pressure
P_{nm}	:	partial pressure of a compound of molecular weight "nm"
S_c	:	B.E.T. area of the catalyst sample (m^2)
S_m	:	metal area of the catalyst sample (m^2)
S_M	:	intrinsic metal surface area ($m^2\, atom^{-1}$)
t	:	time of exchange
T	:	temperature of exchange (K)
V_e	:	rate of exchange of $^*X$ with the support (atoms $min^{-1}\, g^{-1}$)
V_q	:	rate of equilibration of $^*X_2 + X_2$
V_R	:	reaction volume (cm^3)
X	:	^{16}O or ^1H atoms or ions
$^*X$	:	^{18}O or ^2H atoms or ions
x_m	:	metal loading (wt.-%)
Z^∞	:	value of a variable Z at $t = \infty$
Z^0	:	value of a variable Z at $t = 0$

1. Introduction

There is increasing evidence in the literature that surface mobility phenomena can play a decisive role in catalysis; both in the catalytic steps as well as during certain treatments of activation and regeneration of the catalysts.

The steam reforming of aromatic hydrocarbons,[1,2] the oxygen storage in three-way catalysts,[3,4] certain reactions of selective oxidation,[5,6] the NEMCA effect[7,8] as well as many reactions involving a "spillover" process of reactive species,[9–11] are well-known examples of reactions controlled by surface migration of active species.

The phenomena of surface mobility are often invoked by many authors in the explanation of some specific observations, such as cooperation effects between phases, support effects, resistance to coking and to poisoning, and so forth. They are also considered as elementary steps of certain bifunctional mechanisms. Nevertheless, their influence on the reaction kinetics are rarely discussed, owing to the lack of coherent data on these mobilities.

Kapoor *et al.*[12] have reviewed the surface diffusion of physically adsorbed species and of chemisorbed species. However, most of the data collected in this review concerned those that are obtained by means of the diffusion cell technique, well-adapted to surface diffusion measurements of physically adsorbed species. Spectroscopic techniques such as the field emission microscopy or the field ion microscopy, allowing direct observations of surface migration, were also examined by Kapoor *et al.*

The present review deals with recently developed techniques to measure surface mobility in supported metal catalysts. Special attention will be paid to isotopic exchange techniques, well-adapted to diffusion measurements on metal/support catalysts, and to complementary techniques such as IR spectroscopy, having given a comprehensive view of both the nature and the rate of diffusing species. Kinetic equations of isotopic exchange developed in Sec. 2 of this review will be restricted to the measurements conducted in a closed reactor.

2. Measurement of Surface Diffusion by Isotopic Exchange

2.1. *Overview*

To account for the kinetics of the isotopic exchange of polyatomic molecules on heterogeneous catalysts, several models were proposed, including the following steps:

(i) adsorption and dissociation of the parent molecule, with formation of adsorbed atoms or ions
(ii) exchange of these entities with those present on or in the catalyst
(iii) desorption of the exchanged molecules.

Surface diffusion can intervene in step (ii), if the chemical species to be exchanged are not located in the same catalyst region after their primary step of adsorption. This is the case for:

- multisite catalysts, with heterogeneous distribution of these sites,
- multiface catalysts, where the surface species migrate from one crystallographic plane to another, for the exchange to occur,
- multiphase catalysts, where the exchange occurs between chemically different phases.

Let us consider the isotopic exchange between X (normal) and *X (labeled) atoms in diatomic molecules X_2, where $X = {}^1H$ or ${}^{16}O$ and $^*X = {}^2H$ or ${}^{18}O$. The measurement of the change in $^*X_2$, *XX and X_2 concentrations, by mass spectrometry, as a function of kinetic parameters (e.g., time, temperature, partial pressure) allows the determination of:

- the rate of exchange between the different isotopic species
- the coefficient of surface diffusion $\boldsymbol{D}_s$
- the coefficient of bulk diffusion $\boldsymbol{D}_r$, particularly for oxygen in oxide catalysts
- the numbers N_s of exchangeable atoms.

Since numerous studies have been devoted to the isotopic exchange of oxygen on oxide catalysts, a brief review of the theoretical models developed will first be presented, before examining in detail the isotopic exchange in supported metal catalysts.

2.2. *Mechanisms of isotopic exchange of oxygen on oxides*

2.2.1. *Definitions*

According to Boreskov[13] and Novakova,[14] three types of exchange of dioxygen on oxide catalysts can occur:

(a) *the homoexchange* (or Type I exchange) between adsorbed atoms

$$^{18}O_2(g) + {}^{16}O_2(g) \rightarrow 2\ {}^{18}O^{16}O(g),$$

where (g) refers to the gas phase. The rates of adsorption and of desorption of O_2 are significantly higher than the rate of exchange with the oxygens of the solid not involved in the exchange of the gaseous oxygen. As a result, the atomic concentration of ${}^{18}O$ in gas phase (α_g) does not vary during exchange.

(b) *the simple heteroexchange* (or Type II exchange) between a molecule of dioxygen and an atom of the solid:

$$^{18}O_2(g) + {}^{16}O(s) \rightarrow {}^{18}O^{16}O(g) + {}^{18}O(s)$$

and

$$^{18}O^{16}O(g) + {}^{16}O(s) \rightarrow {}^{16}O_2(g) + {}^{18}O(s),$$

where (s) denotes atoms of the solid.

(c) *the multiple heteroexchange* (or Type III exchange) between a molecule of dioxygen and two atoms of the solid:

$$^{18}O_2(g) + {}^{16}O\ldots{}^{16}O(s) \rightarrow {}^{18}O^{16}O(g) + {}^{18}O\ldots{}^{16}O(s)$$

or

$$^{18}O_2(g) + {}^{16}O\ldots{}^{16}O(s) \rightarrow {}^{16}O_2(g) + {}^{18}O\ldots{}^{18}O(s),$$

as well as the corresponding equations written with $^{18}O^{16}O(g)$. In both cases (simple or multiple), heteroexchange leads to significant changes in the atomic concentration of ^{18}O in gas phase.

2.2.2. *Kinetics of the exchange between a gas and a monophasic solid*

The kinetic equation developed by Boreskov[13] and Winter[15] for the exchange of $^{18}O_2$ on oxides can be extended to the exchange of any molecule $^*X_2$ with a solid containing the element X, and particularly to the exchange of $^2H_2$ with the hydroxyl groups of oxides. Following the hypotheses of Boreskov[13] and Winter,[15] the gas-solid exchange reaction obeys first-order kinetics

$$-N_g \frac{d\alpha_g}{dt} = K(\alpha_g - \alpha_s) \tag{1}$$

with

$$\alpha_g = \frac{P_{36} + \frac{1}{2}P_{34}}{P_{36} + P_{34} + P_{32}} \qquad \text{for } {}^{18}O/{}^{16}O \text{ exchange}$$

and

$$\alpha_g = \frac{P_4 + \frac{1}{2}P_3}{P_4 + P_3 + P_2} \qquad \text{for } {}^{2}H/{}^{1}H \text{ exchange}$$

If the number N_s of exchangeable atoms in the solid remains constant during the exchange (i.e., there is no structural nor morphological change in the solid due to the exchange itself), the mass balance of species $^*X$ at time zero and at time t, yields

$$\alpha_g^0 N_g + \alpha_s^0 N_s = \alpha_g N_g + \alpha_s N_s \tag{2}$$

Introducing $\lambda = N_g/N_s$, Eq. (2) becomes:

$$\lambda\alpha_g^0 + \alpha_s^0 = \lambda\alpha_g + \alpha_s \tag{3}$$

The mass balance of species *X at zero time and at t^∞, yields ($\alpha^* = \alpha_g^\infty = \alpha_s^\infty$):

$$\lambda\alpha_g^0 + \alpha_s^0 = (\lambda + 1)\alpha^* \tag{4}$$

Combining Eqs. (1), (3) and (4), the following equation is obtained

$$-N_g\frac{d\alpha_g}{dt} = K(\lambda + 1)(\alpha^* - \alpha_g) \tag{5}$$

leading to the integral expression

$$-\mathrm{Ln}\,\Gamma = K\left(\frac{\lambda+1}{N_g}\right)t \tag{6}$$

with

$$\Gamma = \frac{\alpha_g - \alpha^*}{\alpha_g^0 - \alpha^*}$$

Eq. (6) implies that α^* be determined with sufficient accuracy, which requires in practice that the equilibrium be reached after a moderate amount of time. The number N_s of exchangeable atoms is calculated via λ by Eq. (4) and the rate constant K is obtained by plotting $-\mathrm{Ln}\Gamma$ vs. t [Eq. (6)].

2.3. *Mechanisms of isotopic exchange of oxygen on oxide-supported catalysts*

On oxide-supported metals, multistep exchange can occur. In the case of $^{18}O/^{16}O$ exchange, the different steps represented in Fig. 1 are:

1. dissociative adsorption of a molecule of $^{18}O_2$
2. transfer of ^{18}O atoms from the metal to the support
3. surface migration of ^{18}O atoms (or ions) on the support
4. exchange of ^{18}O atoms (or ions) with ^{16}O atoms (or ions) of the surface
5. internal migration and exchange of ^{18}O with ^{16}O ions in the support
6. direct exchange of $^{18}O_2(g)$ with the oxygen atoms (or ions) of the support

Every step "i" is coupled with the step "$-i$", corresponding to the reverse route for the exchanged species. For instance, the dissociative adsorption of a molecule $^{18}O_2$ (step 1) is coupled with the desorption of a molecule of $^{18}O^{16}O$.

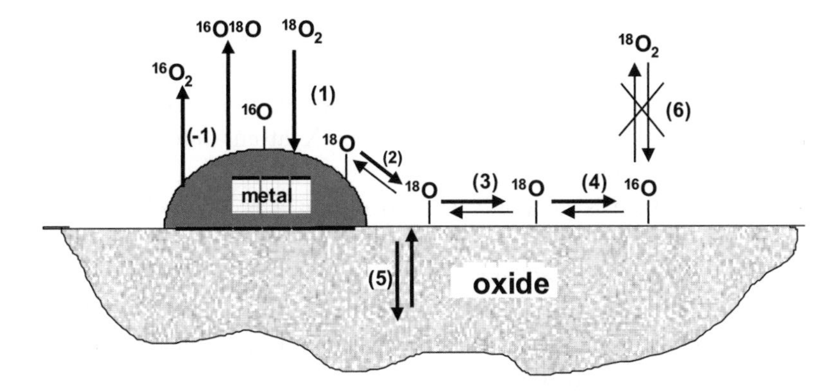

Fig. 1. Isotopic exchange of $^{18}O_2$ with ^{16}O of support via metal particles.

Given that the rates of the steps 5 and 6 are negligible, two types of exchange can be observed:

- The heteroexchange, denoted as "isotopic exchange", is when steps 1–4 occur simultaneously. Experiments are generally carried out with non-equimolar, gaseous mixtures of X_2 and $^*X_2$, as in the study by Schwank *et al.*[16] in the case of isotopic exchange of O_2 on Au and Ru catalysts; or with pure $^*X_2$ as in the study by Duprez *et al.*,[17] of the exchange of $^{18}O_2$ and of $^2H_2$ on Rh/Al_2O_3 catalysts.
- The homoexchange, denoted as "isotopic equilibration", is when only steps 1 and -1 occur at a significant rate. Isotopic equilibration gives useful information on the nature and surface area of metallic particles. Experiments are generally carried out with equimolar mixtures of X_2 and $^*X_2$.

In the following, these two types of exchange will be detailed in the case of $X = {}^{16}O$ or ^1H and $^*X = {}^{18}O$ or ^2H. In Secs. 2.4 and 2.5, kinetic equations for heteroexchange will be given for $\alpha_g^o = 1$ (pure $^*X_2$ at t = 0) and α_s^o, while the natural isotopic abundance of $^*X$ in the support will be neglected.

2.4. *Isotopic heteroexchange on supported metal catalysts — general parameters*

2.4.1. *Mechanism and kinetic conditions*

The exchange obeys a three-step mechanism represented in Fig. 1, namely, the dissociative adsorption of $^*X_2$ on the metal particle (step 1), the transfer of $^*X$ species to the support (step 2) and the surface migration of $^*X$ species to the sites of exchange (step 3). Likewise, the reverse steps such as the desorption of $^*XX$ or of X_2 (step -1), the transfer from the metal to the support (step -2) and the back diffusion of X atoms (step -3) are also involved in this mechanism.

Isotopic exchange is generally carried out in a recycle, closed reactor coupled to a mass spectrometer, as represented on the Fig. 2. The recycle pump is necessary to avoid any diffusion effect in the gas phase, limiting the changes of partial pressure measured by the mass spectrometer. Typical curves of exchange are represented in Fig. 3 which shows the changes, with time, of the partial pressures of $^{16}O_2$, $^{16}O^{18}O$ and $^{18}O_2$ during the exchange of $^{18}O_2$ with the ^{16}O species of an alumina-supported rhodium catalyst. To obtain information on the surface mobility of oxygen and hydrogen on supported metal catalysts, three conditions must be fulfilled:

- Exchange must occur via the metal particle (i.e., the rate of direct exchange — step 6 in Fig. 1 — is negligible)
- Surface migration must be the rate-determining step
- Exchange must occur essentially with the surface atoms.

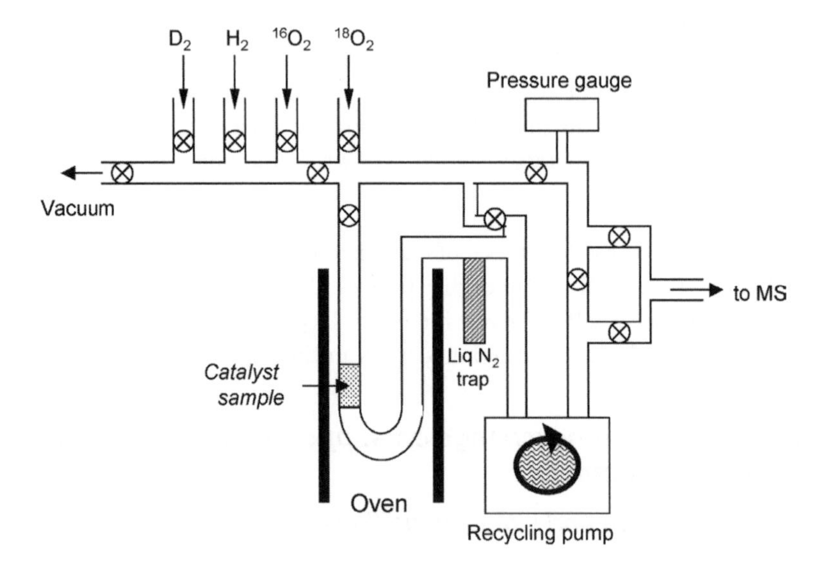

Fig. 2. Closed-loop reactor for exchange experiments.

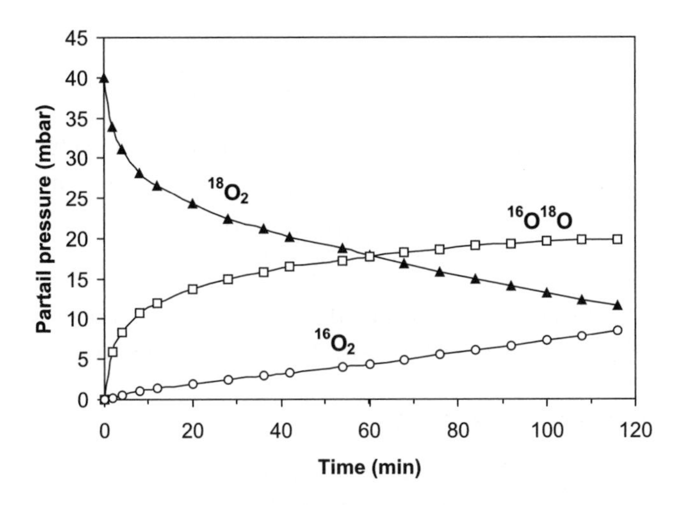

Fig. 3. Exchange of $^{18}O_2$ over Rh/Al_2O_3 at 400°C.

2.4.2. *Determination of the rate of exchange*

The mass balance of *X atoms in the metal particles gives

$$-N_g \frac{d\alpha_g}{dt} = N_m \frac{d\alpha_m}{dt} + N_s \frac{d\alpha_s}{dt} \tag{7}$$

Rate of disappearance of *X atoms from gas phase (net adsorption)	Rate of accumulation of *X in the metal particles	Rate of exchange (net transfer of *X atoms to support)	or

$$r_e = -\left[N_g \frac{d\alpha_g}{dt} + N_m \frac{d\alpha_m}{dt}\right] \tag{8}$$

In most cases, compared with N_s and N_g, N_m can be neglected so that Eq. (8) reduces to

$$r_e = -N_g \frac{d\alpha_g}{dt} \tag{9}$$

From the definition of α_g given in Eq. (1), it becomes

$$\frac{d\alpha_g}{dt} = \frac{1}{P_0}\left[2\frac{dP_{*X*X}}{dt} + \frac{dP_{*XX}}{dt}\right] \tag{10}$$

with

$$P_0 = P_{*X*X} + P_{*XX} + P_{XX}$$

During a pure exchange process (reaction or adsorption excluded), P_0 is constant. In this case, $dP_0/dt = 0$ and Eq. (10) can be written

$$\frac{d\alpha_g}{dt} = -\frac{1}{P_0}\left[\frac{dP_{*XX}}{dt} + \frac{dP_{XX}}{dt}\right] \quad (11)$$

The rate of exchange can be calculated from the slopes of the curves P_{*X*X}, P_{*XX} and P_{XX} vs. time either by Eqs. (9) and (10) or by Eqs. (9) and (11).

2.4.3. *Initial rates of exchange*

In most cases, diffusivities are deduced from the initial rates of exchange. In the very beginning of exchange, curves P_{*X*X} and P_{*XX} vs. time can be assimilated to straight lines

$$P_{*X*X} = P_0(1 - at) \quad \text{and} \quad P_{*XX} = P_0 bt \quad (12)$$

Assuming that the gas phase is in equilibrium with the metal particles, we have

$$\left.\begin{aligned} P_{*X*X} &= P_0\alpha_m^2 \\ P_{*XX} &= 2P_0\alpha_m(1 - \alpha_m) \\ P_{XX} &= P_0(1 - \alpha_m)^2 \end{aligned}\right\} \quad (13)$$

or

$$P_{XX} = \frac{P_{*XX}^2}{4\,P_{*X*X}} = \frac{P_0 b^2 t^2}{4(1 - at)} \quad (14)$$

Derivation of Eq. (14) leads to

$$\frac{dP_{XX}}{dt} = \frac{P_0 b^2}{4}\left[\frac{t\,(2 - at)}{(1 - at)^2}\right] \quad (15)$$

which shows that the initial slope of P_{XX} should be nil. Accordingly, the initial rate of exchange can be written as

$$r_e(t \to 0) = \frac{N_g}{P_0}\left[\frac{dP_{*XX}}{dt}\right] \quad (16)$$

This is well-verified for slow exchanges. In the case of fast exchange, however, $\frac{dP_{XX}}{dt}$ increases so rapidly that its apparent, initial value seems to be positive. Therefore, we recommend using Eqs. (9)–(10) in every case to calculate the initial rates of exchange.

2.4.4. *Numbers of exchanged and of exchangeable atoms*

The number of *exchanged* atoms corresponds to the number of $^*X$ lost from the gas phase

$$N_e = N_g(\alpha_g^0 - \alpha_g)$$

or

$$N_e = N_g(1 - \alpha_g) \tag{17}$$

for experiments carried out with pure $^*X_2$. If the number of exchanged atoms on the metal can be neglected, the values of N_e calculated by Eq. (17) correspond to the number of exchanged atoms at the support surface.

The number of *exchangeable* atoms is given by Eq. (2) which can be written for $\alpha_g^\circ = 1$ and $\alpha_s \approx 0$ as

$$\alpha_s N_s = N_e = N_g(1 - \alpha_g) \tag{18}$$

When the gas/solid equilibrium is reached, we have

$$N_s = \frac{N_e}{\alpha^*} = N_g \left[\frac{1 - \alpha^*}{\alpha^*}\right] \tag{19}$$

As far as the surface atoms (or ions) of the support are concerned, three cases can occur (Fig. 4):

- All the atoms exchange rapidly and homogeneously (with the same rate constant): α^* is easily determined at the asymptotic value of α_g, when $d\alpha_g/dt = 0$ [Fig. 4(a)].
- All the atoms exchange homogeneously, but the equilibrium is not rapidly reached (Fig. 4b). In this case, it is necessary to determine the best value α_i^* of α^*, fitting the kinetic law for exchange. This method, represented in Fig. 4(c) for rate Eq. (6), was used by Boreskov and Muzykantov[18] for $^{18}O_2$/oxide exchanges, and by Schwank *et al.*[16] for $^{18}O_2$ exchange with Ru/SiO_2 and Au/SiO_2 catalysts.
- There is a bimodal distribution of exchangeable species, the first of them exchanging rapidly, while the second exchanges much more slowly. Two values of α^* can be defined: α_r^* for "rapid" species and α_{sl}^* for "slow" species. Figure 4(d) shows how α_r^* and α_{sl}^* can be determined in Boreskov's coordinates. A strong curvature can be observed at time t_r,

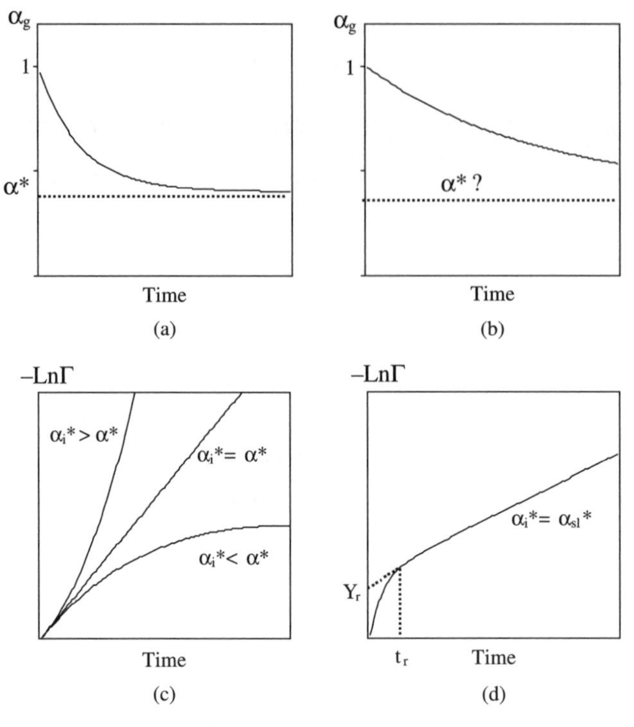

Fig. 4. Determination of the value α^* of α_g corresponding to exchange equilibrium. (a) fast exchange; (b) slow exchange (α^* is not reached); (c) determination of α^* by Eq. 6; (d) bimodal distribution of sites: rapid (r) and slow (sl).

when "rapid" species are totally exchanged: α_r^* is obtained from the ordinate value y_r of $-\mathrm{Ln}\Gamma$. Plotting $y - y_r$ vs. t allows α_{sl}^* to be determined after adequate linearisation as in Fig. 4(c).

2.5. *Isotopic heteroexchange on supported metal catalysts kinetics*

We will now examine the kinetic expressions derived for the different rate-controlling steps of exchange.

2.5.1. *Exchange is controlled by adsorption-desorption of X_2 on the metal*

When exchange experiments are carried out to determine surface mobility of oxygen or hydrogen at the surface of oxides, metal particles are considered as portholes for O_2 and H_2 on the support. Nevertheless, under certain conditions, adsorption/desorption of the gas molecules at the metal surface can be the rate-determining step of exchange. In this case, Eq. (1) can be written

$$-N_g \frac{d\alpha_g}{dt} = K_1 \, S_m(\alpha_g - \alpha_s) \tag{20}$$

When adsorption/desorption is the rate-determining step, the rate of exchange becomes very close to the rate of equilibration (homoexchange) between X_2 and $^*X_2$ on the metal. In this case, no information relative to the support can be obtained.

2.5.2. *Exchange is controlled by the metal → support transfer*

The metal particles are in equilibrium with the gas phase ($\alpha_g = \alpha_m$). Moreover, surface diffusion being rapid, the concentration of $^*X$ species at the support surface changes with time, but does not depend on x and y abscissa parallel to the surface (Fig. 5).

The driving force for exchange is the gradient of concentration across the metal/support interface. The rate of exchange is then given by the rate of mass transfer across this interface

$$-N_g \frac{d\alpha_g}{dt} = K_2 \, I_0(C_m^* - C_s^*) \tag{21}$$

I_0 is the specific perimeter of the metal particles at the metal/support interface

$$I_0 = N \cdot l$$

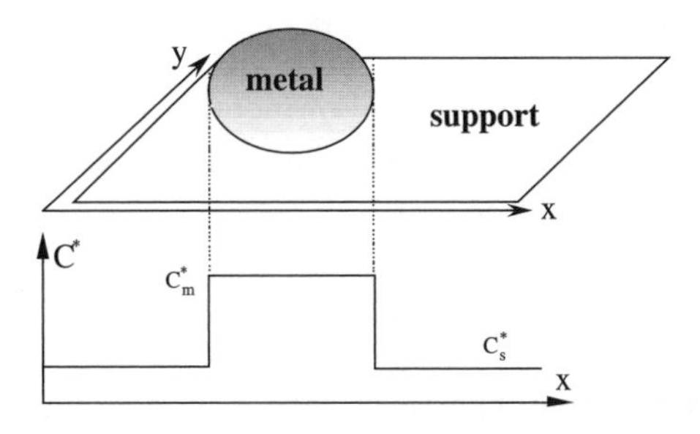

Fig. 5. Schematic view of the concentration profiles when exchange is controlled by the metal-to-support transfer.

with N, number of particles per gram of catalyst and l, perimeter of each particle. I_0 can be calculated as a function of the metal loading x_m, the particle diameter d and the metal density ρ[19]:

$$I_0 = 0.12\,\frac{x_m}{\rho d^2} \tag{22}$$

for hemispherical particles of diameter d, and

$$I_0 = 0.04\,\frac{x_m}{\rho d^2} \tag{23}$$

for cubic particles of length d.

Introducing the metal dispersion D% in Eqs. (22) and (23), it becomes

$$I_0 = \beta x_m D^2 \tag{24}$$

where β is a parameter depending on the nature of metal and particle shape. For Rh particles, $\beta = 8.9 \times 10^5$ (hemispherical) or $\beta = 4.3 \times 10^5$ (cubic), while for Pt particles, $\beta = 5.4 \times 10^5$ (hemispherical) or $\beta = 2.6 \times 10^5$ (cubic). Two remarks can be made about I_0:

- It represents an exceptionally long frontier between two regions of the catalyst (the metal and the support). For instance, in a 0.5 wt.% Rh catalyst with a 50% dispersion, $I_0 = 5.37 \times 10^8\,\mathrm{m\,g^{-1}}$ that is more than the earth-moon distance!
- I_0 depends very much on the metal dispersion (term in D^2), but relatively little on the shape of particles (factor 2 between hemispherical and cubic particles).

The concentrations C_m^* and C_s^* in Eq. (21) are linked to the total site concentrations on the metal and on the support

$$C_m^* = C_{m0}\alpha_g \quad \text{and} \quad C_s^* = C_{s0}\alpha_s \tag{25}$$

C_{m0} and C_{s0} being the concentration of metal sites chemisorbing X or *X atoms and C_{s0}, the concentration of exchangeable sites of support. C_{m0} and C_{s0} can easily be calculated by:

$$C_{m0} = \frac{\theta}{S_M} \quad \text{and} \quad C_{s0} = \frac{N_s}{S_c - S_m} \tag{26}$$

θ being the coverage of metal particles with X + *X atoms during exchange, and $S_c - S_m$, the surface area of support. If one assumes that the number of X + *X atoms on the metal is negligible with respect to the number of

X + *X atoms in the gas phase, the number of exchanged atoms is given by Eq. (18)

$$N_e = \alpha_s N_s = N_g(1 - \alpha_g)$$

Letting $\lambda C_{s0} = \xi C_{m0}$, it becomes

$$-N_g \frac{d\alpha_g}{dt} = K_2\, I_0\, C_{m0}\, [(1+\xi)\alpha_g - \xi] \quad (27)$$

whose integral form is

$$\frac{1}{1+\xi}\, \mathrm{Ln}\, [(1+\xi)\alpha_g - \xi] = \frac{K_2\, I_0\, C_{m0}}{N_g} t \quad (28)$$

with $\frac{\xi}{1+\xi} < \alpha_g < 1$.

For most metals, oxygen coverage is close to unity over a wide temperature range. Accordingly, in oxygen exchange, $C_{m0} \approx 12$–15 atoms nm^{-2}.[20] Moreover, C_{s0} is generally close to C_{m0} or slightly lower.[21] Therefore, the value of ξ depends mainly on the value of λ. In the case of hydrogen exchange, the situation is more complex since θ and thus C_{m0} depend very much on the temperature, while C_{s0} is directly linked to the hydroxyl group coverage.

2.5.3. *Exchange is controlled by the surface diffusion on the support*

It is the situation normally expected of a correct measurement of the surface mobility. In this case, the metal particles are in equilibrium with the gas phase ($\alpha_g = \alpha_m$) and there is no gradient of concentration across the metal/support interface. The driving force is the gradient of concentration at the support surface along x axis (Fig. 6).

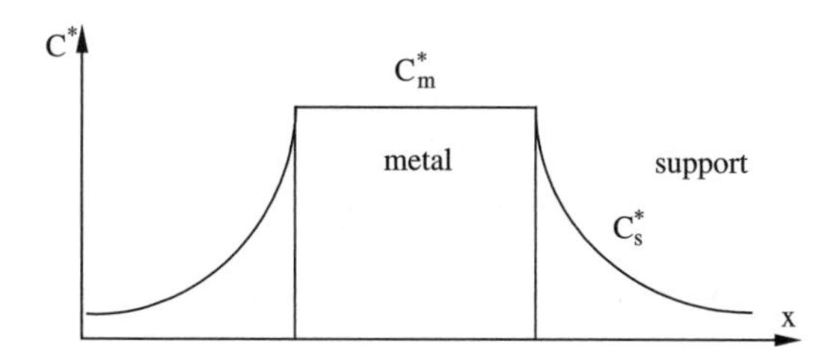

Fig. 6. Profile of surface concentrations when exchange is controlled by surface diffusion on the support.

Surface diffusivity can be determined in this case by the model developed by Kramer and Andre[22] for the data processing of hydrogen spillover on Ni and Pt catalysts.

Let us consider metal particles as circular sources of diffusing species randomly distributed at the support surface, whose area is infinitely great with regard to that of the metal particles. The amount of species having diffused on the support between 0 and t is given by

$$\mathrm{N_e} = 2\,\mathrm{NdC_m^*}\sqrt{\pi D_\mathrm{s}\mathrm{t}} \quad (\text{atoms g}^{-1}) \tag{29}$$

or

$$\mathrm{N_e} = \frac{2}{\sqrt{\pi}}\,\mathrm{C_m^*}\,I_0\sqrt{D_\mathrm{s}\mathrm{t}} \quad (\text{atoms g}^{-1}) \tag{30}$$

N_e can easily be calculated by Eq. (18) and C_m^* by Eq. (25). However, C_m^* is time-dependent so that the determination of D_s by Eq. (29) or (30) should normally be restricted to the initial part of the curve N_e vs. $\sqrt{t}$, when $C_m^* \approx C_{m0}$. Nevertheless, by substituting Eqs. (25) and (18) in Eq. (30), a more general equation can be obtained

$$\frac{\mathrm{N_e}}{\alpha_\mathrm{g}} = \frac{2}{\pi}\,\mathrm{C_{m0}}\,\mathrm{I_0}\,\sqrt{D_\mathrm{s}\mathrm{t}} \tag{31}$$

An example is given in Fig. 7 which shows the curves N_e vs. $\sqrt{t}$ and N_e/α_g vs. $\sqrt{t}$ corresponding to an $^{18}O/^{16}O$ exchange experiment over Rh/ZrO_2 at 400°C.[23]

2.5.4. *Exchange is controlled by bulk diffusion into the support*

When the rate of diffusion in the support network becomes significant, Eq. (31) can no longer describe the kinetics of exchange. Kakioka *et al.*[24,25] have developed a model based on the following assumptions:

- Surface exchange is very fast so that isotopic equilibrium between gas phase and surface atoms is rapidly reached,
- Bulk diffusion occurs in spherical particles.

In this case, the kinetics of diffusion in the solid follows Eq. (32)

$$-\mathrm{Ln}\left[\frac{\alpha_\mathrm{g} - \alpha_\mathrm{s}^0}{\alpha^* - \alpha_\mathrm{s}^0}\right] = \frac{2}{\pi}\,\frac{\rho\mathrm{A}}{\mathrm{Ng}}\,\sqrt{D_\mathrm{r}\mathrm{t}} \tag{32}$$

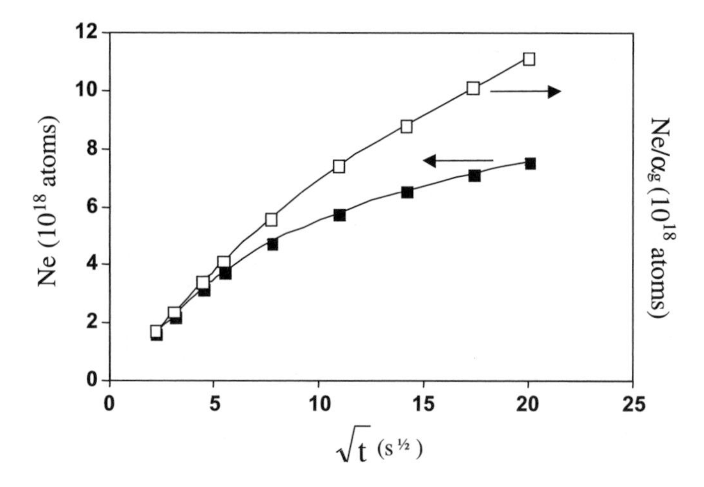

Fig. 7. Curves Ne vs. $\sqrt{t}$ Eq. (30) and Ne/αg vs. $\sqrt{t}$ Eq. (31) for a $^{18}O/^{16}O$ exchange experiment over a Rh/ZrO_2 catalyst at 400°. After Ref. 23.

which reduces to

$$-\mathrm{Ln}\frac{\alpha_g}{\alpha^*} = \frac{2}{\pi}\frac{\rho A}{\mathrm{Ng}}\sqrt{D_r t} \tag{33}$$

for α^* and $\alpha_g \gg \alpha_s^0$. A fact worth noticing is that α^* corresponds to the value of α_g, after surface exchange is completed and when bulk diffusion starts. The coefficient of bulk diffusion can be evaluated from the slope "a" of $\mathrm{Ln}\,\alpha_g$ vs. $\sqrt{t}$, for $t > t_A$ (Fig. 8).

2.6. *Isotopic homoexchange on supported metal catalysts*

This type of exchange corresponds to an *equilibration* of two isotopomers at the catalyst surface according to the reaction

$$^*X_2\ (\mathrm{gas}) + X_2\ (\mathrm{gas}) \Leftrightarrow 2^*XX\ (\mathrm{gas}) \tag{34}$$

In most cases, this reaction is much more rapid on the metal than on the support. Those metals which cannot catalyse equilibration are poor $^*X$ donors for exchange with the support. In what follows, we will consider the case where homoexchange occurs at a significant rate only on the metal.

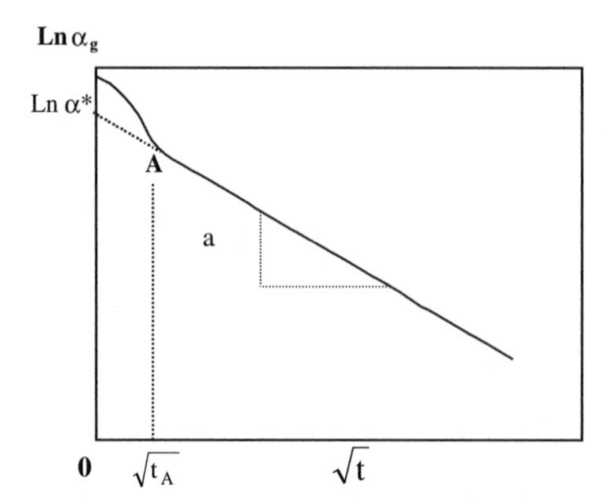

Fig. 8. Determination of the coefficient of bulk diffusion. Extrapolation of the linear part of the curve $\text{Ln}\,\alpha_g$ vs. $\sqrt{t}$ gives the isotopic ratio α^* of the surface when bulk diffusion starts.

2.6.1. *Mechanisms*

The mechanisms of homoexchange are similar to those of heteroexchange, where the atoms of the solid in the latter case are being replaced by adsorbed atoms in homoexchange. Three mechanisms can occur according to the number of these adsorbed atoms involved in the reaction: zero, one or two atoms.[13,26–28]

- $$^*X_2(g) + X_2(g) \rightarrow [^*X^*XXX] \rightarrow 2^*XX(g) \tag{35}$$

X atoms exchange without the participation of adsorbed species. This is the mechanism proposed by Schwab to explain certain results of $^1H/^2H$ exchange.

- $$\left.\begin{aligned} ^*X_2(g) \Leftrightarrow 2^*X_a \text{ and } X_2(g) \Leftrightarrow 2X_a \\ X_2(g) + {}^*X_a \rightarrow {}^*XX(g) + X_a \\ ^*X_2(g) + X_a \rightarrow {}^*XX(g) + {}^*X_a \end{aligned}\right\} \tag{36}$$

The reaction obeys a Rideal-Eley mechanism between a gas phase molecule and an adsorbed atom.

- $$\left.\begin{aligned} ^*X_2(g) \Leftrightarrow 2^*X_a \text{ and } X_2(g) \Leftrightarrow 2X_a \\ 2^*X_a + 2X_a \rightarrow 2^*XX \end{aligned}\right] \tag{37}$$

This is a Bonhoeffer-Farkas mechanism between two pairs of adsorbed atoms.

2.6.2. *Kinetics*

The initial rate of equilibration (in exchanged *atoms*) is given by

$$r_q(t = 0) = \frac{2N_g}{P_0}\left[\frac{dP_{*XX}}{dt}\right]_{t=0} \tag{38}$$

When homoexchange obeys one of the first two mechanisms (zero or one adsorbed atoms), the general rate equation is given by[13,26]:

$$N_g\frac{dP_{*XX}}{dt} = K_q\left(P^{\infty}_{*XX} - P_{*XX}\right) \tag{39}$$

whose integral form is

$$-\mathrm{Ln}\left[\frac{P^{\infty}_{*XX} - P_{*XX}}{P_{*XX} - P^{0}_{*XX}}\right] = \frac{K_q t}{N_g} \tag{40}$$

P^0_{*XX} and P^{∞}_{*XX} are the partial pressures of *XX at $t = 0$ and at equilibrium respectively. During homoexchange, P_0 and α_g remain constant and P^{∞}_{*XX} is given by

$$P^{\infty}_{*XX} = 2\alpha_g(1 - \alpha_g) \tag{41}$$

When homoexchange obeys the third mechanism (two adsorbed atoms), the rate equation is more complex and a second term must be added to the right-hand part of Eq. (40).[27] As Schwab's mechanism is rarely encountered, distinction should be made between the mechanisms involving one or two adsorbed atoms. A simple method for discriminating these two mechanisms was proposed by Klier *et al.*[27] It is based on the variation of the equilibrium constant K^* during exchange experiments

$$K^* = \frac{(P_{*XX})^2}{P_{XX} \cdot P_{*X*X}}$$

Starting with a pre-equilibrated mixture ($K^* = 4$), K^* remains constant in the case of a one-atom mechanism, while K^* passes through a minimum before reaching the value of 4 again in the case of a two-atoms mechanism. This method was used by Klier *et al.*[27] to show that the exchange of ^{18}O on MgO followed, in part, a two-atoms mechanism. A similar approach was developed by Winter[29] to investigate oxygen exchange on a great number of oxides and to determine the respective rates via each mechanism of exchange. Another method for distinguishing the two mechanisms was reported by Boreskov.[13] In the case of the one-atom mechanism, the P_{*XX}/P_{*X*X} ratio is a linear function of $(P_{*X*X})^{-1/2}$

$$\frac{P_{*XX}}{P_{*X*X}} = \left[\left(\frac{P^0_{*XX}}{P^0_{*X*X}} + 2\right)\left(\frac{P^0_{*X*X}}{P_{*X*X}}\right)^{1/2} - 2\right] \tag{42}$$

By contrast, P_{*XX}/P_{*X*X} is constant when exchange obeys a two-atoms mechanism.[13] This method was used by Boreskov to show that K_2SO_4-promoted V_2O_5 contained a significant number of sites exchanging oxygen via the two-atoms mechanism. It was also used by Pichat and colleagues to investigate the kinetics and the mechanism of exchange of $^{18}O_2$ with different oxides under irradiation: SnO_2, ZnO, ZrO_2[30] and TiO_2.[31,32]

The homoexchange is generally performed with 50%-50% mixtures of X_2 and $^*X_2$. An example of experiment is shown on Fig. 9 for a Rh/Al_2O_3 catalyst. The initial slope of the curve P_{*XX} versus time allows the calculation of the rate of equilibration r_q by Eq. (38).

Equilibration corresponds to the steps 1 and −1 (adsorption/desorption on the metal particles) of the mechanism of exchange depicted in Fig. 1. It can be used to determine in what kinetic domain, r_e (rate of exchange of $^*X_2$ with X species of the support) is measured

$r_q > r_e$: exchange is not limited by adsorption/desorption on the metal which acts as an "open" porthole for $^*X_2$ on the support

$r_q \approx r_e$: the rate of exchange is determined by the rate of adsorption/desorption on the metal (see Sec. 2.5.1).

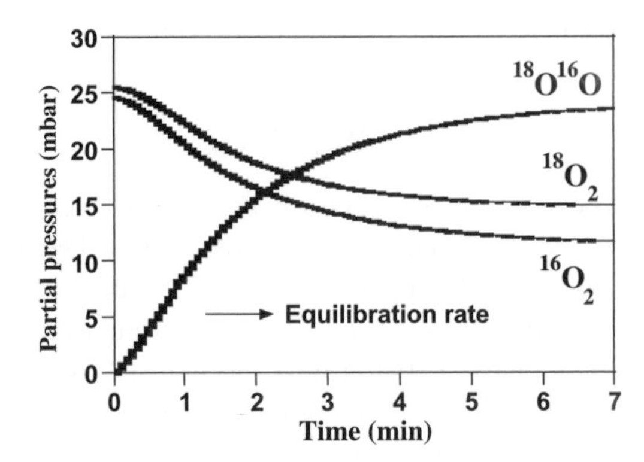

Fig. 9. $^{16}O_2 + ^{18}O_2$ equilibration over a 0.5 wt-% Rh/Al_2O_3 at 300°C.

The change of α_g with time gives useful information on the relative values of r_q and r_e. For $r_q > r_e$, α_g remains approximately constant during equilibration ($\alpha_g = 0.5$ for an equimolar mixture of $^*X_2$ and X_2). On the contrary, if $r_q \approx r_e$, α_g decreases rapidly with time, where the exchange of the support occur during equilibration.

3. Surface Diffusion of Hydrogen and Oxygen

3.1. *Overview*

The first observations of mobility phenomena in catalysis made in the fifties concerned essentially hydrogen surface diffusion. Kuriacose,[33] who was studying the decomposition of germanium hydride at the surface of Ge films, noticed that the presence of platinum electrodes, employed for measuring the electrical conductivity of the films, increased the rate of GeH_4 decomposition significantly. These results were published only in 1967, after having been interpreted by Taylor as a manifestation of hydrogen mobility.[34,35] In the absence of platinum, the hydride decomposition is controlled by H_2 desorption from the surface of GeH_4, while the presence of the electrodes provides a new way of desorption via the Pt surface.

In 1964, Khoobiar showed that a mechanical mixture of Pt/Al_2O_3 and WO_3, treated in H_2, led to the formation of tungsten bronzes H_xWO_3 at room temperature.[36] Apparently, hydrogen species formed by dissociative adsorption of dihydrogen on Pt can migrate and react with WO_3 to form a hydrogen bronze. The same year, Robell *et al.*[37] reported that Pt/C catalysts could store large quantities of hydrogen, much higher than those corresponding merely to H_2 chemisorption on the metal. This was explained by a slow diffusion of H species on the carbon support with a diffusivity of $3.4 \times 10^{-23}\,m^2\,s^{-1}$ at 300°C and of $5.8 \times 10^{-21}\,m^2\,s^{-1}$ at 392°C. The possible role of oxygen spillover in catalysis was invoked more recently by Bond *et al.*,[38] who evidenced a strong synergy effect between Pd and SnO_2 in CO oxidation. The reaction occurs both on Pd *and* on SnO_2, after the migration of CO and O_2 from the metal to the support. Ducarme and Védrine[39] showed that oxygen spillover could occur during exchange of $C^{18}O_2$ with the ^{16}O atoms (or ions) present at the surface of Pt/Al_2O_3 catalysts. They found that all the oxygen atoms chemisorbed on Pt, plus one support atom in five were exchangeable at 180°C. Moreover, the excess of oxygen exchanged on Pt/Al_2O_3 was significantly greater than the number of oxygen atoms exchanged by alumina in the absence of platinum.

Since the pioneering work of Kuriakose and Taylor on hydrogen spillover, and of Bond on oxygen spillover, several hundreds of papers directly connected to this phenomenon and probably several thousands suggesting spillover effects to explain results, were published in the literature. We will now examine those papers which are directly concerned with the measurement of relative or absolute coefficients of surface diffusion.

3.2. $^1H_2/^2H_2$ and $^{16}O_2/^{18}O_2$ homoexchange on metals

In the following, we will consider the case where the rate of the direct exchange with the support is negligible. In this case, metal particles act as portholes for the heteroexchange of hydrogen or of oxygen with surface species of the support (Fig. 1) and the steps of adsorption/desorption of H_2 and O_2 on these metal particles should be fast under the conditions of exchange. $^1H_2/^2H_2$ and $^{16}O_2/^{18}O_2$ homoexchange (or equilibration) is a measurement of the overall rate of adsorption/desorption (steps 1 and −1 in Fig. 1). A fast equilibration being a prerequisite to a correct measurement of surface mobility, we will now examine the literature data concerning these rates of equilibration over various Group 8 metals and copper.

3.2.1. $^1H_2/^2H_2$ equilibration

This reaction is generally very fast and can occur at sub-ambient temperature on most group 8 metals. Table 1 gives typical values of the intrinsic rate of $^1H_2/^2H_2$ equilibration on Group 8 and Cu metal catalysts.

Noble metals are the most active catalysts, while copper and iron are the least active at low temperature (<20°C). However, Schuit and van Reijen[40] showed that the activation energy varied in the reverse order of the intrinsic activity at 0°C

$$\{\text{Cu}, \text{Fe}\}, 31 > \text{Ni}, 13 > \text{Co}, 6 > \{\text{Pt}, \text{Rh}, \text{Ru}\}, 4\,\text{kJ}\,\text{mol}^{-1}$$

so that Cu, Fe, Ni and Co have virtually the same activity by 200°C. The reaction seems to be sensitive to the particle size of metal and to the nature of support. Except for well-dispersed Rh on ceria-alumina, intrinsic activities of Pt and of Rh increase when their dispersion decreases, and are higher on alumina than on ceria-alumina.[45] Contrary to what was reported by Schuit and van Reijen,[40] Taha and Duprez found Pt much more active than Rh at 0°C. It seems that certain sites of Pt surfaces (steps, kinks ...) are extremely active in $^1H_2/^2H_2$ equilibration.[46] For

Table 1. Rates of $^1H_2/^2H_2$ equilibration on Group 8 and Cu catalysts (metal dispersion is given in parentheses).

Catalysts	T°C	P mbar	Intrinsic rates at.H m_{met}^{-2} s^{-1}	Ref.
Group 8 and Cu on SiO_2	0	500	{Pt,Rh,Ru}:Co:Ni:{Fe,Cu} = 10^4:10^2:10:1 (relative rates)	[40]
Ni films	20	1.6	2×10^{20}	[41]
Ni foil	0	0.3	6×10^{20}	[42]
Ni/SiO_2	100	50	4×10^{17} (15% dispersion)	[43, 44]
Cu/SiO_2	100	50	1×10^{16} (15% dispersion)	[43, 44]
Cu/SiO_2	350	50	5×10^{17} (15% dispersion)	[43, 44]
Pt/Al_2O_3	0	50	4.3×10^{20} (57% dispersion) 4.6×10^{21} (4% dispersion)	[45]
Pt/CeO_2-Al_2O_3	0	50	1.5×10^{20} (84% dispersion) 1.2×10^{21} (4% dispersion)	[45]
Rh/Al_2O_3	0	50	3.3×10^{18} (87% dispersion) 1.3×10^{19} (8% dispersion)	[45]
Rh/CeO_2-Al_2O_3	0	50	8×10^{18} (84% dispersion) 1×10^{18} (14% dispersion)	[45]

instance, Bernasek and Somorjai found that stepped platinum surfaces were about three to four orders of magnitude more active than flat surfaces.[47] In the same way, Nishiyama *et al.*[48–50] showed that surface roughening obtained by ultra-high vacuum treatment at 800°C strongly increased the reaction rate on Pt and parallely decreased its activation energy. This is in agreement with previous work of Christmann *et al.*[51] who showed that the activation energy of $^1H_2/^2H_2$ equilibration on Pt could vary between 4 and 24 kJ mol^{-1} depending on the nature of the exposed face and of the sample pretreatment.

It is worth noticing that the hydrogen equilibration reaction normally does not occur on metal surfaces covered with carbon deposits. This property was used by Andersson *et al.*[53] to measure the active surface area of nickel in fresh[52] and coked Ni/SiO_2 catalysts. A similar procedure was employed by the same group to study the deactivation of Pt and Pt-Sn/Al_2O_3 catalysts during propane dehydrogenation.[54] By contrast, it was shown that sulfur, when present at low coverage, could promote the reaction on Pt(111), a poisoning effect of sulfur being observed only for $\theta_S > 0.33$.[55]

3.2.2. $^{16}O_2/^{18}O_2$ equilibration

This reaction over Group 8 metals and Cu was investigated by Duprez *et al.*[17,43–45,56–59] The main results are given in Table 2. $^{16}O_2/^{18}O_2$ equilibration is much slower than the corresponding reaction with hydrogen. Therefore, oxygen equilibration is generally carried out at higher temperatures, typically within the 250–500°C temperature range. Under such conditions, the metals are partially or totally oxidised.

Table 2. Rates of $^{16}O_2/^{18}O_2$ equilibration on Group 8 and Cu catalysts (metal dispersion is given in parentheses).

Nr	Catalysts	Cl content*	T°C	P mbar	Intrinsic rates at.O $m_{met}^{-2}\,s^{-1}$	Ref.
1	Rh/Al_2O_3	DCl	400	50	3.5×10^{18} (50% dispersion)	[56]
2	Rh/Al_2O_3	DCl	400	17	1.3×10^{18} (80% dispersion)	[17, 57]
			400	50	3.9×10^{18} (80% dispersion)	[17, 57]
			320	50	0.74×10^{18} (80% dispersion)	[57]
			260	50	0.17×10^{18} (80% dispersion)	[57]
3	Rh/Al_2O_3	Cl	300	50	0.03×10^{18} (95% dispersion)	[58]
		DCl	300	50	0.52×10^{18} (80% dispersion)	[58]
		DCl	300	50	0.38×10^{18} (15% dispersion)	[58]
4	Rh/Al_2O_3	ClF	300	50	6.0×10^{18} (87% dispersion)	[45]
		ClF	300	50	2.0×10^{18} (8% dispersion)	[45]
5	Pt/Al_2O_3	DCl	500	50	1.7×10^{18} (66% dispersion)	[57]
6	Pt/Al_2O_3	DCl	300	50	0.026×10^{18} (47% dispersion)	[58]
7	Pt/Al_2O_3	ClF	300	50	0.017×10^{18} (57% dispersion)	[45]
		ClF	300	50	0.183×10^{18} (7% dispersion)	[45]
		ClF	300	50	0.75×10^{18} (4% dispersion)	[45]
8	Pd/Al_2O_3	DCl	500	50	0.10×10^{18} (30% dispersion)	[57]
9	Pd/Al_2O_3	ClF	300	50	7.5×10^{14} (38% dispersion)	[59]
10	Ni/Al_2O_3	ClF	500	50	$<10^{15}$ (15% dispersion)	[57]
11	Ni/SiO_2	ClF	450	50	$<10^{14}$ (15% dispersion)	[43, 44]
12	Cu/SiO_2	ClF	300	50	4.5×10^{14} (13% dispersion)	[43, 44]
		ClF	450	50	2.4×10^{16} (13% dispersion)	[43, 44]
		ClF	300	50	9.8×10^{14} (15% dispersion)	[43, 44]
		ClF	400	50	7.8×10^{15} (15% dispersion)	[43, 44]

*Cl = catalyst prepared with a chlorinated compound (≈0.5–0.6 wt.% Cl in the catalyst).
*DCl = catalyst prepared with a chlorinated compound and dechlorinated under a flow of $H_2O + H_2$ at 350°C (residual content: ≈0.1–0.15 wt.% Cl).
*ClF = catalyst prepared with a chlorine-free precursor.

From the data of Table 2, the relative activities in $^{16}O_2/^{18}O_2$ equilibration at 300°C (chlorine-free precursors) would be

$$\text{Rh}, 10^4 > \text{Pt}, 10^2 - 10^3 > \text{Pd}, 1 - 10 \approx \text{Cu}, 1 - 10 > \text{Ni} \approx 0$$

Chloride ions behave as a poison for the equilibration reaction (see example 3 in Table 2). Most of these Cl ions are located on the support in the vicinity of the metal particles. However, it was shown by EXAFS[60] and by SIMS[61] that some Cl atoms were still bound to the metals after high-temperature oxidation or reduction, which can explain the inhibiting role of chlorine in equilibration.

$^{16}O_2/^{18}O_2$ equilibration is a structure-sensitive reaction. However, changing the particle size affects Pt much more than Rh catalysts. Moreover, if we examine the results obtained on fresh and sintered catalysts, an inverse tendency can be observed that the reaction is faster on the largest particles of platinum (see example 7 in Table 2), while the reverse seems to occur with the rhodium (examples 3 and 4).

The $^{16}O_2/^{18}O_2$ isotopic exchange that occurs during thermal desorption of oxygen from polycrystalline rhodium was investigated by Matsushima.[62] O_2 chemisorbed at very low temperature (−148°C) desorbs in two peaks: (i) a low-temperature peak at −110°C resulting from a molecular adsorption of O_2, and (ii) a high-temperature peak with a maximum varying from 450 (at high oxygen coverage) and 850°C (at low coverage), and corresponding to an atomic adsorption of O_2. When $^{16}O_2$ and then $^{18}O_2$ are adsorbed at −148°C, no exchange is found in the LT peak, while a significant exchange can be observed in the HT peak. This means that O_2 dissociation is a prerequisite for isotopic equilibration to occur on rhodium. Moreover, the $^{18}O/^{16}O$ ratio is higher in the molecular than in the atomic peak, which indicates that most of the molecularly adsorbed oxygen is composed of ^{16}O (the first to be adsorbed). Similar LT and HT peaks were observed by Gland[63] and by Sobyanin *et al.*[64] on Pt. However, by contrast to Rh, the isotopic ratio in the molecular and atomic peaks are almost equal on Pt, which means that the proportion of dissociated O_2 depends little on the oxygen coverage, while it is significantly greater on Rh at low coverage. The behaviour of platinum monocrystals was investigated by Sobyanin.[63] Pt (111) is much more active for oxygen equilibration than Pt (110), Pt (100) being the least active. This structure sensitivity can explain the differences of activity found for Pt/Al_2O_3 catalysts as a function of the particle size (Table 2).

The $^{16}O_2/^{18}O_2$ isotopic equilibration on metals is support-sensitive, particularly with reducible supports which can lead to SMSI effects. Shpiro *et al.*[65] showed that the rate of equilibration at 20°C over Rh/TiO_2 and Pt/TiO_2 depended strongly on the temperature of catalyst reduction. The rate is maximum for catalysts reduced at 300°C, while it decreases abruptly after reduction above 500°C. Moreover, all the catalysts show an induction period during which the oxygen equilibration occurs at a slow rate. After this induction period, the reaction rate increases by a factor of 3 to 8. Apparently, for the catalysts reduced at 300°C, the reaction seems to be slightly faster on Pt (5×10^{16} at. $m_{met}^{-2}\,s^{-1}$) than on Rh (2.5×10^{16} at. $m_{met}^{-2}\,s^{-1}$). These figures differ from the relative values obtained with the alumina-supported metals at 300°C, a temperature at which the rhodium is much more active than the platinum (Table 2). Two reasons can be proposed to explain these differences. Firstly, it is likely that the activation energy of O_2 equilibration on Pt/TiO_2 is lower than on Rh/TiO_2. On alumina, the values are of $62 \pm 4\,kJ\,mol^{-1}$ for Pt and of $75 \pm 5\,kJ\,mol^{-1}$ for Rh,[56] while the differences are still greater for the zirconia-supported metals ($37\,kJ\,mol^{-1}$ for Pt and $80\,kJ\,mol^{-1}$ for Rh).[66] If we suppose that these activation energies do not change throughout the entire temperature range, the A_{Rh}/A_{Pt} activity ratio should increase between 20 and 300°C by a factor 14 for Al_2O_3 catalysts, and by a factor of 4500 for ZrO_2 catalysts. The relative high activity of Pt at low temperature is thus consistent with the differences in activation energies between the two metals. The second reason concerns a likely effect of TiO_2 on the $^{16}O_2/^{18}O_2$ equilibration. Shpiro *et al.*[65] have not determined the activation energies for their titania catalysts. However, we can reasonably accept that they are higher than $30\,kJ\,mol^{-1}$, which leads to intrinsic rates at 300°C, comprised between 10^{19} and 10^{20} at. $m_{met}^{-2}\,s^{-1}$ for Pt/TiO_2 and Rh/TiO_2. These rates are of one order of magnitude higher than those found for the alumina-supported catalysts, which suggest that specific sites localised at the metal/support interface could be active sites for $^{16}O_2/^{18}O_2$ homoexchange. Reducing the catalysts at elevated temperatures would lead to both an incorporation of metal cluster boundary atoms into the oxygen vacancies sites (geometric effects), and to metal-Ti^{3+} interactions (electronic effects). As discussed by Shpiro *et al.*,[65] filling the oxygen vacancies and enhancing the electron density on metal would cause a decline in the rate of $^{16}O_2/^{18}O_2$ homoexchange. Significant equilibration rates were also obtained by Cunningham *et al.*[67,68] on RhO_x/CeO_2 at ambient temperature. In these experiments, the active sites for exchange were Rh cations in close association with cerium oxide, according to the equation (written with Kroger-Vink notation)

$$Rh_2O_3 + 2Ce_{Ce}^{x} + O_O^{x} \Leftrightarrow 2Rh'_{Ce} + V_O^{**} + 2CeO_2 \tag{43}$$

Although no clear indication about the sample weight and the rhodium dispersion was given by Cunningham *et al.*, a minimum rate of 10^{16} at. m_{met}^{-2} s^{-1} can be estimated for their 2%RhO_x/CeO_2 sample at 18°C, which is close to the values found by Shpiro *et al.*[65] for Rh/TiO_2 catalysts.

The $^{16}O_2/^{18}O_2$ equilibration is also sensitive to adatoms present in the supported metal catalyst. Masai *et al.*[69–71] investigated the effect of tin on the oxygen homoexchange over Rh/SiO_2 and Ru/SiO_2 catalysts at very low temperatures (−160 to −80°C). Under these conditions, the noble metals remained essentially in a zero-valence state. The presence of tin increases the rate of equilibration of Rh and of Ru by two orders of magnitude. It is assumed that oxygen molecules are dissociatively adsorbed on Rh or Ru. Adsorbed oxygen atoms then spill over to adjacent tin atoms where their recombination can occur more easily than on the noble metals.

3.3. $^2H_2/^1H_S$ and $^{18}O_2/^{16}O_S$ heteroexchange on supports

For the measurements of surface diffusion by $^2H_2/^1H_S$ and $^{18}O_2/^{16}O_S$ heteroexchange to be valid, it is necessary (Sec. 2.4.1) that the rate of the direct exchange with the support be very slow, compared with the rate obtained in the presence of metal particles. Thus, it seems important to examine the literature data dealing with $^2H_2/^1H_S$ and $^{18}O_2/^{16}O_S$ exchange, with the oxides which are often used as supports. We will restrict our review to the following materials Al_2O_3, SiO_2, ZrO_2, MgO, ZnO and CeO_2.

3.3.1. $^2H_2$ exchange with ^1H_S species of oxides

Most $^2H_2/^1H_S$ exchange experiments were carried out by using IR and mass spectrometry. IR absorption bands of surface deuteroxyl groups occur at much lower frequencies than those of hydroxyl groups by about 1000 cm^{-1}. On Al_2O_3, O^1H absorption bands occur between 3550 and 3850 cm^{-1} while O^2H bands can be observed in the 2550–2850 cm^{-1} frequency range.[72–79] On SiO_2, an intense, sharp band at 3755 cm^{-1} assigned to free O^1H groups can be observed, while associated hydroxyls give a weak, broad band between 3300 and 3700 cm^{-1}. The respective deuteroxyl bands are found at 2755 cm^{-1} (free O^2H), and between 2300 and 2700 cm^{-1} (associated O^2H).[80–84] IR spectroscopy was also employed by Davidov *et al.*[85] to monitor the heteroexchange of silanol groups with ^2H_2O. IR spectroscopy can thus be used for kinetic measurements of hydrogen exchange, either directly or in association with mass spectrometry.

Peri and Hannan[72] investigated the kinetics of $^2H_2/O^1H$ heteroexchange over a *gamma*-alumina pretreated at 650°C by IR. At 250°C, the equilibrium between the gas phase and the surface is not reached, even after an 18-hour contact time. By contrast, the equilibrium is rapidly reached above 500°C. Three different hydroxyl bands are present on the alumina sample which are exchanged at different rates. This point will be discussed in Sec. 4. Keith Hall and Lutinski[86] studied the exchange of the hydroxyl groups of an alumina sample (ex-Al isopropoxide, calcined at 640°C) which is constituted of a mixture of *eta* and *gamma* phases. Temperature-programmed isotopic exchange was carried out between 20 and 500°C at $2°C\,min^{-1}$, and the hydrogen isotopomers were analysed by a special technique based on a thermal conductivity measurement. The exchange begins at 80°C with two maxima at 180 (weak) and at 250°C (main peak). Depending on the pretreatment (oxidation, reduction under dry or wet H_2), the number of exchanged hydrogen species may vary between 2.5 and $3.1\,at.\,H\,nm^{-2}$. The presence of fluorine on alumina (1.22 wt.-%) increases the activation energy of exchange which starts at 220°C and decreases the maximal rates of exchange. Only one peak of exchange centered around 320°C can be observed, while the number of exchanged atoms is decreased by a factor of 2 ($1.7\,at\,H\,nm^{-2}$). Carter *et al.*[74] used an *eta* alumina pretreated at 500°C to investigate, by IR spectroscopy, the exchange of O^1H groups with $^2H_2$ between 75 and 250°C. The kinetics of the exchange follow an exponential rate law [Eq. (44)], which is typical of many chemisorption processes

$$r_e = \frac{dx}{dt} = b\,\exp(-\alpha x) \tag{44}$$

where x is the fraction of O^1H groups exchanged. The parameter b varies between $2.8 \times 10^{-5}\,s^{-1}$ at 75°C and $2.9 \times 10^{-3}\,s^{-1}$ at 250°C, while $\alpha = 28.6$ at 75°C and 12.9 at 250°C. The apparent activation energy decreases as the temperature increases from $12\,kJ\,mol^{-1}$ between 75 and 150°C, to $7\,kJ\,mol^{-1}$ between 150 and 250°C.

Temperature-programmed isotopic exchanges (TPIE) of $^2H_2$ with hydroxyl groups of several oxides were carried out by Hall *et al.*,[87] and more recently, by Martin and Duprez.[88,89] In both studies, the temperature was raised at $2°C\,min^{-1}$ during TPIE. The activation energies were calculated by Keith Hall *et al.* according to Eq. (45)

$$E = \frac{RT_{max}^2 (dx/dt)_{max}}{(x_\infty - x)_{max}(dT/dt)_{max}} \tag{45}$$

where subscripts, max, indicate that the values were taken at the inflection point of the curve "x (fraction of O^1H exchanged) versus T", corresponding to the maximal rate of exchange. Equation (46) supposes that there is a first-order kinetics for the istopic exchange, where only one form of hydrogen is exchanged

$$r_e = k_0 N_{OH} \exp(-E/RT) \times (x_\infty - x) \tag{46}$$

The values of E obtained by Eq. (46) depend essentially on T_{max}. For the determination of the activation energies, Martin and Duprez directly used the beginning of the curves of the exchange by plotting the values of x at different temperatures ($T < T_{max}$) in Arrhenius coordinates. The values of E obtained by this procedure are more dependent on the shape of the curves at low temperatures, than on the position of the rate maximum.

Table 3 gives the main conclusions of these studies. The results of Hall *et al.*[87] lead to the following order for the maximal rates of exchange

$$Al_2O_3, 250°C > Al_2O_3\text{-}SiO_2, \approx 400°C > SiO_2, 610\text{–}625°C.$$

The behaviour of the alumina-silica sample is specific of a mixed oxide in which silica and alumina are intimately associated. TPIE of a mechanical mixture of silica and alumina gives two peaks corresponding to the independent contributions of the two oxides. Martin and Duprez have confirmed that the exchange of $^2H_2$ on silica is very difficult. Although a very small amount of hydrogen ($0.08\,H\,nm^{-2}$) can be exchanged at relatively low temperature on this silica sample, hydrogen exchange starts at 250°C for most of the hydroxyl groups. It is worth noticing that a very pure silica pretreated at 320°C is able to equilibrate a $^1H_2/^2H_2$ at 150°C,[90] which confirms that specific sites of silica could possess a significant activity for hydrogen exchange. On the basis of the maximal rate of exchange, the following order is obtained for the various oxides studied by Martin and Duprez

$$CeO_2(red), 100°C > MgO, 120°C > ZrO_2, 145°C > CeO_2(ox),$$
$$160°C > Al_2O_3, 190°C > Cl/Al_2O_3, 200°C > SiO_2, 540°C.$$

Table 3. Temperature-Programmed Isotopic Exchange of $^2H_2$ with O^1H groups of oxides.

Oxides	Pretreatment	A_{BET} $m^2\,g^{-1}$	$T_0(°C)^a$	$T_{max}(°C)^a$	$N_{OH}{}^b$ at. H nm^{-2}	Ea kJ mol^{-1}	Ref.
SiO_2	525°C; O_2, 2h + vacuum, 16h	565	450	625	1.6	96	[87]
SiO_2	"	285	450	610	2.6	92	[87]
γ-Al_2O_3	"	190	130	250 (175, 350)	4.4	25	[87]
12%Al_2O_3 $-SiO_2$	"	327	275	395	1.5	42	[87]
SiO_2	450°C; O_2, 15 min. + vacuum, 30 min.	200	250	540	1.2	30	[88, 89]
γ-Al_2O_3	"	100	(80) 100	190	2.0	72	[88, 89]
Cl/Al_2O_3 (0.25%Cl)	"	100	100	200	1.6	62	[88, 89]
ZrO_2	"	40	50	145	1.4	48	[88, 89]
MgO	"	150	25	120	3.6	47	[88, 89]
CeO_2(ox)	"	60	80	160	1.4	59	[88, 89]
CeO_2(red)	450°C; H_2,15 min. + vacuum, 30 min	60	25	100	2.8	26	[88, 89]

$^a T_0$ is the temperature at which the exchange can be detected while T_{max} is the temperature of the maximal rate of exchange. Figures in parentheses correspond to minor peaks or shoulders.
$^b N_{OH}$ is the number of hydroxyl groups exchanged at the end of the TPIE experiment.

Hydrogen exchange is much easier on alumina than on silica. Chlorine has a moderate effect on the rate of exchange, even though it decreases the amount of exchangeable hydrogen. However, $^2H_2/O^1H$ exchange occurs at still lower temperatures on the other oxides, specially on magnesia and on reduced ceria. With these two oxides, a significant rate can be observed at room temperature. As expected, the activation energies reported by Hall *et al.*[87] follow the same order as the T_{max}. The values of E obtained by Martin and Duprez also increase with T_{max}. There are however two noticeable exceptions, Cl/Al_2O_3 and SiO_2, which results from a site heterogeneity on these oxides. Certain sites are easily exchanged, while most of them are difficult (Cl/Al_2O_3) or very difficult (SiO_2) to be exchanged.

A detailed investigation of the hydrogen exchange on magnesia was reported by Shido *et al.*,[91] who studied the exchange of $^1H_2$ with deuteroxyl groups of MgO. The surface of magnesia was deuterated for 1h at 400°C. The exchange reaction was monitored both by IR and MS. On MgO, six different deuteroxyl groups can be observed. Their activation energies for $^1H_2/O^2H$ exchange vary from 34 to 85 kJ mol^{-1}. Deuteroxyl groups beared by magnesium ions exchange more rapidly than deuterium atoms bonded to lattice oxygens. The coordination number of the Mg and of the O atoms in O^2H-Mg-O sites has a marked influence on the rate of exchange, the most active site being the Mg_{3C}-O_{3C} pair. Knözinger *et al.*[92,93] have shown that these sites were responsible for the reversible heterolytic dissociation of H_2 on MgO, while homolytic dissociation could occur on $O^{\bullet -}$ radical anions (irreversible) or on $Mg^{\bullet +}$ radical cations (reversible). It seems that the isotopic exchange of $^2H_2$ with O^1H groups of MgO requires a homolytic dissociation of $^2H_2$ on $O_{3C}^{\bullet -}$ radical anions,[93] rather than a heterolytic dissociation as proposed by Hoq *et al.* for $R^1H/^2H_2$ exchange (RH = methane, neopentane, benzene...).[94]

Agron *et al.*[95] have shown there were essentially two types of hydroxyl groups on ZrO_2: linear OHs (Zr-OH) and bridged OHs (Zr_2-OH).[95] These hydroxyl groups can be easily exchanged by ^2H_2O at 40°C. Kondo *et al.*[96,97] have reported that zirconia was also able to dissociate H_2 at low temperature and to produce $Zr\langle^{H}_{H}$, Zr-H and OH species. Although Kondo *et al.*[88,89] used $^2H_2$ to study isotopic effects, no clear indication about the hydrogen exchange was given in their reports. However, we can expect that this particular behaviour of ZrO_2 in H_2 activation can contribute to the rapid exchange observed by Martin and Duprez.

3.3.2. $^{18}O_2$ *exchange with* $^{16}O_S$ *species of oxides*

In most cases, $^{18}O/^{16}O$ exchange experiments were carried out by using mass spectroscopy analysis. Oxygen exchange cannot be easily monitored by IR spectroscopy because of the very low isotopic shift between ^{18}O and ^{16}O surface species (generally less than 20 cm^{-1}). There is an exception to this rule concerning the oxygen binuclear species (superoxides O_2^- and peroxides O_2^{2-}), which give isotopic shifts of about 50–70 cm^{-1}. However, IR studies are then restricted to those oxides, such as CeO_2, giving rise to binuclear species.[98–100] Raman spectroscopy which gives very narrow metal-oxygen bands could be used for the study of oxygen exchange over oxides like MoO_3, Sb_2O_4 and SnO_2.[101–103]

The $^{18}O/^{16}O$ heteroexchange was investigated by Winter[29] and by Boreskov and Muzykantov[18] over a great variety of oxides. Some of their results are summarised in Table 4 which gives the preexponential factor, the activation energy and the order m with respect to O_2 of the rate equation

$$r_e = P(O_2)^m \times A_0 \exp\left(-\frac{Ea}{RT}\right) \quad (47)$$

The values of the rate at 400°C were calculated from the kinetic parameters (A_0 and Ea) reported by Winter[29] and by Boreskov and Muzykantov.[18] Compared with hydrogen (Table 3), oxygen exchanges with a relatively high activation energy, the differences between Ea (O) and Ea (H) ranging from 50 to 100 kJ mol^{-1}. With the exception of orders with respect to O_2, the kinetic data of Winter and of Boreskov and Muzykantov show a very good agreement. On the basis of the rate of exchange at 400°C, the following order for the oxides listed in Table 4 can be obtained (data of Winter, base 10 for γ-Al_2O_3)

$$\begin{aligned}&Rh_2O_3, 2\times 10^5 \gg CeO_2, 1650 \gg MgO, 470 > ZrO_2,\\&280 \gg ZnO, 47 > TiO_2, 26 > \gamma\text{-}Al_2O_3, 10 > \delta\text{-}Al_2O_3,\\&0.5 \gg SiO_2 \text{ (very low)}\end{aligned}$$

The reaction is very fast on Rh_2O_3. Among the oxides currently used as supports, ceria exhibits the highest rate of oxygen exchange. It also shows a relatively high rate for the multiple exchange (Type III), which can be linked to the formation of binuclear species upon O_2 adsorption.[98,99,104,105] On the opposite side of the scale, alumina shows a modest ability to exchange its oxygen atoms, while silica practically cannot exchange oxygen at all.

TPIE experiments were carried out over some of these oxides by Martin and Duprez.[66,88] The main results reported in Table 5 lead to the following order for the oxides, according to the temperature of their maximal rate of exchange

$$\begin{aligned}&CeO_2(ox), 400°C > CeO_2, 410°C > CeO_2\text{-}Al_2O_3, 480°C > MgO,\\&490°C > ZrO_2, 530°C > Al_2O_3, 620°C \gg SiO_2, 850°C.\end{aligned}$$

This confirms that SiO_2 requires a very high temperature for oxygen exchange and is totally inactive below 600°C. The activation energies

Table 4. Heteroexchange of $^{18}O_2$ with ^{16}O of oxides.

Oxides	A_{BET} m^2g^{-1}	Pretreatment* at T°C	Reaction Temp. °C	Type of Exchange	Preexp. Factor** Molec. O_2 m^{-2} s^{-1}	Ea $kJ\,mol^{-1}$	O_2 Order	r_e at 400°C** Molec. O_2 m^{-2} s^{-1}	Ref.
SiO_2	296	(a); 750	650	II	—	—	—	—	[29]
γ-Al_2O_3	95	(a); 600	480–580	II	2.14×10^{25}	163	0	4.9×10^{12}	[29]
δ-Al_2O_3	41	(a); 730	620–720	II	9.59×10^{23}	163	0	2.2×10^{11}	[29]
ZrO_2	10.5	(b)	350–450	II	8.13×10^{23}	126	1.0	1.4×10^{14}	[29]
MgO	61	(b)	370–450	II	4.90×10^{26}	159	0	2.3×10^{14}	[29]
CeO_2	1.7	(b)	450–520	II, III	1.91×10^{23}	108	0.8	8.1×10^{14}	[29]
TiO_2 rut.	17	(b)	500–580	II	2.75×10^{24}	146	1.0	1.3×10^{13}	[29]
ZnO	5.3	(b)	350–410	II	1.0×10^{25}	150	1.0	2.3×10^{13}	[29]
Rh_2O_3	1.5	(b)	290–370	II, III	1.70×10^{19}	42	0	9.4×10^{16}	[29]
Al_2O_3	124	(c); 600	350–475	II	5.0×10^{19}	155	1	1.4×10^{13}	[18]
MgO	9	(c); 600	480–530	II	1.1×10^{18}	104	1	9.4×10^{13}	[18]
TiO_2	67	(c); 600	530–600	II, III	—	≈210	—	$\approx 5 \times 10^{12}$	[18]
ZnO	1	(c); 550	425–525	II, III	1.0×10^{27}	167	0.9	1.1×10^{14}	[18]

*Pretreatments: (a), outgassed at T°C, then pretreated with O_2 (2000 Pa) for at least 18 hr at the highest reaction temperature.
(b), only pretreated with O_2 at the highest reaction temperature.
(c), pretreated with O_2 (1330 Pa) at T°C.
**Given for $P(O_2) = 1330$ Pa.

Table 5. Temperature-Programmed Isotopic Exchange of $^{18}O_2$ with ^{16}O atoms of oxides.[66,88]

Oxides[a]	A_{BET} $m^2\,g^{-1}$	T_0[b] °C	T_{max}[b] °C	T_{end}[b] °C	$N_{O(max)}$[c] at. O nm^{-2}	$N_{O(end)}$[c] at. O nm^{-2}	Ea $kJ\,mol^{-1}$
SiO_2	200	650	850	900	15.0	22.7	111
γ-Al_2O_3	100	460	620	800	11.5	27.5	125
ZrO_2	40	380	530	780	21.5	48.1	111
ZrO_2 (ox)	40	380	530	780	18.6	42.5	105
MgO	150	390	490	800	7.8	31.4	166
CeO_2	60	310	410	480	19.6	37.7	110
CeO_2 (ox)	60	310	400	480	15.7	35.0	115
CeO_2-Al_2O_3 (12 wt.-% CeO_2)	100	370	480	800	9.9	27.7	101

[a]Pretreatment at 450°C: O_2, 15 min + H_2, 15 min + vacuum, 30 min; except (ox): only O_2 + vacuum.
[b]T_0 is the temperature at which the exchange can be detected, T_{max}, the temperature of the maximal rate of exchange and T_{end}, the temperature at the end of the TPIE experiment.
[c]$N_{O(max)}$ is the number of oxygen atoms exchanged at T_{max} and $N_{O(end)}$ at the end of the TPIE experiment.

measured in these TPIE experiments are not very different from those found by Winter (Table 4), where all are greater than 100 $kJ\,mol^{-1}$, confirming that O_2 exchange is a strongly activated process. The density of surface oxygen atoms depends on the nature of the oxide and on the hydroxyl coverage. However, it lies between 8 and 13 at. O nm^{-2},[21] which shows that the whole *surface* can be exchanged before, T_{max}, while at T_{end}, a significant fraction of the *solid* is exchanged (between two and four layers).

3.4. *Surface diffusion of hydrogen species on supported metal catalysts*

3.4.1. *$^2H_2$ heteroexchange with ^1H species of alumina in the presence of metals*

TPIE of $^2H_2$ with ^1H species of a 0.75 wt.% Pt/Al_2O_3 were carried out by Hall and Lutinski.[86] In parallel, $^2H/^1H$ TPIE of the same alumina (a mixture of η- and γ-Al_2O_3, 136 $m^2\,g^{-1}$) not impregnated with Pt had been investigated (see Sec. 3.3.1.). The presence of multiple peaks of exchange assigned to different hydroxyl groups can be noticed on Pt/Al_2O_3, as on Al_2O_3. However, depending on the nature of pretreatment (oxidation, reduction or outgassing), the different maxima of the rate of exchange can be observed on Pt/Al_2O_3 for temperatures ranging from 180 to 350°C, while a main peak centered around 250°C is observed on Al_2O_3. Apparently, the presence of Pt does not decrease the activation energy for the $^2H_2/O^1H$ exchange, and could even have a negative effect on the reaction in certain cases (reduced catalysts). Hall and Lutinski assumed that this inhibition could result from the presence of chloride ions (0.62 wt.-% Cl) introduced in the catalyst during metal impregnation (H_2PtCl_6). However, it seems that chlorine has little effect on the number of exchangeable H atoms while fluorine decreases both the rate of exchange and the number of exchangeable atoms.

The inhibiting effect of chlorine on the $^2H_2/O^1H$ exchange was confirmed by Carter *et al.*,[74] who studied the reaction between 75 and 250°C over variously loaded Pt/Al_2O_3 catalysts (0.001 to 2% Pt) with different chlorine contents (chlorine and chlorine-free precursors). The exchange was monitored by IR spectroscopy. The presence of very small amounts of Pt (0.001%) on chlorine-free catalysts has a definite promoter effect on the rate of hydrogen exchange of alumina. When the catalyst is prepared by impregnation with chloroplatinic acid, there is a weak effect of Pt, regardless of the metal loading (0.1 to 2%). In this case, the effect of adding platinum is less marked than that with the Cl-free 0.001% Pt catalyst. However, contrary to what was observed by Hall and Lutinski,[86] Pt keeps a promoter effect in the presence of chlorine. Carter *et al.*[74] explained the poisoning effect of chlorine by electronic interactions which would strengthen the O-H bond, and by the blockage of certain sites of exchange at the alumina surface. Chlorine would then be viewed as an inhibitor of surface diffusion. Moreover, there is no effect of the pretreatment (oxidised, reduced...) on the rate of exchange over chlorine-free catalysts. The observations reported by Hall and Lutinski[86] could result from a dechlorination of the catalyst sample during its pretreatment.

Since the pioneering work of Hall and Lutinski[86] and of Carter *et al.*,[74] many authors have shown that there was a promoter effect of the noble metals (Pt, Rh or Pd) on the $^2H_2$ exchange with the O^1H groups of alumina.[77,88,89,106–109] Cavanagh and Yates[77] have used IR spectroscopy to monitor the hydrogen exchange over a 2.2% Rh/Al_2O_3 catalyst. In the presence of Rh, the exchange is noticeable at 37°C, while there is no exchange on the bare alumina. CO chemisorption on the supported Rh leads to a significant reduction of the rate of exchange. Cavanagh and Yates assume that the rate-determining step of exchange is the surface migration of H species on alumina. The dissociation of dihydrogen on the metal particles is normally a very fast process. Nevertheless, CO chemisorption would be effective in preventing this process from occurring, which blocks the step of transfer on the support, and thus the hydrogen exchange on alumina. A similar effect was obtained by Dmitriev *et al.*[107] by chemisorbing hydrocarbons on a Pt/alumina catalyst. However, the exchange of ^2H atoms (generated in gas phase) with O^1H groups of alumina is less rapid than the exchange of $^2H_2$ with O^1H groups of Pt/Al_2O_3, which shows that there is a specific function of the metal/support interface.[106]

Martin and Duprez[88,89] have investigated the effect of Rh on the rate of $^2H_2/O^1H$ exchange over several supports at temperatures ranging from 50 to 100°C. At 75°C, the rate of exchange on Rh/oxide is *two to three orders* of magnitude higher than the rate measured on the bare oxide. The reaction is never limited by the rate of adsorption/desorption of H_2 on the metal, since the $^2H_2/^1H_2$ homoexchange process is extremely fast even at low temperatures (Table 1). However, a change in the rate-determining step of exchange was found around 75°C. Martin and Duprez assume that at $T < 75°C$, the $^2H_2/O^1H$ exchange is limited by the rate of transfer of H species from metal to support, while at $T > 75°C$, the reaction would be limited by the rate of migration of H species on alumina. The activation energy of this process ($7.8\,kJ\,mol^{-1}$) is much lower than the activation energy found for the exchange on the bare support (Table 3).

Most studies on the $^2H_2/O^1H$ exchange were performed on supported noble metal catalysts. However, non-noble metals can also promote hydrogen exchange with hydroxyl groups of alumina. Chen and Falconer[110,111] carried out $^2H_2$ TPIE by heating a 5.1 wt-% Ni/Al_2O_3 catalyst in a 10% $^2H_2$ flow. Hydrogen exchange started at 170°C with a maximal rate at 220°C. The large amount of ^1H that exchanged ($2900\,\mu$mol g-1) showed that $^2H_2$ was exchanged not only with ^1H on the metal, but also with O^1H groups of the support. The OH density estimated by TPIE was 7.9 $OH\,nm^{-2}$, which corresponds to an exchange of the alumina surface that would be almost total at ≈280°C.

Water was shown to promote hydrogen spillover on numerous oxides including alumina.[108,112–117] This could be due to the fact that $^2H_2/^1H_2O$ (or $^1H_2/^2H_2O$) exchange over metal catalysts is more rapid than the exchange of deuterium with the OH groups of oxides.[26] The order of activity found by McNaught *et al.* for $^1H_2/^2H_2O$ exchange is Pt > Rh > Pd ≫ Ni.[118] For the hydrogen/deuterium oxide exchange at 0°C, they report turnover frequencies over 10^{18} molec. $s^{-1}\,m^{-2}$ for Pt and over 10^{17} molec. $s^{-1}\,m^{-2}$ for Rh, while rates of 10^{16}-10^{17} molec. $s^{-1}\,m^{-2}$ are observed for the $^2H_2/O^1H$ exchange on Rh/Al_2O_3 above 50°C. It seems therefore necessary to carry out deuterium/hydroxyl exchange in a controlled atmosphere by trapping the traces of water vapor.[88,89] Sermon and Bond assume that the main effect of H_2O is to transport H species from metal to support or between catalyst grains *via the gas phase.*[113] However, *adsorbed* water molecules are certainly able to exchange and to transport H species in accordance with the model proposed by Peri for the $^2H_2/O^1H$ exchange.[119]

3.4.2. *Hydrogen exchange on other supports*

The effect of metals on the $^2H_2$ exchange with O^1H groups of silica was investigated by many authors. Conner *et al.*[81,83,84] have developed a spatially- and time-resolved FTIR technique in order to monitor the hydroxyl exchange of various silicas. The point source of deuterium species was either a small grain of $4\%Pt/SiO_2$[81] or of $0.05\%Pt/Al_2O_3$[83,84] pressed together with the silica in the centre of the wafer that was 2.5 cm in diameter. Exchange of silanols with deuterium occurred at 200°C and was relatively fast at 240°C. Blank experiments showed that $O^1H \rightarrow O^2H$ exchange of silica was very slow in the absence of the metal catalyst, in agreement with other work.[87,89] This demonstrates the crucial role of the point source of Pt in providing hydrogen exchangeable species. Conner *et al.* showed that the O^1H exchange increased with time and decreased with distance from the point source. Inexpectedly, however, O^2H gradients measured along a radius of the silica wafer were extremely flat. Exchange occurred over distances of several millimeters from the point source, suggesting that H diffusion was very rapid. As adsorption/desorption of hydrogen on Pt is an extremely fast process above 200°C, Conner *et al.* concluded that the exchange was controlled by the transfer of deuterium species from the metal to the support as well as by the reverse process.

Similar experiments were carried out by Bianchi *et al.*[82] who studied the $^2H_2/O^1H$ exchange between 25 and 150°C on a silica wafer (20 mm diameter, 50–80 mg, $200\,m^2\,g^{-1}$), compressed with a small grain of a $0.4\%Pt/Al_2O_3$ catalyst (2 mm diameter, 3 mg) located at the edge of the wafer. Bianchi *et al.* focused their attention on the possible effect of water on the deuterium exchange. Preliminary experiments were carried out in the absence of the Pt catalyst. Silica alone is not able to exchange $^2H_2$ in a dry atmosphere. Introducing a partial pressure of water (between 10^{-4} and 1 torr) allows silica to exchange deuterium. When Pt is present, a significant rate of exchange is obtained in dry atmosphere. Traces of water cannot account for this exchange promoted by platinum, which confirms the role of spillover and surface diffusion in this phenomenon. IR spectra of silica were taken both in the upper part of the wafer, far from the Pt catalyst "spot", and in its lower half, close to the spillover source. A significant gradient of deuteroxyl species is found for $t < 200$ minutes. After this exchange time, O^2H concentration tends to become uniform on the whole pellet. From these experiments, Bianchi *et al.* claimed that the migration of deuterium was the slow step in the exchange.

Ravi and Sheppard have investigated the $^2H_2/O^1H$ exchange at 25°C over a 10% Ir/SiO_2 catalyst.[120] The reaction did not proceed on silica alone, while the presence of iridium gave rise to a measurable exchange (52% of O^1H exchanged after 1200 minutes). However, the exchange proceeded more slowly over "hydrogen-precovered" catalyst than over bare iridium. It seems that hydrogen can eliminate traces of oxygen which promotes $^2H_2/O^1H$ exchange by formation of deuterium oxide.

Recently, Basallote *et al.*[121] have compared $^2H_2$ and $^1H_2$ TPD with $^2H_2 + {}^1H_2$ homoexchange rates on Pt/SiO_2 and Pt/CeO_2. TPD curves revealed that all the hydrogen chemisorbed on the metal desorbed between 25 and 280°C, while a significant amount of hydrogen was stored on ceria and desorbed above 300°C. The hydrogen stored on the support did not really participate in the $^2H_2 + {}^1H_2$ homoexchange carried out at 200°C. This reaction occurred essentially on the platinum surface. However, TPD carried after $^2H_2 + {}^1H_2$ homoexchange, revealed that the high temperature peak on Pt/CeO_2 was mainly composed of $^2H^1H$ and $^2H_2$, which showed that H species stored on the support was exchanged in the course of the homoexchange reaction.

Copper can also promote hydrogen spillover on oxide support. A hydrogen diffusion process was reported by Jung and Bell[122] to occur in methanol synthesis over Cu/ZrO_2 catalysts. FTIR showed that formate species adsorbed on ZrO_2 are formed by the reaction of CO with OH group of zirconia. Exchange of $^2H_2$ with these ^1HCOO-Zr species occurred at a low rate and formate species were shown to inhibit $^2H/^1H$ of O^1H groups of ZrO_2.

3.5. *Surface diffusion of oxygen species on supported metal catalysts*

Rhodium is the metal having the highest $^{16}O_2 + {}^{18}O_2$ equilibration rate on alumina (Table 2). There are many indications in the literature that this remains true for most supports, although Ru and Ir were shown recently to be slightly more active than Rh on CeO_2.[123] Rhodium nanoparticles were then currently used as porthole for oxygen in oxide-supported metal catalysts to measure oxygen surface diffusivity on these oxides. The results obtained by Martin and Duprez on zirconia-supported Rh and Pt catalysts are shown in Fig. 10 (BET area of the support: $40\,m^2\,g^{-1}$; Rh dispersion: 66%; Pt dispersion: 24%).

For rhodium catalysts, there is a breakpoint on the curve of exchange at about $T_0 = 280°C$. At low temperature ($T < T_0$), exchange is limited by adsorption/desorption of oxygen on the metal (steps 1 and -1 of Fig. 1). In this temperature range, exchange occurs at the same rate as equilibration. At high temperatures ($T > T_0$), exchange is limited by the surface diffusion process. The measurement of the coefficient of diffusion should be restricted to this temperature range. Platinum did not show the same behaviour as rhodium. There is no apparent breakpoint in the curves of exchange probably because the process is limited by the adsorption/desorption of oxygen on Pt within an extended range of temperature.

Similar results were obtained with other supports such as silica, alumina, magnesia ceria. It was also verified that the direct exchange between the gas phase and the support could not occur at a significant rate on most oxide supports. $^{16}O/^{18}O$ TPIE were carried out on the bare supports and on the corresponding Rh catalysts. An example is represented on Fig. 11

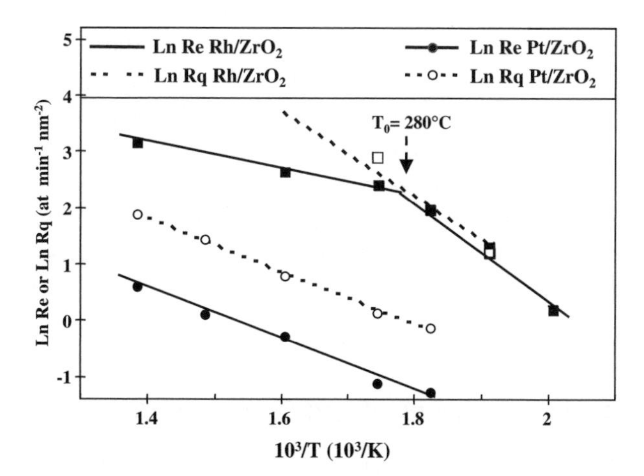

Fig. 10. Arrhenius plots of the rates of exchange (Re) and of equilibration (Rq) of oxygen over zirconia-supported catalysts (after Ref. 66).

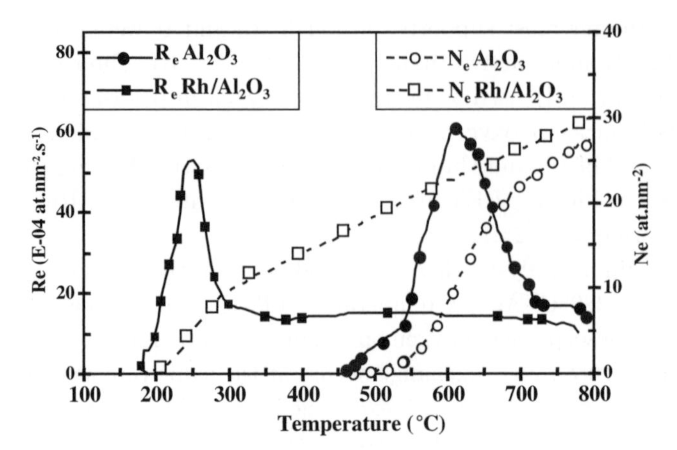

Fig. 11. $^{16}O/^{18}O$ TPIE over bare Al_2O_3 and Rh/Al_2O_3 catalysts. Rates of exchange Re and number of exchanged atoms Ne (after Ref. 66).

for Al_2O_3 and Rh/Al_2O_3 catalysts (BET area of the support: $100\,m^2\,g^{-1}$; Rh dispersion: 87%).

Oxygen exchange starts at 450°C on bare alumina, while it can be detected below 200°C in the presence of rhodium. There is more than 350°C between the two maxima of exchange, which confirms that there is virtually no direct exchange within the temperature range (300–450°C) in which the coefficient of diffusion have been currently measured on alumina. TPIE over Rh/Al_2O_3 shows a sharp peak in between 180°C and 300°C, followed by a slow exchange up to the highest temperature (800°C). Interestingly, the number of exchanged atoms at 300°C is close to 10 atoms nm^{-2}, which means that all the surface of alumina has already been exchanged at this temperature. The slow rate of exchange observed above 300°C in the presence of rhodium would correspond to a bulk exchange in alumina. Similar differences between the starting temperatures of exchange on bare ceria and Rh/CeO_2 were reported by Cunningham *et al.*[124] Onset changes in the isotopomer composition can be observed at 530°C, while in the presence of rhodium, exchange starts at 260°C.

3.6. *Comparison between H and O surface diffusion*

On most oxides, O surface diffusion can be observed and measured in the 250–500°C temperature range. Above 500°C, rates of exchange on supported metal catalysts are extremely high and the coefficient of surface

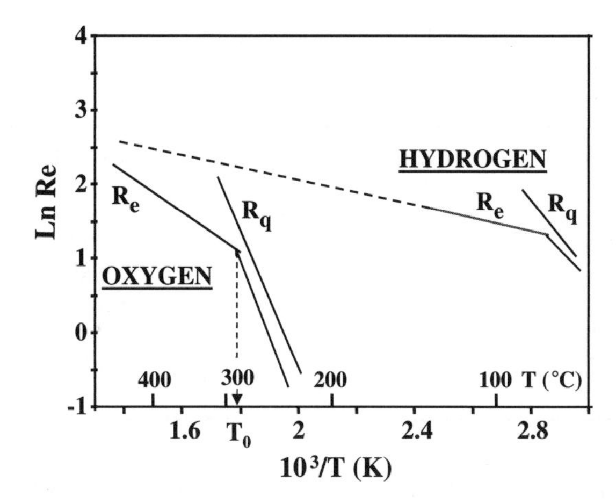

Fig. 12. Arrhenius plots of oxygen and hydrogen exchange (Re) and equilibration (Rq) over Rh/Al_2O_3 (after Refs. 43, 66 and 89).

diffusion cannot be measured with sufficient accuracy by this technique. H surface diffusion occurs at much lower temperatures, typically between 0 and 100°C (Fig. 12).

Changes in the rate determining steps of exchange are observed both for oxygen (around 300°C) and hydrogen (around 75°C). At low temperature, the mechanism of hydrogen migration cannot involve oxygen mobility. It is likely to imply jumps of H atoms (or ions) between two adjacent OH groups. However, H diffusion is virtually not activated ($Ea < 10\,kJ\,mol^{-1}$), while O diffusion is clearly an activated process (Ea between 20 and $40\,kJ\,mol^{-1}$).[43,66,89] At high temperatures (typically around 450°C), H and O diffusion occur practically at the same rate, which can be viewed as a collective migration of OH groups at the alumina surface.

Relative values of the coefficient of diffusion are given in Table 6.

Ceria exhibits both the highest oxygen and the highest hydrogen mobility, in connection with well-known properties of this oxide: (i) high Oxygen Storage Capacity, OSC[125–132] and (ii) high spillover rates of hydrogen on metal-ceria catalysts.[133–135] Mathematical modeling of the elementary steps involved in OSC allowed Costa *et al.* to evaluate the back-spillover rate constant of oxygen on Pd/CeO_2.[136] By contrast, silica has the lowest oxygen and hydrogen mobility, which allows to conclude that silanols groups are immobile at the silica surface within a wide range of temperature. Acid-base properties of these oxides were measured[137]: it was shown that oxygen mobility could be correlated satisfactorily with basic properties

Table 6. Relative O and H mobility over various oxides (base 10 for alumina). The absolute coefficients of diffusion on alumina are $2 \times 10^{-18}\,m^2\,s^{-1}$ for oxygen at 400°C and $0.43 \times 10^{-18}\,m^2\,s^{-1}$ for hydrogen at 75°C.

Oxides (BET area)	Relative Oxygen Mobility at 400°C		Relative Hydrogen Mobility at 75°C	
γ-Al_2O_3, $100\,m^2\,g^{-1}$	CeO_2	2300	CeO_2	80
12%CeO_2/Al_2O_3, $100\,m^2\,g^{-1}$	MgO	50	MgO	22
CeO_2, $60\,m^2\,g^{-1}$	ZrO_2	28	CeO_2/Al_2O_3	16
MgO, $150\,m^2\,g^{-1}$	CeO_2/Al_2O_3	18	Al_2O_3	10
SiO_2, $200\,m^2\,g^{-1}$	Al_2O_3	10	ZrO_2	2.3
ZrO_2, $40\,m^2\,g^{-1}$	SiO_2	0.1	SiO_2	Very low

of oxides while no correlation was found between hydrogen mobility and acid-base properties. It seems as if a very weak acidity was sufficient to promote hydrogen migration.

4. Recent Developments of $^{16}O/^{18}O$ Isotopic Exchange: Case of Oxides with Very High Oxygen Mobility

The model developed in Sec. 2 can be applied to oxides having a moderate oxygen mobility (see Sec. 3). In this case, initial rates of exchange can be measured with sufficient accuracy and surface and bulk diffusion can be distinguished as shown in Fig. 8. Cerium-zirconium oxides were developed in the 90's to incorporate materials with very high OSC in exhaust gas catalysts.[138–150] It was proven that surface and bulk oxygen mobility was so fast that the models based on the initial rates of exchange can no longer be used. It was thus necessary to develop new models adapted for $^{16}O/^{18}O$ isotopic exchange on these mixed oxides. A simplified model was first developed by Holmgren *et al.*,[130] while a more complete version of the model including a computer program (called PLATINA) was developed by Galdikas *et al.*[151–152] Metal particles were assimilated to cubes of size "d". Surface and volume were divided into elementary units which are represented in Fig. 13(a) (top view for the surface) and Fig. 13(b) (perpendicular view for the volume).

The flux of species "j" arriving at the surface (adsorption step) was calculated by

$$I_j = \frac{P_j N_A}{\sqrt{2\pi\, M_j\, RT}} \tag{48}$$

assuming a sticking coefficient of 1 for oxygen on metals (Rh or Pt).

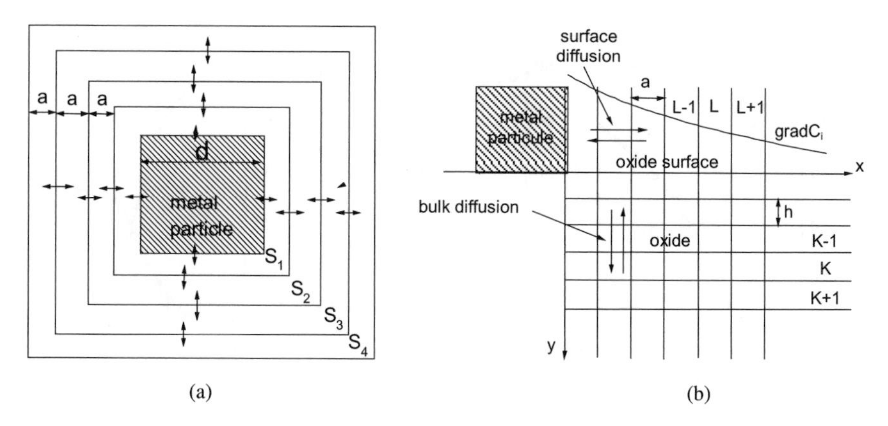

Fig. 13. Square-shaped layers around the cubic metal particles: (a) at the surface (top view) and (b) in the bulk (view perpendicular to the surface).

The desorption was considered as recombination process between two atoms. For instance, the rate of $^{16}O_2$ desorption is given by:

$$\frac{dn_{32}}{dt} = \frac{S_m}{2V} k C_{16}^2 \tag{49}$$

where Sm is the metal surface area, V, the reactor volume, C_{16}, the concentration of ^{16}O atoms on the metal at time t, and k, the desorption rate constant:

$$k = A \ \exp\left(-\frac{Q}{RT}\right) \tag{50}$$

Assuming that the diffusion fluxes follow the Fick's law, C_{16} and C_{18} were calculated in each elementary surface and in each elementary volume as a function of time, which allows one to estimate the concentration of each isotopomer in the gas phase. Values of Q, D_s and D_b were adjusted to minimise the squares of the differences between calculated and measured concentrations in gas phase. A refinement was recently brought to this model by including a dual site population on the support. Certain sites exchange oxygen via a simple exchange mechanism, while other sites would exchange oxygen via a multiple exchange mechanism (see Sec. 2.2.1). This modification in the model was required because preliminary investigations showed that the multiple exchange mechanism was predominant on CeZrOx oxides (Fig. 14).

The initial formation of $^{16}O_2$ is not observed on ZrO_2, which showed that zirconia exchange oxygen via a simple mechanism exclusively. The appearance of $^{16}O_2$ together with $^{16}O^{18}O$ is characteristic of a multiple

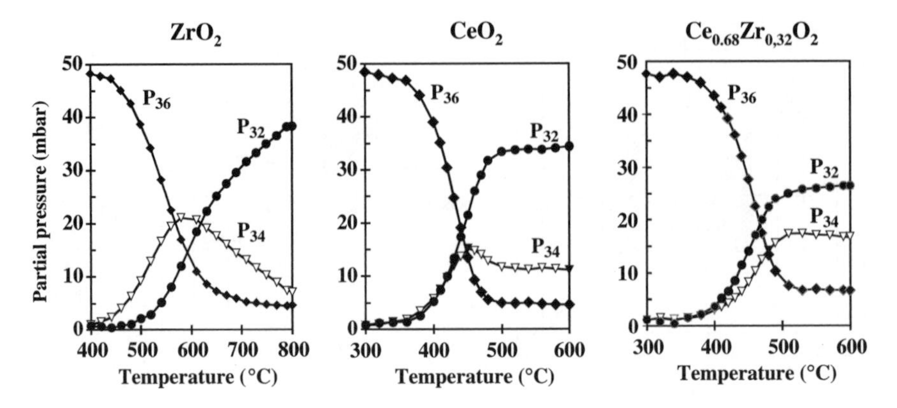

Fig. 14. $^{16}O/^{18}O$ TPIE over ZrO_2, CeO_2 and $Ce_{0.68}Zr_{0.32}O_2$.

Table 7. Coefficients of diffusion determined with the PLATINA model specially developed for ceria-zirconia supported catalysts (after Ref. 151).

Catalyst	Pt/CZ-D	Pt/CZ-O	Pt/CZ-R
BET area ($m^2\,g^{-1}$)	37	104	3
Metal particle size (nm)	2	5.2	59
Temperature (°C) of exchange	334	323	332
Surface diffusion D_s ($\times 10^{-20}\,m^2\,s^{-1}$)	1.7	109	1450
Bulk diffusion D_b ($\times 10^{-23}\,m^2\,s^{-1}$)	0.34	4.5	36

mechanism. This feature is clearly observed on ceria and ceria-zirconia, but the P_{32}/P_{34} ratio is higher than 1 over the entire range of temperature for ceria-zirconia, which denotes a predominant multiple exchange mechanism on this oxide.

Galdikas *et al.*[151] have reported the results obtained on three 1%Pt/ $Ce_{0.5}Zr_{0.5}O_2$ catalysts (Table 7). CZ-O is a conventional mixed oxide prepared by hydrolysis of $ZrO(NO_3)_2$ with an aqueous ammonia solution in the presence of a fine ceria powder. CZ-R was obtained by a reducing treatment in CO at 1200°C, while CZ-D was prepared by high energy ball milling of CeO_2 and ZrO_2 powders. After platinum impregnation ($Pt(NH_3)_2(NO_2)_2$ as a precursor), all the samples were calcined at 500°C. In spite of its very low BET area and metal dispersion, Pt/CZ-R exhibits the highest oxygen diffusivity both at the surface and in the bulk.

Compared with Pt/CZ-R, oxygen diffusion is slow both at the surface and in the bulk of catalysts prepared by high energy ball milling. It is very likely that grain boundaries play an important role in the diffusion process. This was recently confirmed in the case of perovskite catalysts for methane combustion.[153]

This study emphasises the special behaviour of ceria-zirconia catalysts and related compounds exhibiting very high oxygen diffusivity. The use of refined mathematical models and computer-assisted calculations are required for processing the data of isotopic exchange with the highest efficiency and accuracy.

References

[1] Grenoble DC, *J Catal* **51**: 212, 1978.
[2] Duprez D, *Appl Catal* **82**: 111, 1992.
[3] Yao HC, Yu Yao YF, *J Catal* **86**: 254, 1984.
[4] Oh SH, Heickel CC, *J Catal* **112**: 543, 1988.
[5] Weng LT, Ruiz P, Delmon B, Duprez D, *J Mol Cat* **52**: 349, 1989.
[6] Weng LT, Delmon B, *Appl Catal* **81**: 141, 1992.
[7] Bebelis S, Vayenas CG, *J Catal* **118**: 125, 1989.
[8] Vayenas CG, Bebelis S, Ladas S, *Nature* **343**: 625, 1990.
[9] Sermon PA, Bond GC, *Catal Rev* **8**: 211, 1973.
[10] Conner WC Jr, Pajonk GM, Teichner SJ, *Adv Cata* **34**: 1, 1986.
[11] Inui T, Fujimoto K, Uchijima T, Masai M (eds.), *New Aspects of Spillover Effects in Catalysis*, Stud Surf Sci Catal, Vol. **77** (1993).
[12] Kapoor A, Yang RT, Wong C, *Catal Rev-Sci Eng* **31**: 129, 1989.
[13] Boreskov GK, *Adv Catal* **15**: 285, 1964.
[14] Novakova J, *Catal Rev* **4**: 77, 1970.
[15] Winter ERS, *Adv Catal* **10**: 196, 1958.
[16] Schwank J, Galvano S, Parravano G, *J Catal* **63**: 415, 1980.
[17] Duprez D, Abderrahim H, Kacimi S, Rivière J, in *Proc. 2nd Int. Conf. Spillover*, Steinberg KH (ed.), Leipzig Universität, 1989, p. 127.
[18] Boreskov GK, Muzykantov VS, *Ann New York Acad Sci* **213**: 137, 1973.
[19] Duprez D, Pereira P, Miloudi A, Maurel R, *J Catal* **75**: 151, 1982.
[20] Anderson JR, *Structure of Metallic Catalysts*, Academic Press, New York, 1975.
[21] Linsen BG, *Physical and Chemical Aspects of Adsorbents and Catalysts*, Academic Press, London and New York, 1970.
[22] Kramer R, Andre M, *J Catal* **58**: 287, 1979.
[23] Martin D, Duprez D, in *New Aspects of Spillover Effects in Catalysis*, Stud Surf Sci Catal **77**: 201, 1993.
[24] Kakioka H, Ducarme V, Teichner SJ, *J Chim Phys* **68**: 1715, 1971.
[25] Kakioka H, Ducarme V, Teichner SJ, *J Chim Phys* **68**: 1722, 1971.
[26] Ozaki A, *Isotopic Studies of Heterogeneous Catalysis*, Kodansha, Tokyo and Academic Press, New York, 1977.
[27] Klier K, Novakova J, Jiru P, *J Catal* **2**: 479, 1963.

[28] Muzykantov VS, Popovski VV, Boreskov GK, *Kin Katal* **5**: 624, 1964.
[29] Winter ERS, *J Chem Soc A* 2889, 1968.
[30] Courbon H, Pichat P, *C R Acad Sci Pari* **285C**: 171, 1977.
[31] Courbon H, Formenti M, Pichat P, *J Phys Chem* **81**: 550, 1977.
[32] Courbon H, Herrmann J-M, Pichat P, *J Phys Chem* **88**: 5210, 1984.
[33] Kuriacose JC, *Indian J Chem* **5**: 646, 1967.
[34] Taylor HS, *Ann Rev Phys Chem* **12**: 147, 1961.
[35] Taylor HS, *Proc 2nd Int Cong Catalysis, Paris (1960)*, Editions Technip, Paris, 1961, p. 157.
[36] Khoobiar S, *J Phys Chem* **68**: 411, 1964.
[37] Robell AJ, Ballou EV, Boudart M, *J Phys Chem* **68**: 2748, 1964.
[38] Bond GC, Fuller MJ, Molloy LR, *Proceedings of the 6th International Congress in Catalysis*, The Chemical Society, London, p. 356, 1977.
[39] Ducarme V, Védrine JC, *J Chem Soc Farad Trans 1* **74**: 506, 1978.
[40] Schuit GCA, van Reijen LL, *Adv Catal* **10**: 242, 1958.
[41] Eley DD, Shooter D, *J Catal* **2**: 259, 1963.
[42] Takasu Y, Yashamina T, *J Catal* **28**: 174, 1973.
[43] Duprez D, *Spillover and Migration of Surface Species on Catalysts*, Can Li *et al.* (eds.), Stud Surf Sci Catal **112**: 13, 1997.
[44] Kacimi S, Duprez D, Dalmon JA, *J Chim Phys* **94**: 535, 1997.
[45] Taha R, Duprez D, *J Chim Phys* **92**: 1506, 1995.
[46] Paal Z, Menon PG, *Catal Rev-Sci Eng* **25**: 229, 1983.
[47] Bernasek SL, Somorjai GA, *J Chem Phys* **62**: 3149, 1975.
[48] Nishiyama S, Yoshioka K, Matsuura S, Tsuruya S, Masai M, *React Kinet Catal Lett* **33**: 405, 1987.
[49] Nishiyama S, Yoshioka K, Yoshida T, Tsuruya S, Masai M, *Appl Surf Sci* **33/34**: 1081, 1988.
[50] Nishiyama S, Tsuruya S, Masai M, *J Vac Sci Technol A* **8**: 868, 1990.
[51] Christmann K, Ertl G, Pignet T, *Surf Sci* **54**: 365, 1976.
[52] Niklasson C, Andersson B, *Ind Eng Chem Res* **27**: 1370, 1988.
[53] Wrammerfors Å, Andersson B, *J Catal* **146**: 34, 1994.
[54] Larsson M, Andersson B, Bariås OA, Holmen A, in *Catalyst Deactivation 1994*, Delmon B, Froment GF (eds.), Stud Surf Sci Catal **88**: 233, 1994.
[55] Oudar J, Berthier Y, Pradier C-M, *C R Acad Sci* **292**: 577, 1981.
[56] Abderrahim H, Duprez D, in *Catalysis and Automotive Pollution Control I*, Crucq A, Frennet A (eds.), Stud Surf Sci Catal **30**: 359, 1987.
[57] Kacimi S, Duprez D, in *Catalysis and Automotive Pollution Control II*, Crucq A (ed.), Stud Surf Sci Catal **71**: 581, 1991.
[58] Descorme C, Duprez D, *Appl Catal A* **202**: 231, 2000.
[59] Maillet T, Barbier J Jr, Gelin P, Praliaud H, Duprez D, *J Catal* **202**: 367, 2001.
[60] Bazin D, Dexpert H, Lagarde P, Bournonville JP, *J Catal* **110**: 209, 1988.
[61] Zhou Y, Wood MC, Winograd N, *J Catal* **146**: 82, 1994.
[62] Matsushima T, *J Catal* **85**: 98, 1984.
[63] Gland JL, *Surf Sci* **93**: 487, 1980.

[64] Sobyanin VA, Boreskov GK, Cholach AR, Losev AP, *React Kinet Catal Lett* **27**: 299, 1985.
[65] Shpiro ES, Dysenbina BB, Tkachenko OP, Antoshin GV, Minachev Kh M, *J Catal* **110**: 262, 1988.
[66] Martin D, Duprez D, *J Phys Chem* **100**: 9429, 1996.
[67] Cunningham J, Cullinane D, Farell F, O'Driscoll JP, Morris MA, *J Mater Chem* **5**: 1027, 1995.
[68] Cunningham J, Cullinane D, Farell F, Morris MA, Datye A, Kalakkad D, in *Catalysis and Automotive Pollution Control III*, Frennet A, Bastin JM (eds.), Stud Surf Sci Catal **96**: 237, 1995.
[69] Masai M, Murata K, Yabashi M, *Chem Lett* 989, 1979.
[70] Masai M, Nakahara K, Yabashi M, Murata K, Nishiyama S, Tsuruya S, in *Spillover of Adsorbed Species*, Pajonk GM *et al.* (eds.), Stud Surf Sci Catal **17**: 89, 1983.
[71] Nishiyama S, Yanagi H, Tsuruya S, Masai M, in Steinberg K-H (ed.), *Proc 2nd Int Conf Spillover*, Leipzig Universität, p. 116, 1989.
[72] Peri JB, Hannan RB, *J Phys Chem* **64**: 1526, 1960.
[73] Peri JB, *J Phys Chem* **69**: 231, 1965.
[74] Carter JL, Lucchesi PJ, Corneil P, Yates DJC, Sinfelt JH, *J Phys Chem* **69**: 3070, 1965.
[75] Knözinger H, *Adv Catal* **25**: 184, 1976.
[76] Knözinger H, Ratnasamy P, *Catal Rev-Sci Eng* **17**: 31, 1978.
[77] Cavanagh RR, Yates JT Jr, *J Catal* **68**: 22, 1981.
[78] Baumgarten E, Zachos A, *J Catal* **69**: 121, 1981.
[79] Baumgarten E, Zachos A, *Z Phys Chem* **130**: 211, 1982.
[80] Bianchi D, Lacroix M, Pajonk G, Teichner SJ, *J Catal* **59**: 467, 1979.
[81] Conner WC, Cevallos-Candau JF, Shah N, Haensel V, in *Spillover of Adsorbed Species*, Pajonk GM *et al.* (eds.), Stud Surf Sci Catal **17**: 31, 1983.
[82] Bianchi D, Maret D, Pajonk GM, Teichner SJ, *ibid*, p. 45.
[83] Cevallos-Candau JF, Conner WC, *J Catal* **106**: 378, 1987.
[84] Cevallos-Candau JF, Conner WC, in Steinberg K-H (ed.), *Proc 2nd Int Conf Spillover*, Leipzig Universität, p. 18, 1989.
[85] Davidov V Ya, Kiselev AV, Zhuravlev LT, *Trans Farad Soc* **60**: 2254, 1964.
[86] Hall KW, Lutinski FE, *J Catal* **2**: 518, 1963.
[87] Hall KW, Leftin HP, Cheselske FJ, O'Reilly DE, *J Catal* **2**: 506, 1963.
[88] Martin D, *Thesis*, Poitiers, 1994.
[89] Martin D, Duprez D, *J Phys Chem* **101**: 4428, 1997.
[90] Bittner EW, Bockrath BC, Solar JM, *J Catal* **149**: 206, 1994.
[91] Shido T, Asakura K, Iwasawa Y, *J Chem Soc Faraday Trans 1* **85**: 441, 1989.
[92] Knözinger E, Jacob K-H, Singh S, Hofmann P, *Surf Sci* **290**: 388, 1993.
[93] Knözinger E, Jacob K-H, Hofmann P, *J Chem Soc Farad Soc* **87**: 1101, 1993.
[94] Hoq MF, Nieves I, Klabunde KJ, *J Catal* **123**: 349, 1990.

[95] Agron PA, Fuller EL Jr, Holmes HF, *J Coll Int Sci* **52**: 553, 1975.
[96] Kondo J, Sakata Y, Domen K, Maruya K-I, Onishi T, *J Chem Soc Faraday Trans* **86**: 397, 1990.
[97] Kondo J, Domen K, Maruya K-I, Onishi T, *Chem Phys Lett* **188**: 443, 1992.
[98] Li C, Domen K, Maruya K-I, Onishi T, *J Am Chem Soc* **111**: 7683, 1989.
[99] Li C, Domen K, Maruya K-I, Onishi T, *J Catal* **123**: 436, 1990.
[100] Li C, Chen Y, Li W, Xin Q, in *Ref. 11*, p. 217.
[101] Mestl G, Ruiz P, Delmon B, Knözinger H, *J Phys Chem* **98**: 11269, 1994.
[102] Mestl G, Ruiz P, Delmon B, Knözinger H, *J Phys Chem* **98**: 11276, 1994.
[103] Mestl G, Ruiz P, Delmon B, Knözinger H, *J Phys Chem* **98**: 11283, 1994.
[104] Gideoni M, Steinberg M, *J Solid State Chem* **4**: 370, 1972.
[105] Che M, Kibblewhite JFJ, Tench AJ, Dufaux M, Naccache C, *J Chem Soc Farad Trans 1* **69**: 857, 1973.
[106] Mukherji P, Gadgil K, Gonzalez RD, *J Indian Chem Soc* **55**: 943, 1979.
[107] Dimitriev RV, Krasavin SA, Bragin OV, Minachev Kh M, *Izv Akad Nauk SSSR Ser Khim* **3**: 501, 1982.
[108] Ambs WJ, Mitchell MM Jr, *J Catal* **82**: 226, 1983.
[109] Chen H-W, White JM, *J Mol Catal* **35**: 355, 1986.
[110] Chen B, Falconer JL, unpublished results.
[111] Conner WC, Falconer JL, *Chem Rev* **95**: 759, 1995.
[112] Levy RB, Boudart J, *J Catal* **32**: 304, 1974.
[113] Sermon PA, Bond GC, *J Chem Soc Faraday Trans 1* **76**: 889, 1980.
[114] Lobashina NE, Solovieva ES, Savvin NN, Miasnikov IA, *Kinet Katal* **25**: 499, 1984.
[115] Baumgarten E, Denecke E, *J Catal* **95**: 296, 1985.
[116] Baumgarten E, Denecke E, *J Catal* **100**: 377, 1986.
[117] Conner WC Jr, in Paal Z, Menon PG (eds.), Hydrogen Effects in Catalysis, *Spillover of hydrogen*, M Dekker, New York, p. 311, 1988.
[118] McNaught WG, Kemball C, Leach HF, *J Catal* **34**: 93, 1974.
[119] Peri JB, *J Phys Chem* **69**: 220, 1965.
[120] Ravi A, Sheppard N, *J Catal* **22**: 389, 1971.
[121] Basallote MG, Bernal S, Gatica JM, Pozo M, *Appl Catal A* **232**: 39, 2002.
[122] Jung K-D, Bell A, *J Catal* **193**: 207, 2000.
[123] Bedrane S, Descorme C, Duprez D, *Appl Catal A* **289**: 90, 2005.
[124] Cunningham J, Farell F, Gibson C, McCarthy J, *Catal Today* **50**: 429, 1999.
[125] Yao HC, Yu Yao YF, *J Catal* **86**: 254, 1984.
[126] Harrison B, Diwell AF, Hallett C, *Platinum Metals Rev* **32**: 73, 1988.
[127] Kacimi S, Barbier J Jr, Taha R, Duprez D, *Catal Lett* **22**: 343, 1993.
[128] Zafiris GS, Gorte RJ, *J Catal* **139**: 561, 1993.
[129] Trovarelli A, *Catal Rev-Sci Eng* **38**: 439, 1996.
[130] Holmgren A, Duprez D, Andersson B, *J Catal* **182**: 441, 1999.
[131] Holmgren A, Andersson B, Duprez D, *Appl Catal B* **22**: 215, 1999.
[132] Duprez D, Descorme C, Catalysis by Ceria and Related Materials, in Trovarelli A (ed.), Catalysis Science Series (Vol. 2), Imperial College Press, p. 243, 2002.

[133] Bernal S, Calvino JJ, Cauqui MA, Cifredo GA, Jobacho A, Pintado JM, Rodriguez-Izquierdo JM, *J Phys Chem* **97**: 4118, 1993.
[134] Bensalem A, Bozon-Verduraz F, Perrichon V, *J Chem Soc Faraday Trans* **91**: 2185, 1995.
[135] Salasc S, Perrichon V, Primet M, Chevrier M, Mathis F, Moral N, *Catal Today* **50**: 227, 1999.
[136] Costa CN, Christou SY, Georgiou G, Efstathiou AM, *J Catal* **219**: 259, 2003.
[137] Martin D, Duprez D, *J Mol Catal* **118**: 113, 1997.
[138] Bartley GJJ, Shady PJ, D'Anielo MJ, Chandler JGR, Brisley RJ, Webster DE, SAE Paper # 93076, 1993.
[139] Fornasiero P, Di Monte R, Ranga Rao G, Kaspar J, Meriani S, Trovarelli A, Graziani M, *J Catal* **151**: 168, 1995.
[140] Trovarelli A, Zamar F, Llorca J, de Leitenburg C, Dolcetti G, Kiss J, *J Catal* **169**: 490, 1997.
[141] Cuif JP, Blanchard G, Touret O, Seigneurin A, Marczi M, Quemere E, SAE paper #970463, 1997.
[142] Ozawa M, *J Alloys Comp* **275–277**: 886, 1998.
[143] Fornasiero P, Fonda E, Di Monte R, Vlaic G, Kaspar J, Graziani M, *J Catal* **187**: 177, 1999.
[144] Rossignol S, Gérard F, Duprez D, *J Mater Chem* **9**: 1615, 1999.
[145] Madier Y, Descorme C, Le Govic AM, Duprez D, *J Phys Chem B* **103**: 10999, 1999.
[146] Martinez-Arias A, Fernandez-Garcia M, Belver C, Conesa JC, Soria J, *Catal Lett* **65**: 197, 2000.
[147] Duprez D, Descorme C, Birchem T, Rohart E, *Topics Catal* **16/17**: 49, 2001.
[148] Suda A, Ukyo Y, Sobukawa H, Sugiura M, *J Ceram Soc Japan* **110**: 126, 2002.
[149] Bedrane S, Descorme C, Duprez D, *Catal Today* **75**: 401, 2002.
[150] Mamontov E, Brezny R, Koranne M, Egami T, *J Phys Chem B* **107**: 13007, 2003.
[151] Galdikas A, Descorme C, Duprez D, Dong F, Shinjoh H, Proc. CaPoC6, Kruse N, Frennet A, Bastin J-M, Visart de Bocarmé T (eds.), *Topics Catal* **30/31**: 405, 2004.
[152] Galdikas A, Descorme C, Duprez D, *Solid State Ionics* **166**: 147, 2004.
[153] Royer S, Duprez D, Kaliaguine S, *J Catal* **234**: 364, 2005.

Chapter 7

Investigation of Reaction at the Site Level Using SSITKA

Sourabh Pansare, Amornmart Sirijaruphan and
James G Goodwin, Jr

1. Introduction

Steady state isotopic transient kinetic analysis (SSITKA), initially developed by Happel,[1] Bennett[2] and Biloen,[3] is a powerful technique for the kinetic study of heterogeneous catalytic reactions.[4] Sometimes, this technique is referred to as isotopic transient kinetic analysis (ITKA) for measurements that are not at steady state. Under isothermal and isobaric reaction conditions, the technique employs a switch in the isotopic labelling of one of the reactant species generating an isotopic transient response in the isotopic labelling of the product species that is detected using a mass spectrometer. The major use of SSITKA is for the determination of average surface residence time for reaction (τ_p) and surface concentration of the most active reaction intermediates (N_p), at actual reaction conditions. It can also be used to estimate a "site" turnover frequency (TOF), to determine the surface coverage of the intermediates, and to delineate the site heterogeneity.

SSITKA has contributed much to our understanding of many surface catalyzed reactions. To date, the reactions that have been studied by SSITKA are summarized in Table 1. Although most of the published work is presented in Table 1, not all the references for different reactions utilizing SSITKA have been reported, due to space limitations. Effects of temperature,[5] reactant partial pressure,[5] and promotion[6] have also been explored using SSITKA. SSITKA has been used to determine the cause of deactivation for n-butane isomerisation on sulfated zirconia,[7] and the

Table 1. Reactions studied by SSITKA.

Reaction	References
Ammonia synthesis	[6, 25–27]
Ammonia oxidation	[28, 29]
Butane isomerisation	[7, 30–32]
Benzene hydrogenation	[33, 34]
CO oxidation	[8, 18, 35]
CO hydrogenation	[5, 36–43]
Ethane hydrogenolysis	[44–46]
Ethylene hydroformylation	[47–49]
Isobutane hydrogenation and dehydrogenation	[50]
Methane combustion	[51]
Methanation and Fischer-Tropsch synthesis	[3, 5, 11, 37, 41, 42, 52–93]
Methanol synthesis	[16, 94–97]
Methane coupling	[14, 17, 98–106]
Methane partial oxidation	[17, 107–113]
NOx reduction	[114–119]
NO-CO/ NO-H_2 reactions	[35, 118, 120, 121]
Propylene partial oxidation	[122]
Propane dehydrogenation	[123]
Vinyl acetate synthesis	[124]

selective oxidation of CO on $Pt/\gamma\text{-}Al_2O_3$.[8] These various applications have demonstrated the versatility of the SSITKA technique.

2. Theory

As stated earlier, SSITKA is carried out by tracking the change in the isotopic labelled product(s), following a step change in the isotopic label on a reactant. To maintain isothermal and isobaric reaction conditions, flow rate and pressure of the two isotopic labelled reactant flows are adjusted to be identical, prior to switching. A typical reaction system for isotopic transient studies is shown in Fig. 1.

Consider a reversible steady-state heterogeneous reaction taking place in a catalytic bed, in which the reactant species R gives the product species P via formation of adsorbed intermediate species[4]:

$$R_{(g)} + \text{other reactants} \rightleftarrows \text{intermediates (ads)} \rightleftarrows P_{(g)} + \text{other products}$$

The isotopic transient experiment begins with a standard reaction similar to the one above using a reactant with a particular isotopic label, typically

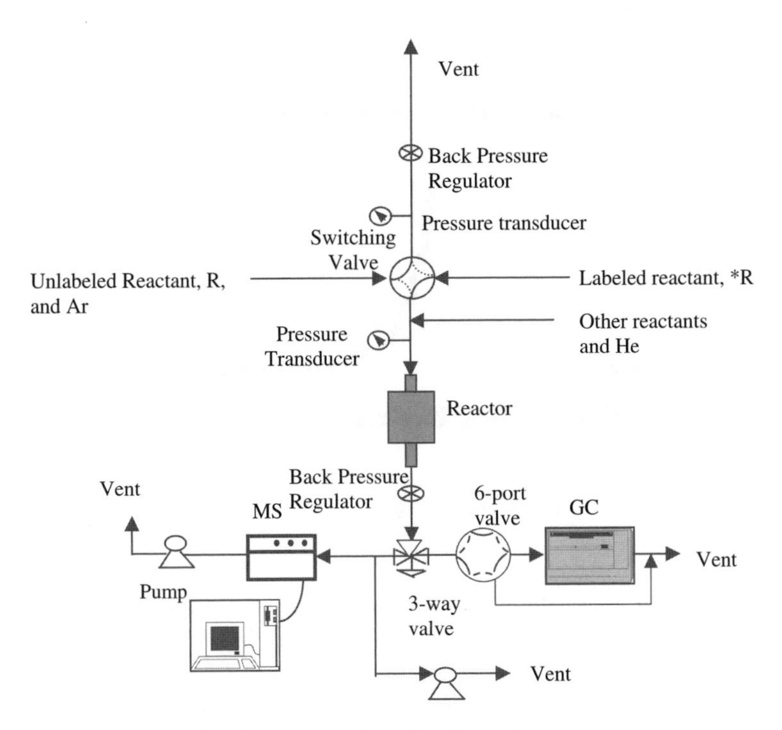

Fig. 1. Typical system for isotopic transient kinetic study.

the naturally occurring one. At a given time-on-stream (TOS), the inlet feed stream is switched to another feed stream of the same reactant containing a different isotopic label. As the reactant progresses through the reactor and reacts on the catalyst surface to form the product, the old isotopic label reactant is replaced by the new isotopic label. An example of the experimentally obtained isotopic transients is shown in Fig. 2.

In SSITKA, it is assumed that the catalyst surface is a system made up of a number of interconnected compartments. Each compartment is a homogeneous or well-mixed system like a continuous stirred tank reactor (CSTR). A separate compartment is assumed to be present for each uniquely adsorbed reaction intermediate species.[4] Based on this assumption, the response of a single unidirectional reaction step to a step change in the isotopic concentration of one of the reactants, which is made at time $t = 0$, is given as

$$\frac{\mathrm{dN_A}}{\mathrm{dt}} = -\frac{\mathrm{N_A}}{\tau_\mathrm{A}} \tag{1}$$

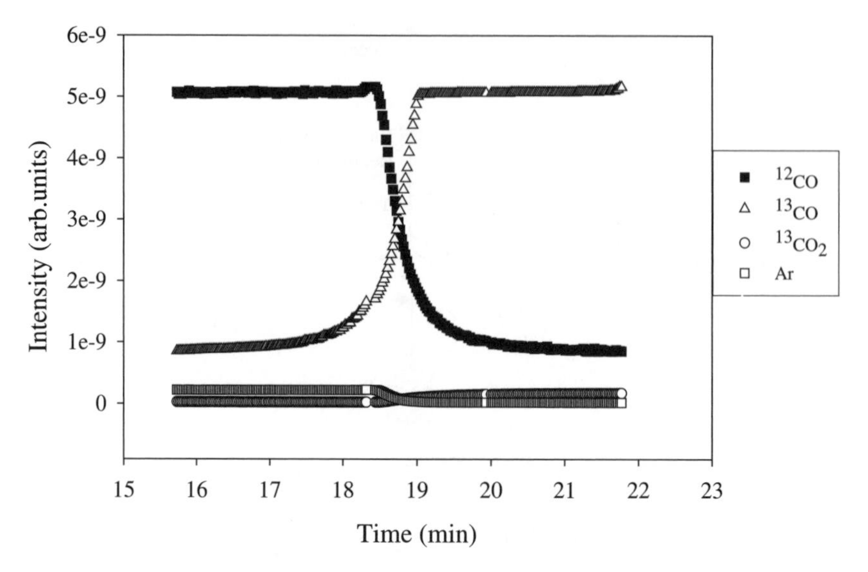

Fig. 2. Typical experimentally obtained isotopic transients for the selective CO oxidation on Pt/γ-Al_2O_3. A switch from ^{12}CO to ^{13}CO was made during the reaction run.

where N_A is the amount of a surface intermediate A containing a particular isotope, and τ_A is the lifetime of that intermediate. Decay in N_A is observed by decay in the rate of flow of the labelled species P from the reactor (r_P). Since A is not measured directly but via P and assuming no readsorption of P after formation, we will refer to N_A as N_P, and τ_A as τ_P. If an assumption is made that there is no isotopic effect, then the total surface concentration of this species remains unchanged with the isotopic switch (R/*R) resulting in P/*P, and τ_P remains constant as N_P changes. The integrated form of Eq. (1) in terms of r_P is given as

$$r_P = r \cdot e^{\frac{-t}{\tau}} \tag{2}$$

where r is a constant and is given by the summation of the formation rates of labelled and unlabelled species

$$r = r_P + r_{*P} \tag{3}$$

If P is a product, r is the overall rate of formation. The normalised form of the Eq. (2) is written as

$$e^{\frac{-t}{\tau}} = \frac{r_P}{r} = \frac{r_P}{r_P + r_{*P}} = F_P(t) \tag{4}$$

where $F_P(t)$ is the normalized rate of formation of P. The "$e^{\frac{-t}{\tau}}$" response is a characteristic of the emptying of an ideal continuous stirred tank reactor (CSTR). This response is shown in Fig. 3. The typical normalised transient responses of a non-ideal differential-bed plug flow reactor (PFR), the kind most often used in SSITKA studies, following a step change in the isotopic labelling of reactant R, R $\rightarrow$ *R, which subsequently appears in product P as P $\rightarrow$ *P, are shown in Fig. 4. An inert tracer is used in SSITKA for the determination of the gas-phase holdup or non-ideality of flow due to back-mixing in the reactor system, and is introduced simultaneously in a small amount, with either of the isotopic-tracer feed streams during the step change.[4]

The subscript m in Fig. 4 represents a measured transient response and the maximum value of 1.0 represents the steady-state flow rate of the reactant or inert, and the steady state formation rate of the product.

The relation between $F_m^P(t)$ and $F_m^{*P}(t)$ is given as

$$F_m^P(t) = 1 - F_m^{*P}(t) \tag{5}$$

The normalised form of the experimentally obtained data for the selective CO oxidation on Pt/γ-Al_2O_3 is shown in Fig. 5 where argon was the

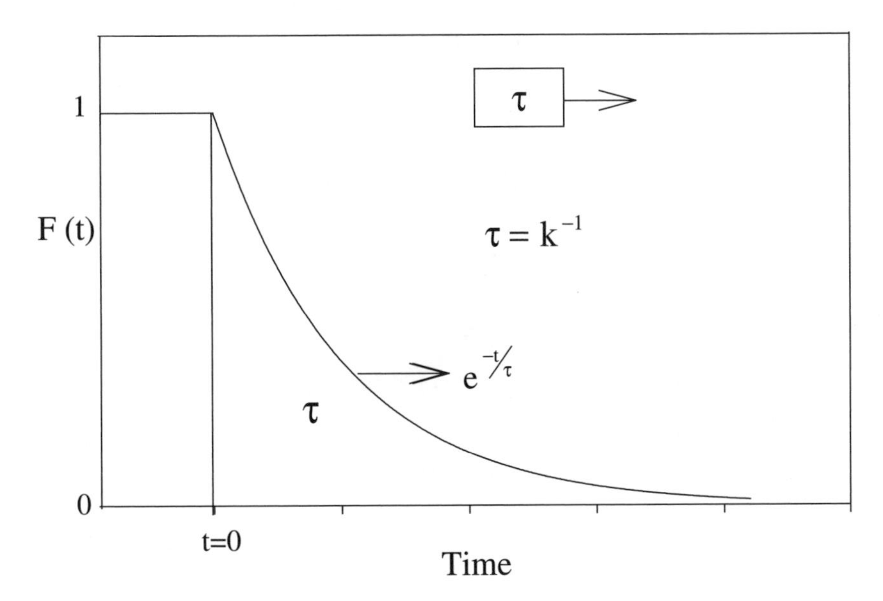

Fig. 3. Response of a single, homogeneous surface pool to a step change in the isotopic concentration of one of the reactants.

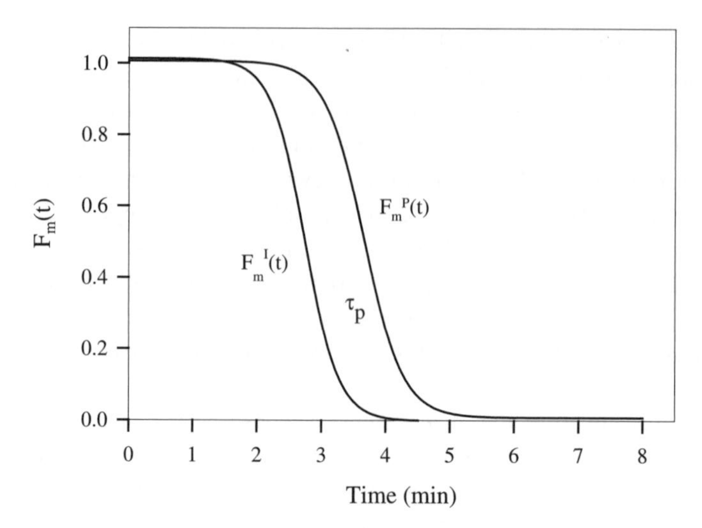

Fig. 4. Typical normalized isotopic transient responses in product species P following an isotopic switch in reactant R → *R. An inert tracer, I, is introduced to determine the gas-phase holdup of the reactor system.

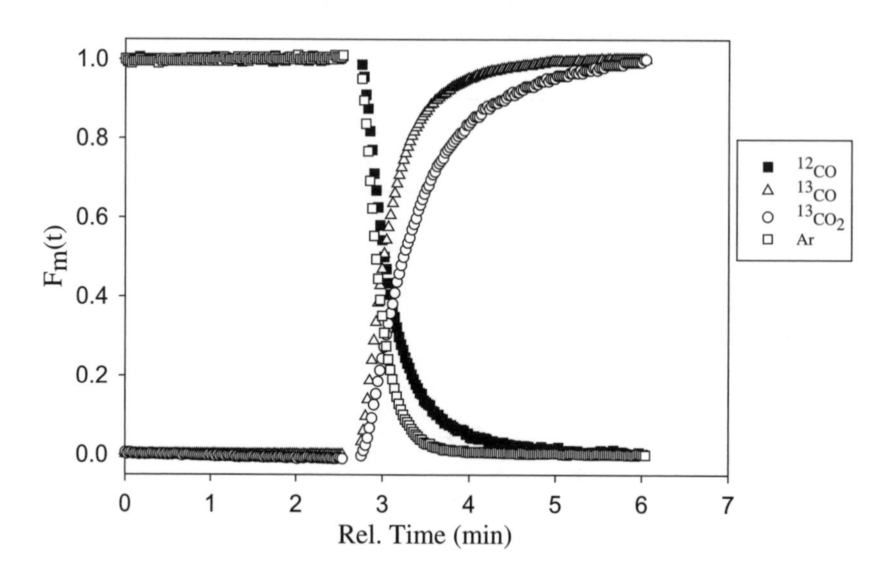

Fig. 5. Normalized isotopic transients for the selective CO oxidation on $Pt/\gamma\text{-}Al_2O_3$.

inert tracer. To determine the mean surface residence time of the most active reaction intermediates (τ_p), the normalised isotopic transient given by Eq. (4) is integrated relative to the inert tracer

$$\tau_P = \int_0^\infty [F_m^P(t) - F_m^I(t)]dt \tag{6}$$

Thus, as shown in Fig. 4, the area under the curve from $t = 0$ to $t = \infty$ gives the value of τ_p, but accounting for non-ideality of the step change in isotopic concentrations.

Based on a simple mass balance

$$N_P = \tau_P \cdot r_P \tag{7}$$

where N_P is the concentration of the most active surface intermediates leading to product P, and r_P is the steady state rate of formation of P. This equation can be rewritten as

$$r_P = \frac{1}{\tau_P} \cdot N_P = k \cdot N_P \tag{8}$$

Thus, $1/\tau_P$ is a pseudo-first order rate constant. This parameter ($k = 1/\tau_P$) is a measure of catalytic site activity. Since $\frac{1}{\tau_P} = \frac{r_P}{N_P} = TOF_{ITK}$, this intrinsic site activity ($1/\tau_p$) is a form of turnover frequency (TOF) which represents the number of molecules reacting per site per second and has units of s^{-1}.

Although N_p can be determined directly from the rate transient of P, and the transient for the step change in inert flow normalised to the total rate of formation of P(r), it is easier to normalise all the transients to their maximum flow rate and integrate to get τ_p first. Like other rate constants, k and hence τ_p is a function of temperature. However, it may also be a function of partial pressure.[9] So k is a form of TOF and a measure of site activity, and like all TOFs,[9] it gives a measure of site activity at a set of reaction conditions.

Another parameter which is often useful to know is the concentration of reversibly adsorbed reactants N_R, i.e., molecules of reactant R that adsorb but desorb without reacting. Although N_R can be meaningful, τ_R is meaningless because (i) the number of readsorptions are unknown, (ii) the extent of bypassing (i.e., reactants that do not adsorb on the catalyst in a differential reactor) is not known. Hence τ_R is only useful for the calculation of N_R.

The values for the surface coverage (θ) of various species (R, P, etc.) can be determined by normalising N_i using the monolayer coverage determined from selective chemisorption or estimated by other means.

Besides the τ for the reaction, the measured τ_P also includes the residence time of additional holdup on the surface due to readsorption.[10]

$$\tau_{\text{measured}} = \tau_{\text{reaction}} + \sum_{i=1}^{n} \tau_{\text{ads,i}} \tag{9}$$

where n is the number of readsorptions each lasting for a time $\tau_{\text{ads,i}}$. It is difficult to obtain the direct values for $\tau_{\text{ads,i}}$, but the readsorption and hence $\tau_{\text{ads,i}}$ can be minimised by (i) decreasing the bed length, (ii) decreasing the holdup time in the reactor by increasing the flow rate, (iii) adding product molecules to the inlet gas which will compete for readsorption sites, (iv) decreasing catalyst granule size, and (v) decreasing catalyst porosity. Usually, multiple experiments are done (e.g., at different flow rates) and the resulting values of τ_P determined are extrapolated to zero readsorption conditions. In this way, it is possible to find out the approximate values for the actual τ of the surface reaction.[10]

Although SSITKA measurements can be useful in studying reaction induction and deactivation, SSITKA transients (which may last from a few seconds to minutes depending on the catalyst, reaction, and temperature) should be used where the rate of reaction during the collection of the transient data varies by only ca. 5%. Otherwise, interpretation may be more difficult.

3. Advanced Mathematical Analysis

3.1. *Deconvolution*

Other useful information that can be extracted from SSITKA data using advanced mathematical analysis includes the reactivity distribution, f(k). On a heterogeneous catalyst surface, the active sites exhibit a non-uniform reactivity which can be characterised by a reactivity distribution function f(k), where k represents a measure of site activity. The determination of f(k) from the isotopic transient of P requires a numerical deconvolution technique.

Two main methods, parametric and nonparametric, have been developed for this deconvolution. The parametric method involves development of a multi-parameter model for obtaining the value of reactivity distribution function. The model is based on the assumption of some form of f(k), e.g., a bimodal Gaussian distribution.[11] However, for a particular catalyst in a particular reaction environment, it is not necessarily clear what form f(k) should have. Hence, nonparametric methods, which do not require the assumption of any specific functional form for f(k), are preferred over parametric techniques. Two nonparametric methods have been developed and applied in SSITKA: the inverse-Laplace-transform technique (ILT) and the Tikhonov-Fredholm method (T-F).

The ILT method is based on the fact that the isotopic transient of the product formation rate $[r_p^*(t)]$ represents the Laplace transform of $N_p \cdot k \cdot f(k)$. For a pseudo-first order reaction, the transient of the product formation rate can be expressed as

$$r_p^*(t) = N_p \int_0^{\infty} ke^{-kt} f(k)dk \tag{10}$$

where N_p is the surface concentration of the most active reaction intermediate producing product P. Thus, f(k) can be determined by applying the inverse-Laplace transform to $r_p^*(t)$. In the T-F method, f(k) can be extracted from $r_p^*(t)$ by the Tikhonov regularisation of Eq. (10). The detailed development of and the analysis based on the T-F method is given elsewhere.[11] Gas phase holdup is important for determination of f(k). Yet, the correction for the gas phase holdup cannot be done mathematically prior to distribution analysis. Hence, the distribution was calculated from the normalised product isotopic transient and then shifted to have the average value calculated from the figure equal to the average value obtained directly by ITKA.

The T-F method is more precise and less subjective than the ILT method; on the other hand, it is more demanding from a computational point of view than the latter.[11] The T-F method, even in the presence of small amounts of random noise, can recover significantly the reactivity distribution.

An original transient for CO_2 in the selective oxidation of CO over Pt/γ-Al_2O_3 is shown in Fig. 6(a). The T-F method deconvolution result for this transient is shown in Fig. 6(b).

3.2. *Convolution*

Another advanced analysis method applicable to the SSITKA studies makes use of convolution, well known for linear systems. Making use of linear convolution in parametric and nonparametric kinetic analyses, provides increased accuracy for the determination of the kinetic parameters

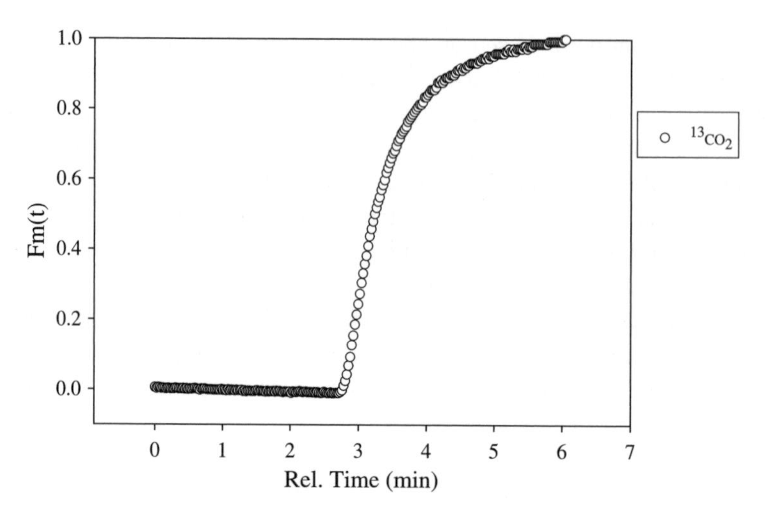

Fig. 6(a). Normalized isotopic transient of $^{13}CO_2$ for the selective CO oxidation on Pt/γ-Al_2O_3.

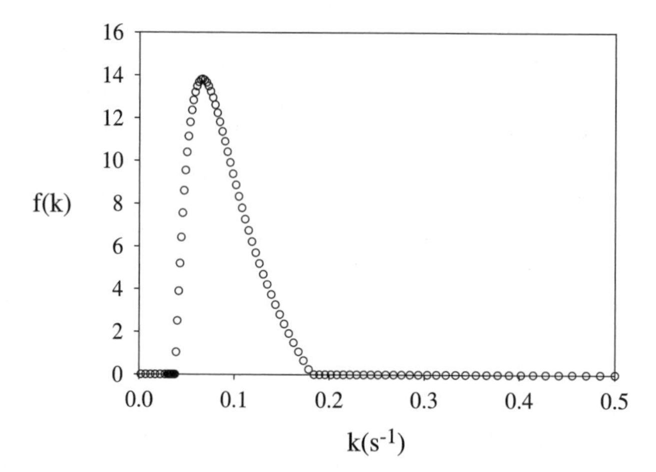

Fig. 6(b). Reactivity distribution obtained by the T-F deconvolution method for the selective oxidation of CO on Pt/γ-Al_2O_3 catalyst.

in SSITKA.[12] For a PFR with a differential length catalyst bed, the relationship between the step input responses is given by the linear convolution integral

$$F_m^{*P}(t) = \int_0^t \left[D_t\{F_c^{*P}(t-\zeta)\} \cdot F_m^{+I}(\zeta)\right] d\zeta \tag{11}$$

where ζ is a dummy convolution time variable; F_m^{+I} is the measured transient response of the inert I, $F_m^{+I}(\zeta)$ is the convolution form of $F_m^{+I}(t)$; and $D_t\left\{F_c^{*P}(t-\zeta)\right\}$ is the convolution form of the first derivative, $D_t = d/dt$, of the catalyst-surface transient response, $F_c^{*P}(t)$. The linear convolution relationship for the step decay response is given by

$$F_m^P(t) = 1 + \int_0^t \left[D_t\{F_c^P(t-\zeta)\} \cdot \left(1 - F_m^{-I}(\zeta)\right)\right] d\zeta \tag{12}$$

Equations (11) and (12) enable the generation of the total isotopic transient responses of a product species given (a) the transient response that characterises hypothesized catalyst-surface behaviour and (b) an inert-tracer transient response that characterises the gas-phase behaviour of the reactor system. Use of the linear-convolution relationships has been suggested as an iterative means to verify a model of the catalyst surface reaction pathway and kinetics.[12] This is attractive since the direct determination of the catalyst-surface transient response is especially problematic for non-ideal PFRs, since a method of complete gas-phase behaviour correction to obtain the catalyst-surface transient response is presently unavailable for such reactor systems.[4] Unfortunately, there are also no corresponding analytical relationships to Eqs. (11) and (12) which permit explicit determination of the catalyst-surface transient response from the measured isotopic and inert-tracer transient responses, and hence, a model has to be assumed and tested. The better the model of the surface reaction pathway, the better the fit of the generated transient to the measured transient.

4. Experimental Limitations and Corrective Techniques

4.1. H_2-D_2

Mechanistic modelling and kinetic parameter determination using SSITKA assume that isotopic substitution does not perturb the steady-state reaction conditions (no isotope effect on reaction). However, during the isotopic exchange of hydrogen/deuterium (H/D) such as H_2/HD, CH_4/CH_3D or C_2H_6/C_2H_5D, the kinetic and thermodynamic differences arising from the relatively large mass differences and the bonding energies between hydrogen isotopes is significantly high such that steady-state reaction is not maintained. The kinetic rates and the concentrations of surface intermediates may be substantially changed following the isotopic switch. However, the application of SSITKA during H_2/D_2 exchange can be useful for the characterization of the active surface. Basallote *et al.*[13] reported the use of SSITKA in characterizing Pt/CeO_2. Since other techniques such as conventional hydrogen chemisorption did not permit straightforward interpretation due to hydrogen spillover, SSITKA results, though flawed, gave better direct information about the number and activity of sites involved in the H_2/D_2 exchange process.

4.2. O_2

Determining the surface kinetic parameters using oxygen isotopic exchange can be more complicated for oxide catalysts or catalysts with oxide supports. Oxygen exchange between the gas phase, the catalyst surface, and the catalyst bulk may result in the failure of the $^{16}O^{18}O$ transient to relax back to zero, following a step change from $^{16}O_2$ to $^{18}O_2$. Bulk oxide in the catalyst is able to act as an oxygen source for exchange. As a result of this complication, oxygen isotopic tracing with SSITKA is sometimes not useful for determining surface reaction kinetic parameters. Low temperature has to be maintained in order to minimise the oxygen exchange rate with the bulk. However, SSITKA has been used to quantify lattice oxygen diffusivity and total oxygen uptake as shown in the high temperature work of Peil *et al.*[14] When combined with other reactant isotopic switches such as methane, the active intermediates and oxygen reaction pathway can be delineated to a larger degree.[10,14,15]

4.3. *Readsorption*

Although SSITKA can be used in determining the average surface residence time and surface concentration of intermediates, it can be complicated by the readsorption of reactant and product molecules. When readsorption occurs, whether at active or non-active sites, the average surface residence time, and thus the surface concentration of intermediates attributed to the reaction, are overestimated. Interparticle readsorption can be minimised by decreasing the bed length or increasing the weight hourly space velocity (WHSV). The presence of readsorption is indicated by an increase in the surface reaction residence time (τ_p) with increasing bed length. If readsorption is problematic, one has to perform a series of experiments at various flow rates or bed lengths and extrapolate to negligible readsorption, as shown in Figs. 7 and 8.

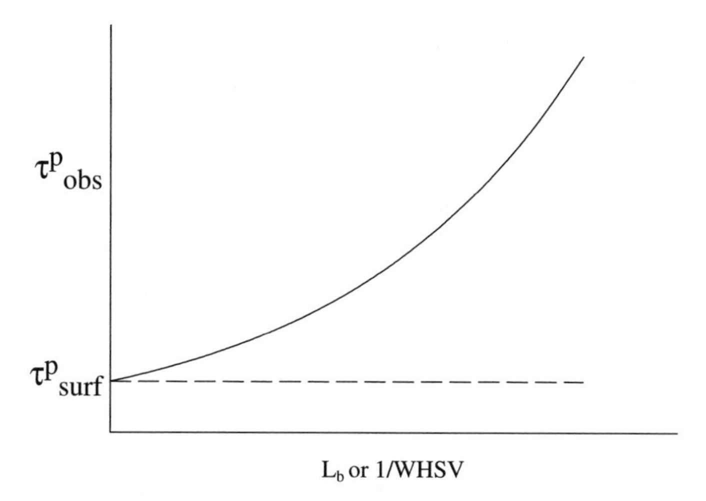

Fig. 7. Experimental determination of the mean surface-residence time in the presence of interparticle readsorption.

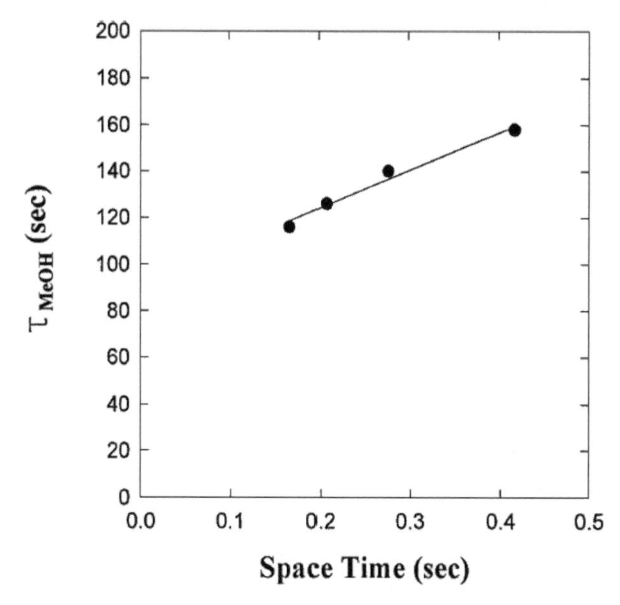

Fig. 8. τ_{MeOH} vs. space time during steady-state MeOH synthesis at 220°C (used to determine τ^0_{MeOH}).[16]

Unfortunately, intraparticle readsorption cannot be corrected using this method. If intraparticle readsorption is significant, it can be detected by adding unlabelled product to the feed stream, to compete for readsorption sites with the labelled product formed during reaction. The observed surface residence time of product will approach the true surface residence time at higher concentrations of product added.

An example of how to determine the true surface residence time, taking into account both interparticle and intraparticle readsorption, was shown by Ali and Goodwin[16] for methanol synthesis on Pd/SiO_2. Surface residence time of intermediates leading to methanol, τ_{MeOH}, was corrected by varying and extrapolating to zero space time holding methanol partial pressure constant, as shown in Fig. 9. Without the correction for methanol readsorption, τ_{MeOH} would have been double the correct value. τ^{00}_{MeOH} represents the extrapolation of data varying both space time and methanol partial pressure.

4.4. *Steady-state versus non-steady-state analysis*

At steady-state reaction conditions, an abrupt switch in the isotopic composition of one of the reactants does not disturb the reaction, provided there is not an isotope effect such as in the case of H_2-D_2. Many studies

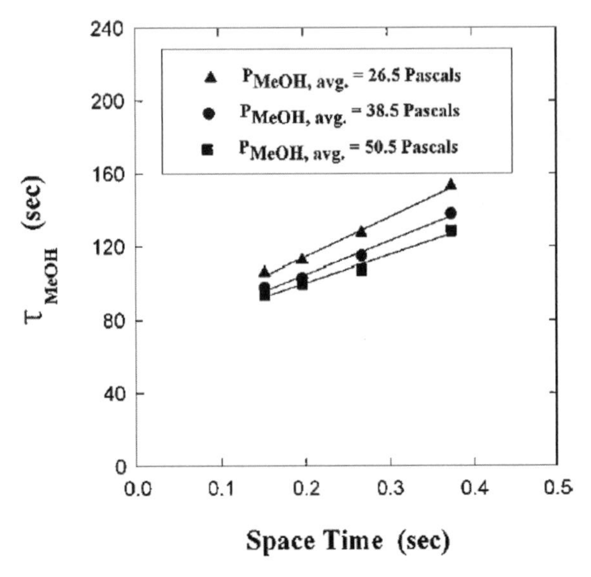

Fig. 9. τ_{MeOH} vs. space time at constant average P_{MeOH} during steady-state MeOH synthesis at 220°C on Pd (used to determine τ^{00}_{MeOH}).[16]

have appeared in literature using non-steady-state transient kinetic analysis (stop-flow, non-isotopic switching) that involves stopping totally of one or more of the reactants. Since the surface experiences major changes, the quantity, types, and kinetics of surface intermediates detected during this transient, may or may not relate to those existing under steady-state reaction condition. This has been clearly shown by Peil *et al.*[17] in the application of SSITKA and of stop-flow measurements during the oxidative coupling of methane. The non-isotopic non-steady-state studies were conducted by stopping the flow of either CH_4 or O_2 in the reactant stream. A step change in the concentration of one of the reactants resulted in transient responses in both the reactant and the product. Carbon surface concentrations of all the surface intermediates, leading to a particular product, were determined by integrating the non-normalized transient. The results were compared with those obtained for the SSITKA study using the switch $^{12}CH_4/^{13}CH_4$. Although the sums of the total carbon on the catalyst surface from the non-steady state CH_4 cutoff and the steady-state SSITKA studies were reasonably close, only the SSITKA results were able to provide an accurate measure of the active intermediates on the catalyst surface during steady-state reaction.

5. Applications

5.1. *TOF on metals*

In order to evaluate the site efficiency of a catalyst compared with other active catalysts for a reaction of interest, many attempts have been made to determine or express the intrinsic activity of a catalyst. This involved dividing the rate of reaction by a parameter related to the number of catalytic sites. With advancement in techniques, this has progressed from dividing the reaction rate by the weight of the catalyst, by the BET surface area, and, in the last 40 years, by the amount of a particular gas chemisorbed (TOF), the latter remains the most popular method from a fundamental standpoint. Although not in the original definition of TOF based on chemisorption, in practice, many researchers apply TOF values determined in a way that assumes that all chemisorption sites, determined in a pure gas usually near room temperature, are active for reaction under reaction conditions. Although, TOF based on chemisorption is certainly useful for comparison purposes for very similar catalysts, the question is raised as to how well TOF, based on chemisorption, relates to a site TOF. In a recent review by Goodwin *et al.*,[9] the use of SSITKA in determining the TOF of metal catalysts based on the number of active intermediates (TOF_{ITK}) was shown and compared with the TOF based on the amount of chemisorption (TOF_{chem}) for a variety of reactions and supported metal catalysts. Rate per number of active intermediates (related to the number of active sites) determined experimentally by SSITKA ($TOF_{ITK} = 1/\tau_P$), is simply the reciprocal of the average surface residence time of product intermediates. It was shown that for a structure insensitive reaction such as methanation, TOF_{ITK} exceeds TOF_{chem} only by an order of magnitude. However, for structure sensitive reactions, TOF_{ITK} exceeds TOF_{chem} by approximately 2–3 orders of magnitude due to the ensemble size and/or geometric arrangement required for active reaction sites.

5.2. *Cause of deactivation*

Since SSITKA can decouple the apparent rate of reaction into the contribution from the intrinsic activity ($\sim$ the reciprocal of surface residence time of intermediates) and the number of active sites ($\sim$ surface concentration of intermediates), the cause of deactivation of a catalyst during reaction can often be revealed. SSITKA has been used in a number of studies for this purpose. Catalyst deactivation during n-butane isomerization and selective CO oxidation are good examples. Deactivation studies are conducted by collecting isotopic transient data at particular times-on-stream as deactivation occurs.

Kim *et al.*[7] successfully showed that for n-butane isomerisation on sulfated zirconia, deactivation was due to a decrease in site activity and a blockage of active sites. The initial loss in activity was caused by the loss of the most active sites. The less active sites deactivated slower and had more of a relative impact on the isomerisation reaction at later TOS.

Deactivation of γ-Al_2O_3 supported Pt and PtFe catalysts for selective CO oxidation was studied by Sirijaruphan *et al.*[8,18] Even though these two catalysts exhibited similar deactivation behaviours in terms of rate of reaction, they behaved differently in terms of surface kinetic parameters. The intrinsic site activity of Pt remained relatively constant during deactivation, while that of PtFe decreased significantly approaching that of Pt with TOS, as shown in Fig. 10. Decrease in surface concentration of CO_2 intermediates indicated site blockage by carbon for both catalysts. However, the reoxidation of Fe in PtFe was suggested to be the probable cause of its decrease in site activity.

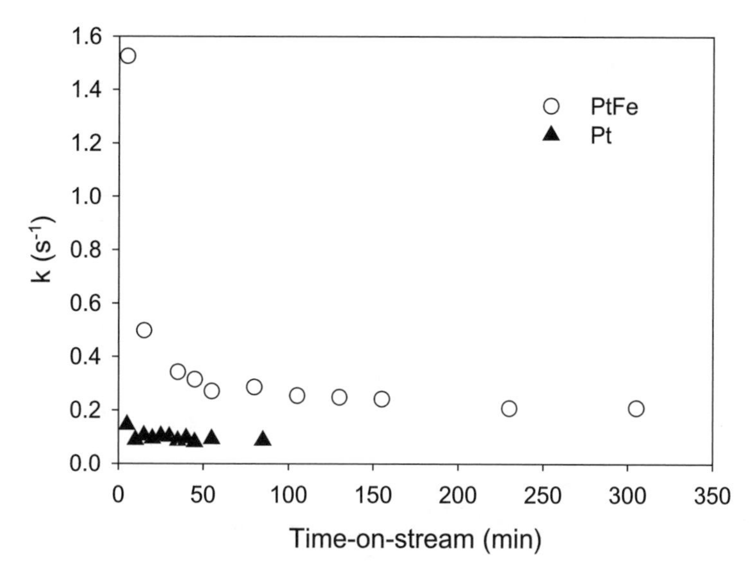

Fig. 10. The time-on-stream behavior of the pseudo-first-order intrinsic rate constant for Pt and PtFe catalysts on alumina.

5.3. *Effects of promotion*

Surface site heterogeneity of Ru/SiO_2 for ammonia synthesis, as well as the effect of K promotion, has been studied using SSITKA.[6] K is a well known activity promoter for ammonia synthesis on Ru and Fe catalysts. The Ru/SiO_2 catalyst was studied at 673 K, 204 kPa, H_2/N_2 ratio of 3 and GHSV between 5 000–23 000. SSITKA experiments were carried out by switching between $^{14}N_2$ and $^{15}N_2$.

The results showed that K promotion resulted in almost 50 times the increase in the reaction rate. TOF based on the amount of hydrogen chemisorption increased by about two orders of magnitude with K-promotion. However, the intrinsic TOF based on SSITKA increased only by a factor of 16. The increase in activity with K promotion was actually due to both a significant increase by a factor of 3 in the surface concentration of intermediates, and an increase by a factor of 16 in the average intrinsic site activity.

Site heterogeneity was investigated using a deconvolution technique based on inverse Laplace transform (ILT) method. As could be seen from the active site distribution shown in Fig. 11, the unpromoted catalyst had essentially only one kind of site (β). K-promotion increased the average

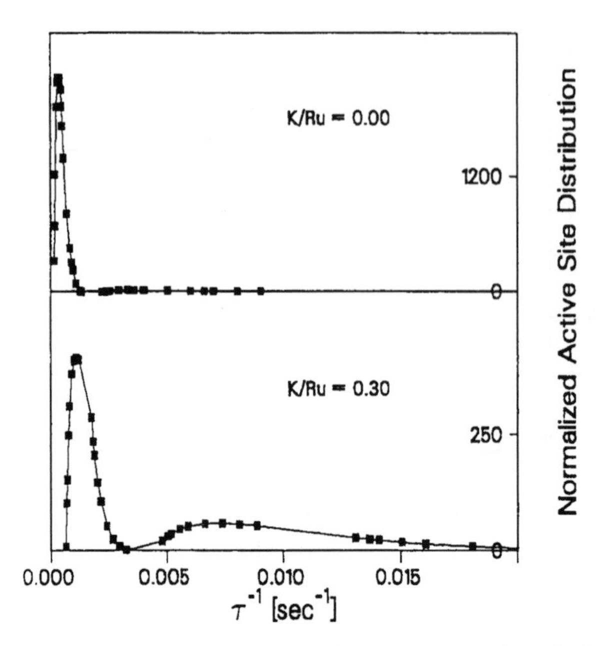

Fig. 11. Active site distribution for unpromoted and K promoted Ru/SiO_2 catalyst for ammonia synthesis.[6]

activity of these sites by four times higher (5.6×10^{-4} to $22 \times 10^{-4}\,s^{-1}$). However, and more importantly, K promotion also resulted in the creation of super active sites (α), having an average activity of $250 \times 10^{-4}\,s^{-1}$. These super active sites accounted for 78% of the reaction rate, even though they only accounted for 20% of the total surface intermediates.

5.4. *Effect of alloying*

The effect of possible alloy formation resulting in changes in surface kinetic parameters for the selective oxidation of CO was investigated for a $PtFe/\gamma$-Al_2O_3 catalyst.[18] At the catalyst composition used (Pt:Fe atomic ratio of 3:1), the intermetallic alloy $FePt_3$ could form during catalyst preparation and pretreatment. The activity, selectivity and surface kinetic parameters were compared between non-promoted and Fe-promoted Pt/γ-Al_2O_3 catalysts. The Fe-promoted catalyst had higher activity and selectivity for selective CO oxidation, consistent with other studies.[19,20] Farrauto and coworkers[19,20] have hypothesised that the promoted catalyst is more active for CO oxidation, due to iron oxide providing more favorable sites for oxygen adsorption than Pt itself, and being located in close contact with surface Pt. This idea was supported by the results from SSITKA, since an increase in intrinsic site activity was seen upon Fe promotion.

Comparing the surface concentration of CO_2 intermediates ($N_{I\text{-}CO_2}$), even though PtFe had a higher concentration of surface intermediates than Pt during most TOS, the maximum concentration, which shows the highest possible number of active sites, was about the same as that for Pt. Despite the fact that $N_{I\text{-}CO_2}$ of PtFe decreased with TOS at a remarkably slower rate, it was concluded that Fe promotion does not increase the maximum concentration of active sites adsorbing CO for CO_2 formation. It is important to note that $N_{I\text{-}CO_2}$ included only intermediates containing carbon, since the isotopic tracing was done with $^{12}C/^{13}C$. Thus, these results suggest that Fe does not promote the catalyst by providing more sites for the adsorption and reaction of CO. The reactivity distribution of PtFe showed that the relative amounts of less active sites to more active sites increased as the reaction progressed, causing the mean value for the distribution to shift to a lower value of k. As shown in Fig. 12, with TOS, site activity of PtFe appeared to asymptotically approach that of Pt, indicating that the surface became more like Pt as deactivation occurred.

5.5. *Acid catalysis*

Sulfated zirconia (SZ) is a well-known solid acid catalyst that has been widely studied in the past 15 years. SZ is a very strong acid catalyst and is active for n-butane isomerisation even at room temperature. Many parameters have been found to impact the catalytic activity of SZ, such as catalyst preparation and pretreatment, which affect the sulfur content, the concentration of Lewis and Bronsted acid sites, and other characteristics. However, the deactivation of SZ during n-butane isomerisation can be severe.

Kim *et al.*[7] studied the fast initial deactivation observed during the first 45 min TOS at 150°C. The SSITKA results, after correction for readsorption, showed an increase in the surface residence time of iso-C_4 intermediates ($\tau^*_{iso\text{-}C_4}$) and a corresponding decrease in site activity, as well as a decrease in the surface concentration of intermediates leading to iso-C_4 with TOS. The high initial activity and rapid deactivation were mainly due to the presence and the deactivation of the more active sites respectively. After calculating the surface coverage based on sulfur content, the surface coverage of iso-C_4 was found to be an order of magnitude smaller than that of reversibly adsorbed n-C_4, at all TOS as given in Table 2. Only a maximum

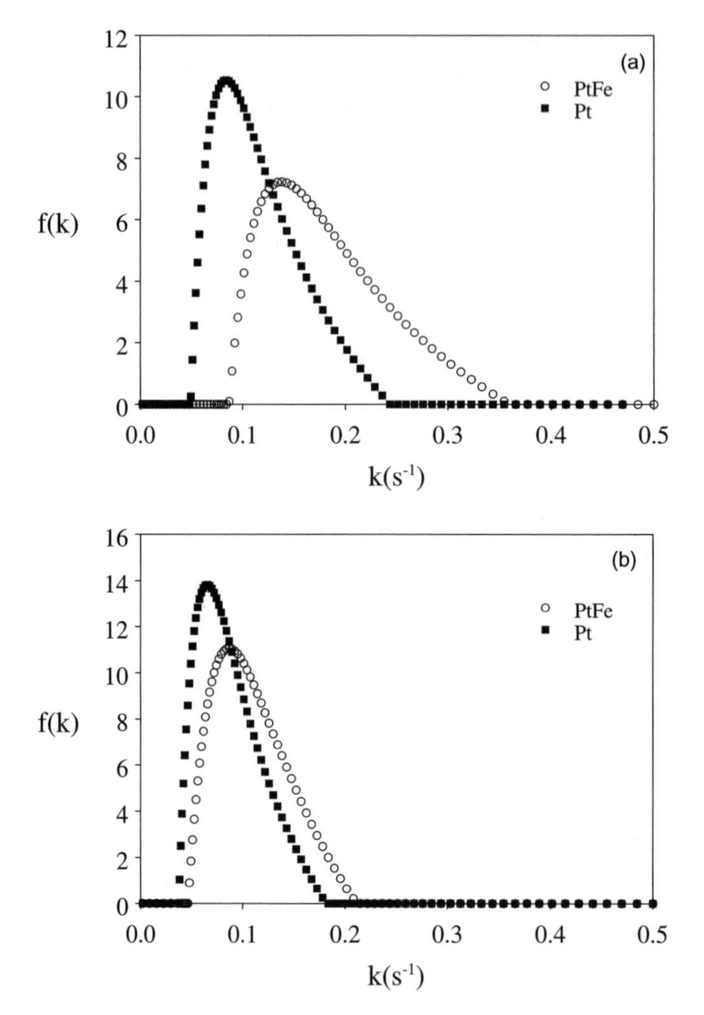

Fig. 12. Rate constant distribution of the Pt and PtFe catalysts at (a) 5 min TOS and (b) steady state.[18]

of 20% of the sulfate species present was found to be n-C_4 adsorption sites and only 1–2% of the sulfated species appeared to adsorb active surface intermediates. The TOF calculated from SSITKA $\left(\mathrm{TOF}^*_{\mathrm{ITK}} \approx 1/\tau^*_{\text{iso-C}_4}\right)$ could be compared with the conventional TOF based on total sulfur concentration ($\mathrm{TOF}_{\mathrm{sulfur}}$). Since $\mathrm{TOF}^*_{\mathrm{ITK}}$ was based on the experimentally determined number of active intermediates rather than on all hypothetical sites, it was suggested to be more accurate. $\mathrm{TOF}_{\mathrm{ITK}}$ was approximately two orders of magnitude higher than $\mathrm{TOF}_{\mathrm{sulfur}}$.

Table 2. Surface coverage of SZ based on sulfur content[a].[7]

TOS (min)	$\theta_{n\text{-}C_4}$ [b]	$\theta_{iso\text{-}C_4}$ [c]
30	0.20	0.022
100	0.19	0.013
150	0.21	0.013
250	0.15	0.008

[a] $\theta_i = N_i/N_s = N_i/(493\,\mu mol\ sulfur\ g_{cat})$.
[b] Max error = ±0.006.
[c] Max error = ±0.0001.

5.6. *High temperature reaction*

Oxidative coupling of methane to ethylene and ethane occurs at high temperature on both MgO and lithium-doped MgO. Of all the reactions studied by SSITKA, this reaction presented the greatest challenge to study with SSITKA because of the high temperature.[4] At high temperatures, the surface residence times of reactive intermediates can be very low, possibly even below the detectability limit of SSITKA. However, by studying the reaction at 645°C rather than at higher temperature, SSITKA was able to be utilised.[10,14,21]

SSITKA gave information about the mechanistic events taking place on the catalyst, leading to formation of the coupled products. It was able to help in determining the reaction pathways during the oxidative coupling reaction, and also in revealing the role of lattice oxygen during production of ethane and ethylene.[10] Peil *et al.*[14] quantified the active intermediates along both the carbon and oxygen reaction pathways by using isotopic switches of oxygen and methane under steady state conditions. They suggested that lattice oxygen plays an important role in this reaction under steady state conditions. The typical transients obtained in this study are shown in Fig. 13.

As shown by Lunsford and co-workers,[22–24] on Li/MgO catalysts the initial step in the coupling of methane is the abstraction of hydrogen by Li^+O^- centers. This leads to the formation of methyl radicals. SSITKA was able to give quantitative information about reactive intermediates and reaction pathways over the catalyst. According to the SSITKA results, the catalyst contributes more than the methyl radical generator centers. It provides two parallel carbon pathways for methane conversion.[14] Out of the two pathways, one is active in the formation of CO and CO_2, while the other

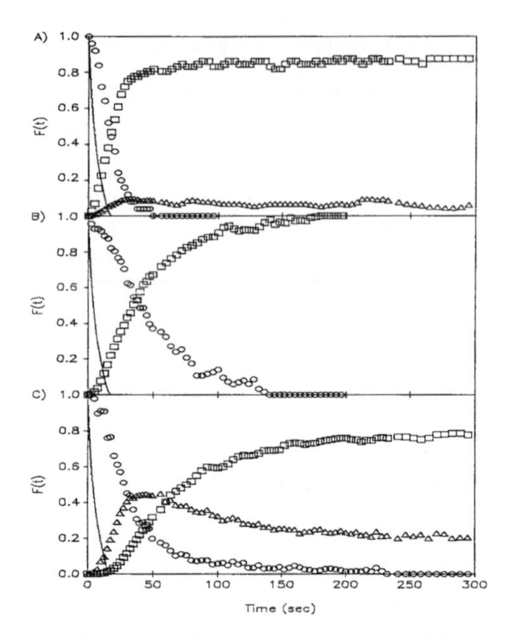

Fig. 13. Molecular oxygen transients during reaction with methane at 645°C over 32.6 mg of Li/MgO. F(t) = $F_{*O_2}(t)$, $F_{C^*O}(t)$, $F_{C^*O_2}(t)$, $F_{He}(t)$. (A) O_2 transients: O,$^{16}O_2$; Δ, $^{16}O^{18}O$; □, $^{18}O_2$. (B) CO transients: O,$C^{16}O$;□,$C^{18}O$. (C) CO_2 transients: O, $C^{16}O_2$; Δ, $C^{16}O^{18}O$; □, $C^{18}O_2$. Other reaction conditions: CH_4/O_2 = 5, CH_4 conversion = 5.3%, CO_2/CO = 2.4, ethane selectivity = 19.3%. Solid lines indicate an inert gas tracer.[14]

is active in the formation of ethane. The surface generated intermediates undergo possible gas phase coupling leading to the formation of ethane.

Peil *et al.*[21] also did another study considering the effect of readsorption of CO_2 formed during the reaction on Li/MgO and Sm_2O_3 catalysts. The SSITKA results showed that all active sites on Li/MgO catalyst can adsorb CO_2. The probability of adsorption is the same for all the sites, and site activity is not affected by this. However, for the Sm_2O_3 catalyst, no interaction between the catalyst and the carbon of gas phase CO_2 was observed at the reaction conditions.

5.7. *Effect of temperature*

Methanation is a relatively simple reaction that can be used for studying surface hydrogenation. The surface reaction mechanism of methanation has been proposed by a number of different authors. For Ru and Co, the hydrogenation of carbonaceous species has been widely concluded to be the rate-limiting step. Bajusz and Goodwin[5] studied the effect of temperature on the surface coverage and the activity of reaction intermediates on Ru/SiO_2 during methanation. The impact of temperature (240°C–270°C) on the overall reaction and the SSITKA surface reaction parameters depended on the hydrogen partial pressure and the H_2/CO ratio. The average surface residence time of intermediates (τ_M) appeared to decrease with increasing temperature for both H_2/CO ratios studied ($H_2/CO = 5$ and 20) as might be expected. Surface concentration of methane intermediates, N_M, increased with temperature at high H_2/CO ratio, but remained fairly constant at low H_2/CO ratio, because of a concomitant partial deactivation of the active sites with increasing temperature due to low hydrogen partial pressure. The activity distributions for methane formation were evaluated numerically using the T-F deconvolution technique as shown in Fig. 14. Two types of carbon pools, more active (α) and less active (β), were detected at 240°C and were able to be hydrogenated to methane. The activity of each pool shifted towards higher values of k with increasing temperature. The

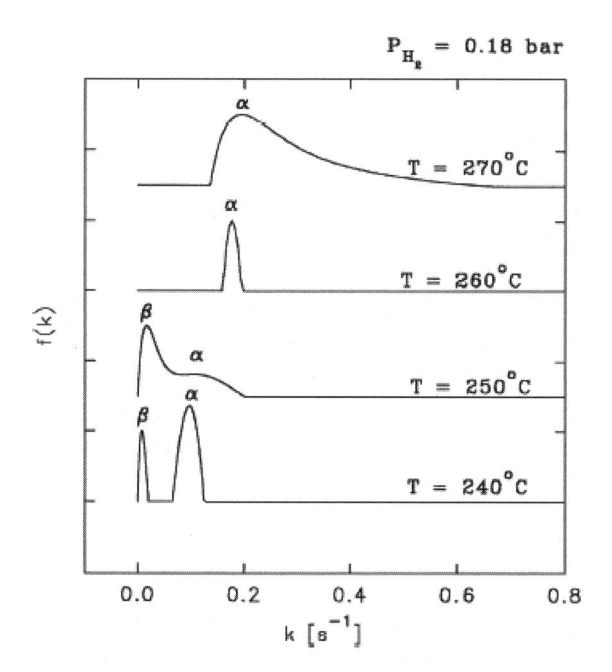

Fig. 14. Site activity distribution for methanation on Ru/SiO_2 at different temperatures ($P_{H_2} = 0.18$ bar, $P_{CO} = 0.0036$ bar, $H_2/CO = 5$).[5]

transformation of the less active β carbon into inactive γ carbon resulted in a single pool of active intermediates detected at 250°C.

5.8. *Effect of partial pressure of reactant*

The effect of partial pressure of hydrogen on the surface kinetic parameters for methanation on Ru/SiO_2 was shown in the study by Bajusz and Goodwin.[5] The rate of reaction can be expressed as

$$r_M = \left(\frac{1}{\tau_M}\right) \cdot N_M \tag{13}$$

where τ_M is the average surface residence time and N_M is the concentration of the active intermediates leading to methane. The reciprocal of the average surface residence time of carbon-containing intermediates leading to methane ($1/\tau_M$), representing a pseudo-first-order rate constant (k), also includes the hydrogen surface concentration (N_H) dependence. By varying the hydrogen partial pressure, the average residence time of the intermediates (τ_M) decreased and the concentration of surface methane intermediates (N_M) increased while the surface concentration of reversibly adsorbed CO remained essentially constant. The decrease in τ_M with an increase in H_2 partial pressure is related to its dependence on N_H. At 240°C it was shown that the relative surface concentration of hydrogen on the catalyst increased with increasing hydrogen partial pressure.

Site activity distributions determined from SSITKA data showed the evolution of the activity with increasing hydrogen partial pressure (Fig. 15). At low hydrogen partial pressure (0.14 bar), only the more active carbon pool (α) was detected. On increasing the hydrogen partial pressure from 0.14 bar to 0.18 bar, the less active carbon pool (β) was able to be detected. Further increase in hydrogen partial pressure resulted in a shift in the site activity peak toward higher activity, due to the increase in N_H and the dependence of k on N_H. The effect of hydrogen partial pressure on the less active β pool was more significant than on the more active α pool.

6. Summary

Steady state isotopic transient kinetic analysis is a powerful technique for *in situ* analysis of heterogeneous catalytic reactions. It provides information

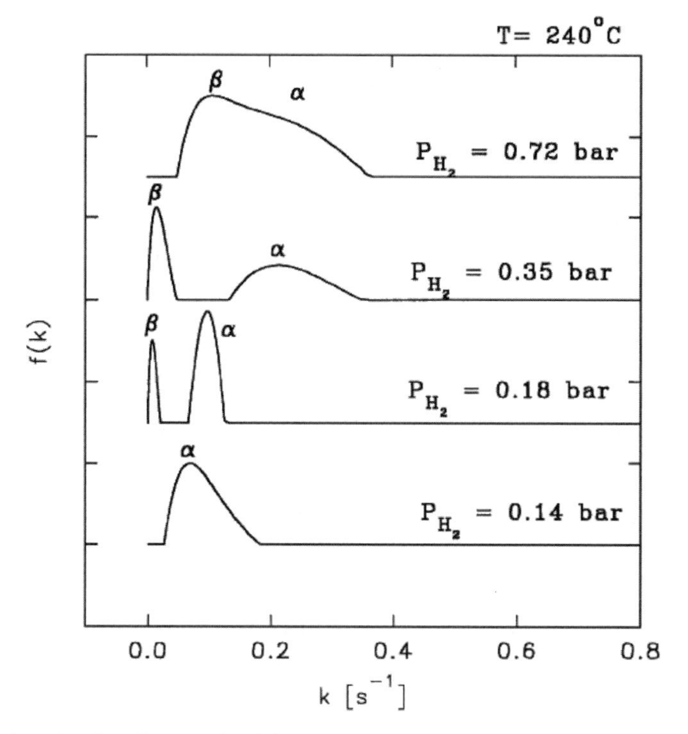

Fig. 15. Activity distributions for different hydrogen partial pressures at 240°C (P_{CO} = 0.036 bar).[5]

about the average residence time (τ_P), concentration (N_P) and surface coverage (θ_P) of the reactive intermediates. Advanced mathematical analysis techniques like deconvolution and convolution give information about reactivity distributions or site heterogeneity. SSITKA is applied easily to most smaller molecule gas phase reactions. SSITKA has helped in enhancing current knowledge about many reactions and catalytic performances.

Acknowledgement

This material is based upon work supported by the National Science Foundation under Grant No. CTS — 0211495. Any opinions, findings, and conclusions or recommendations expressed in this material are those of the authors and do not necessarily reflect the views of the National Science Foundation.

References

[1] Happel J, *Chem Eng Sci* **33**: 1567, 1978.
[2] Bennett CO, Understanding heterogeneous catalysis through the transient method, in Bell AT, Hegedus LL (eds.), *Catalysis Under Transient Conditions, ACS Symposium Series; American Chemical Society*, Washington, DC, pp. 1–32, 1982.
[3] Biloen P, *J Mol Catal* **21**: 17, 1983.
[4] Shannon SL, Goodwin JG, Jr, *Chem Rev* **95**: 677, 1995.
[5] Bajusz I-G, Goodwin JG, Jr, *J Catal* **169**: 157, 1997.
[6] Nwalor JU, Goodwin JG, Jr, *Topic Catal* **1**: 285, 1994.
[7] Kim SY, Goodwin JG, Jr, Galloway D, *Catal Today* **63**: 21, 2000.
[8] Sirijaruphan A, Goodwin JG, Jr, Rice RW, *J Catal* **221**: 288, 2004.
[9] Goodwin JG, Jr, Kim SY, Rhodes WD, Turnover frequencies in metal catalysis: meanings, functionalities and relationships, in Spivey J, Roberts G (eds.), *Catalysis* Vol. 17, Royal Chemistry Society, London, pp. 320–347, 2003.
[10] Peil KP, Marcelin G, Goodwin JG, Jr, The role of lattice oxygen in the oxidative coupling of methane, in Wolf EE (ed.), *Methane Conversion by Oxidative Processes-Fundamental and Engineering Aspects*, Van Nostrand Reinhold, New York, pp. 139–145, 1992.
[11] Hoost TE, Goodwin JG, Jr, *J Catal* **134**: 678, 1992.
[12] Shannon SL, Goodwin JG, Jr, *Appl Catal A Gen* **151**: 3, 1997.
[13] Basallote MG, Bernal S, Garcia JM, Pozo M, *Appl Catal A Gen* **232**: 39, 2002.
[14] Peil KP, Goodwin JG, Jr, Marcelin G, *J Catal* **131**: 143, 1991.
[15] Peil KP, Marcelin G, Goodwin JG, Jr, Kiennemann A, Transient kinetic analysis of the oxidative coupling of methane over Sm_2O_3, in Albright LF (ed.), *Novel Production Methods for Ethylene, Light Hydrocarbons and Aromatics*, Marcel Dekker Inc. New York, pp. 61–74, 1992.
[16] Ali SH, Goodwin JG, Jr, *J Catal* **171**: 339, 1997.
[17] Peil KP, Goodwin JG, Jr, Marcelin G, *J Catal* **132**: 556, 1991.
[18] Sirijaruphan A, Goodwin JG, Jr, Rice RW, *J Catal* **224**: 304, 2004.
[19] Korotkikh O, Farrauto R, *Catal Today* **62**: 249, 2000.
[20] Liu X, Korotkikh O, Farrauto R, *Appl Catal A Gen* **226**: 293, 2002.
[21] Peil KP, Goodwin JG, Jr, Marcelin G, Influence of product CO_2 on the overall reaction network in the oxidative coupling of methane, in Holmen A, Jens KJ, Kolboe S (eds.), *Natural Gas Conversion*, Elsevier Science Publishers BV, Amsterdam, pp. 73, 1991.
[22] Driscoll DJ, Martir W, Wang J-X, Lunsford JH, *J Am Chem Soc* **107**: 58, 1985.
[23] Campbell KD, Morales E, Lunsford JH, *J Am Chem Soc* **109**: 7900, 1987.
[24] Lunsford JH, in *Methane conversion*, Bibby DM, Chang CD, Howe RF, Yurchak S (eds.), 1988, Elsevier: Amsterdam. p. 359.
[25] McClaine BC, Davis RJ, *J Catal* **211**: 379, 2002.
[26] McClaine BC, Davis RJ, *J Catal* **210**: 387, 2002.

[27] Siporin SE, Davis RJ, *J Catal* **222**: 315, 2004.
[28] Nwalor JU, Goodwin JG, Jr, Biloen P, *J Catal* **117**: 121, 1989.
[29] Ozkan US, Cai Y, Kumthekar MW, *J Catal* **149**: 357, 1994.
[30] Hammache S, Goodwin JG, Jr, *J Catal* **218**: 258, 2003.
[31] Kim SY, Goodwin JG, Jr, Hammache S, Auroux A, Galloway D, *J Catal* **201**: 1, 2001.
[32] Yaluris G, Larson RB, Kobe JM, González MR, Fogash KB, Dumesic JA, *J Catal* **158**: 336, 1996.
[33] Mirodatos C, *J Phys Chem* **90**: 481, 1986.
[34] Kazi AM, Chen B, Goodwin JG, Jr, *J Catal* **157**: 1, 1995.
[35] Oukaci R, Blackmond DG, Goodwin JG, Jr, Gallaher G, Steady-state isotopic transient kinetic analysis investigation of CO-O_2 and CO-NO reaction over a commercial automotive catalyst, in Silver RG, Sawyer JE, Summers JC (eds.), *Catalytic Control of Air Pollution: Mobile and Stationary Sources*, Washington, DC, 1992.
[36] Chen B, Goodwin JG, Jr, *J Catal* **158**: 511, 1996.
[37] Haddad GJ, Chen B, Goodwin JG, Jr, *J Catal* **161**: 274, 1996.
[38] Panpranot J, Goodwin JG, Jr, Sayari A, *J Catal* **213**: 78, 2003.
[39] Shen WJ, Ichihashi Y, Ando H, Matsumura Y, Okumura M, Haruta M, *Appl Catal A Gen* **217**: 231, 2001.
[40] Wang SY, Moon SH, Vannice MA, *J Catal* **71**: 167, 1981.
[41] Kogelbauer A, Goodwin JG, Jr, *J Catal* **160**: 125, 1996.
[42] Rothaemel M, Hanssen KF, Blekkan EA, Schanke D, Holmen A, *Catal Today* **38**: 79, 1997.
[43] van Dijk HAJ, Hoebink JHBJ, Schouten JC, *Chem Eng Sci* **56**: 1211, 2001.
[44] Chen B, Goodwin JG, Jr, *J Catal* **154**: 1, 1995.
[45] Chen B, Goodwin JG, Jr, *J Catal* **158**: 228, 1996.
[46] Ozkan US, Cai Y, Kumthekar MW, *J Catal* **149**: 390, 1994.
[47] Balakos MW, Chuang SSC, *J Catal* **151**: 253, 1995.
[48] Balakos MW, Chuang SSC, *J Catal* **151**: 266, 1995.
[49] Hendrick SA, Chuang SSC, Brundage MA, *J Catal* **185**: 73, 1999.
[50] Kao JY, Piet-Lahanier H, Walter E, Happel J, *J Catal* **133**: 383, 1992.
[51] Ocal M, Oukaci R, Marcelin G, Jang BWL, Spivey JJ, *Catal Today* **59**: 205, 2000.
[52] Soong Y, Krishna K, Biloen P, *J Catal* **97**: 330, 1986.
[53] Stockwell MD, Chung JS, Bennett CO, *J Catal* **112**: 135, 1988.
[54] Belambe AR, Oukaci R, Goodwin JG, Jr, *J Catal* **166**: 8, 1997.
[55] Stockwell MD, Bennett CO, *J Catal* **110**: 354, 1988.
[56] Biloen P, Helle JN, Van der Berg FGA, Sachtler WMH, *J Catal* **81**: 450, 1983.
[57] Zhang X, Biloen P, *Chem Eng Commun* **44**: 303, 1986.
[58] Zhang X, Biloen P, *J Catal* **98**: 468, 1986.
[59] Nwalor JU, *Dissertation.* 1988, University of Pittsburgh: Pittsburgh, Pennsylvania.
[60] Bertole CJ, Kiss G, Mims CA, *J Catal* **223**: 309, 2004.

[61] Bertole CJ, Mims CA, Kiss G, *J Catal* **221**: 191, 2004.
[62] Bertole CJ, Mims CA, Kiss G, *J Catal* **210**: 84, 2002.
[63] Mims CA, Krajewski JJ, Rose KD, Melchior MT, *Catal Lett* **7**: 119, 1991.
[64] Mims CA, McCandlish LE, *J Phys Chem* **91**: 929, 1987.
[65] Mims CA, McCandlish LE, *J Am Chem Soc* **107**: 696, 1985.
[66] Mims CA, *Catal Lett* **1**: 293, 1988.
[67] Mims CA, McCandlish LE, Melchior MT, *Catal Lett* **1**: 121, 1988.
[68] Efstathiou AM, *J Mol Catal* **67**: 229, 1991.
[69] Efstathiou AM, Bennett CO, *Chem Eng Commun* **83**: 129, 1989.
[70] Efstathiou AM, Bennett CO, *J Catal* **120**: 137, 1989.
[71] Efstathiou AM, Chafik T, Bianchi D, Bennett CO, *J Catal* **148**: 224, 1994.
[72] Otarod M, Happel J, Walter E, *Appl Catal A Gen* **160**: 3, 1997.
[73] Happel J, Suzuki J, Kokayeff P, Fthenakis V, *J Catal* **65**: 59, 1980.
[74] Happel J, Cheh H, Otarod M, Ozawa S, Severdia AJ, Yoshida T, Fthenakis V, *J Catal* **75**: 314, 1982.
[75] Otarod M, Ozawa S, Yin F, Chew M, Cheh H, Happel J, *J Catal* **84**: 156, 1983.
[76] Siddall JH, Miller ML, Delgass WN, *Chem Eng Commun* **83**: 261, 1989.
[77] Vada S, Hoff A, Adnanes E, Schanke D, Holmen A, *Topic Catal* **1**: 155, 1995.
[78] Vada S, Chen, Goodwin JG, Jr, *J Catal* **153**: 224, 1995.
[79] Ali SH, Chen B, Goodwin JG, Jr, *J Catal* **157** 35, 1995.
[80] Soong Y, Biloen P, *Langmuir* **1**: 768, 1985.
[81] Yang CH, Soong Y, Biloen P, *J Catal* **94**: 306, 1985.
[82] Winslow P, Bell AT, *J Catal* **86**: 158, 1984.
[83] Hoost TE, Goodwin JG, Jr, *J Catal* **137**: 22, 1992.
[84] Krishna KR, Bell AT, *Catal Lett* **14**: 305, 1992.
[85] Koerts T, Welters WJJ, van Santen RA, *J Catal* **134**: 1, 1992.
[86] Hanssen KF, Blekkan EA, Schanke D, Holmen A, *Stud Surf Sci Catal* **109**: 193, 1997.
[87] Bajusz I-G, Kwik DJ, Goodwin JG, Jr, *Catal Lett* **48**: 151, 1997.
[88] Agnelli M, Swaan HM, Marquez-Alvarez C, Martin GA, Mirodatos C, *J Catal* **175**: 117, 1998.
[89] Ali SH, Goodwin JG, Jr, *J Catal* **176**: 3, 1998.
[90] Marquez-Alvarez C, Martin GA, Mirodatos C, *Stud Surf Sci Catal* **119**: 155, 1998.
[91] Rohr F, Holmen A, Barbo KK, Warloe P, Blekkan EA, *Stud Surf Sci Catal* **119**: 107, 1998.
[92] Blekkan EA, Rohr F, Holmen A, *Abstr Pap Am Chem Soc* **217**: U244–U245 Part2, 1999.
[93] Rohr F, Lindvag OA, Holmen, Blekkan EA, *Catal Today* **58**: 247, 2000.
[94] Ali SH, Goodwin JG, Jr, *J Catal* **170**: 265, 1997.
[95] Ali SH, Goodwin JG, Jr, *J Catal* **171**: 333, 1997.
[96] Haddad GJ, Goodwin JG, Jr, *J Catal* **157**: 25, 1995.
[97] Vada S, Goodwin JG, Jr, *J Phys Chem* **23**: 9479, 1995.
[98] Mirodatos C, Dalmon JA, Martin GA, *J Catal* **105**: 405, 1987.

[99] Lacombe S, Sanchez JG, Delichere P, Mozzanega H, Tatibouet JM, Mirodatos C, *Catal Today* **13**: 273, 1992.
[100] Mirodatos C, Holmen A, Mariscal R, Martin GA, *Catal Today* **6**: 601, 1990.
[101] Peil KP, Goodwin JG, Jr, Marcelin G, *J Phys Chem* **93**: 5977, 1989.
[102] Efstathiou AM, Lacombe S, Mirodatos C, Verykios XE, *J Catal* **148**: 639, 1994.
[103] Kalenik Z, Wolf EE, *Catal Lett* **9**: 441, 1991.
[104] Peil KP, Goodwin JG, Jr, Marcelin G, *J Am Chem Soc* **112**: 6129, 1990.
[105] Efstathiou AM, Verykios XE, *Appl Catal A Gen* **151**: 109, 1997.
[106] Mims CA, Hall RB, Jacobson AJ, Lewandowski JT, Myers G, *ACS Symposium Series* **482**: 230, 1992.
[107] Koranne MM, Goodwin JG, Jr, Marcelin G, *J Catal* **148**: 378, 1994.
[108] Koranne MM, Goodwin JG, Jr, Marcelin G, *J Phys Chem* **97**: 673, 1993.
[109] Smith MR, Ozkan US, *J Catal* **142**: 226, 1993.
[110] Mauti R, Mims CA, *Catal Lett* **21**: 201, 1993.
[111] Ekstrom A, Lapszewicz JA, *J Am Chem Soc* **110**: 5226, 1988.
[112] Ekstrom A, Lapszewicz JA, *Chem Commun* 797, 1988.
[113] Ekstrom A, Lapszewicz JA, *J Phys Chem* **93**: 5230, 1989.
[114] Kumthekar MW, Ozkan US, *Appl Catal A Gen* **151**: 289, 1997.
[115] Sadovskaya EM, Suknev AP, Pinaeva LG, Goncharov VB, Bal zhinimaev BS, Chupin C, Mirodatos C, *J Catal* **201**: 159, 2001.
[116] Sadovskaya EM, Suknev AP, Pinaeva LG, Goncharov VB, Bal zhinimaev BS, Chupin C, Perez-Ramirez J, Mirodatos C, *J Catal* **225**: 179, 2004.
[117] Sadovskaya EM, Suknev AP, Goncharov VB, Bal zhinimaev BS, Mirodatos C, *Kinetics Catal* **45**: 436, 2004.
[118] Burch R, Shestov AA, Sullivan JJ, *J Catal* **186**: 353, 1999.
[119] Burch R, Shestov AA, Sullivan JJ, *J Catal* **182**: 497, 1999.
[120] Shestov AA, Burch R, Sullivan JJ, *J Catal* **186**: 362, 1999.
[121] Burch R, Shestov AA, Sullivan JJ, *J Catal* **188**: 69, 1999.
[122] Saleh-Alhamed YA, Hudgins RR, Silveston PL, Peil KP, Goodwin JG, Jr, *Unpublished Work*, University of Pittsburgh, Pittsburgh, Pennsylvania, 1992.
[123] Barias OA, *Dissertation*, University of Trondheim, Trondheim, Norway, 1993.
[124] Crathorne EA, MacGowan D, Morris SR, Rawlinson AP, *J Catal* **149**: 254, 1994.

Chapter 8

Positron Emission Profiling — The Ammonia Oxidation Reaction as a Case Study

A M de Jong, E J M Hensen and R A van Santen

1. Introduction

In the study of the kinetics of heterogeneously catalysed reactions, the chemical nature of reacting species and the rates at which these species undergo transformations is of major interest. A vast quantity of techniques is available which are capable of measuring chemical species in a quantitative manner. Many spectroscopic and surface science techniques have been successfully applied to this problem over the past forty years and have generated a large body of data. However, the experimental conditions under which these measurements are performed (single crystal catalysts and low pressure), often differ considerably from those that exist under normal process conditions. Thus, a question remains as to whether the data obtained can be extrapolated to the real system. As these data have generally been obtained under conditions differing from the actual process, a link must be made back to the original systems of interest occurring in chemical reactors at elevated temperatures, and at atmospheric pressure and above. Mathematical models based on previously determined mechanistic information may provide this link.

Mathematical models of reactor kinetics are used to describe the concentration distributions of various reactants, intermediates and products within the reactor during the course of the reaction. In order to validate these models, concentration distributions must be measured. However, as these chemical reactions, under real process conditions, generally occur within metal reactors enclosed by a heating or pressure mantle, it is not a simple matter to "look inside". Rather, most techniques rely on measurements of concentration distributions measured at the reactor inlet and outlet only.

Ideally, one would like to have a technique which could "image" the reactor in such a way as to be able to determine both the chemical identity and quantity of all reactants, intermediates and products as a function of time and position under identical conditions of temperature, pressure, flow rate etc. In addition, it is desirable that the probe method does not disturb the reaction, i.e., the method should be non-invasive.

Medical research has provided an incentive to develop several sophisticated non-invasive, *in situ*, imaging techniques. Techniques such as magnetic resonance imaging (MRI) and computer-assisted tomography (CT) are now widely used diagnostic tools to study structure within the living human. Positron Emission Tomography (PET) based on the detection of emitted radiation from injected radioactive tracers is used to study biochemical and metabolic functions. In this chapter, we will illustrate the potential of this technique in the field of catalysis research with a study on the ammonia oxidation reaction. Although PET is now well established as a diagnostic tool for *in vivo* imaging of the function of human organs, particularly of the brain and the heart,[1] it is probably unknown to most researchers outside of the medical community. Thus, we will start with a description of the technique and its basic principles.

1.1. *Positron emission and positron-electron annihilation*

The decay of radioactive isotopes via electron emission, known as beta decay, is a well known phenomenon. In this mode, unstable nuclei that have an excessive number of neutrons, for example ^{14}C, can emit fast electrons, β^- particles, in order to attain a stable nuclear configuration. Nuclei with insufficient neutrons, such as ^{11}C, can obtain stability by emitting fast positrons, β^+ particles (the anti-matter equivalents of electrons). Both processes are classified as radioactive β decay. In each case, the mass number of the nucleus remains constant but the atomic number changes.

There exist several positron emitting isotopes of which ^{11}C, ^{13}N and ^{15}O in particular are of interest for catalytic reaction studies. Since the half-life time of these isotopes is only 20, 10 and 2 minutes respectively, they must be produced on-site. Production of such radioactive isotopes is normally done by irradiation of an appropriate target material with protons or deuterons at high energy.

Since the positron is the antiparticle of the electron, an encounter between them can lead to the subsequent annihilation of both particles. Their combined rest mass energy then appears as electromagnetic radiation. Annihilation can occur via several mechanisms: direct transformation into one, two, or three photons; or the formation of an intermediate, hydrogen-like bound state between the positron and the electron, known as a positronium (Ps). The extent to which each annihilation mechanism contributes depends on the kinetic energy of the positron-electron pair.

Positrons emitted during the β^+ decay process, possess a statistical distribution of kinetic energies ranging from zero to a maximum value, T_{max}, dependent on the decaying nucleus ($T_{max} = 0.96$ MeV for ^{11}C). The average kinetic energy is equal to 0.4 T_{max}. The probability of annihilation is negligibly small at high energies. The emitted positrons must therefore be slowed down by inelastic scattering interactions with the nuclei and the bound electrons within the surrounding medium to near thermal values, before annihilation can occur. The lifetime of a positron is of the order of nanoseconds. During its lifetime, the positron will travel a distance known as the stopping distance, which is dependent on the energy of the positron and on the density of the surrounding material. For 0.4 MeV positrons (average kinetic energy of positrons emitted from ^{11}C) in a medium with a density of 0.5 g/ml (such as a zeolite or metal oxide), this corresponds to circa 3 mm.[2]

The predominant annihilation process for thermalised positrons is via the direct production of two photons (Fig. 1). If both the positron and

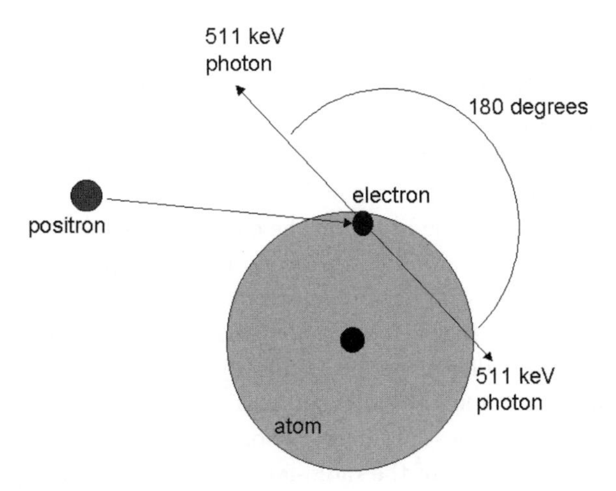

Fig. 1. When a positron and electron annihilate, two 511 keV gamma photons are emitted in opposite directions.

the electron were at rest upon annihilation, conservation of energy dictates that the energy of each emitted photon would be equal to the 511 keV, rest mass energy of the positron or electron. Conservation of momentum implies that the two gamma photons be emitted in opposite directions, since the initial momentum of the positron-electron pair was zero.

1.2. *Positron emission detection methods*

The emitted gamma photons produced by positron-electron annihilation can be detected using scintillation crystal detectors such as sodium iodide (NaI), bismuth germanium oxide (BGO) and cerium doped lutetium oxy-orthosilicate (LSO). The short half-life of most positron emitters leads to high specific activity. Only a very small quantity of radio-labelled molecules is thus required, making positron annihilation detection techniques very non-invasive. In fact, practical catalyst studies can be carried out using less than 37 kBq of carbon-11, corresponding to less than 6.5×10^7 molecules.

The first use of positron-emitting isotopes as tracers in catalysis research was published in 1984 by Ferrieri and Wolf.[3,4] ^{11}C-labelled acetylene and propylene were employed to monitor the alkyne cyclotrimerisation reaction on silica-alumina supported chromium(VI). Single annihilation photons were collimated and detected using a sodium iodide scintillation detector. The aromatic products (benzene, toluene, xylenes) desorbed from the surface were analysed using radio gas chromatography. These authors named their technique, Positron Annihilation Surface Detection (PASD). Baiker and co-workers[5] used ^{13}N-labelled NO to investigate the selective catalytic reduction (SCR) of NO by NH_3 over vanadia/titania at very low reactant concentrations. Concentrations of 5×10^{-9} ppm ^{13}NO were used. Again, only single annihilation photons were detected by NaI scintillation detectors placed after the reactor bed.

For imaging, a technique based on the coincident detection of both photons produced via the annihilation event is often applied. This can be achieved by using two scintillation detectors, each placed on opposite sides of the emitting source. In this mode, only pairs of detected events that occur within a preset coincidence window (typically less than 50 ns) are counted. The position of the annihilation event that gave rise to the two detected photons, can then be located along a chord joining the two detector elements. The concentration of the radio-labelled isotope at that position can also be determined by integrating the number of events detected during a fixed time.

Due to the penetrating power of the emitted 511 keV gamma photons, which can pass through several millimetres of stainless steel, detection is possible from within steel reactors or process vessels. The coincident detection of photons is the principle of techniques such as Positron Emission Tomography (PET), Positron Emission Particle Tracking (PEPT), and Positron Emission Profiling (PEP), which are discussed below.

1.3. *Positron Emission Tomography (PET), Particle Tracking (PEPT) and Profiling (PEP)*

Positron Emission Tomography (PET) is now well established as a diagnostic technique in nuclear medicine, providing 3D images of the distribution of radio-labelled molecules within living human organs. The development of a new breed of small self-shielding cyclotrons in the 1980s and significant improvements in computer hardware and software has led to an explosive growth in the number of PET facilities world wide. Unlike CT and MRI that measure structural information, PET is capable of providing rate information regarding biochemical and metabolic processes. Injected radiotracers such as [^{18}F]-2-fluoro-2-deoxyglucose and [^{13}N]-ammonia are used to make *in vivo* measurements of the cerebral metabolism and myocardial blood flow. Feliu published an excellent overview of this technique.[6]

Medical PET detectors normally employ one or more rings of small scintillation detectors. The NeuroECAT tomograph, for example, consists of eight detector banks arranged in an octagonal pattern. Each bank contains 11 BGO scintillation detectors. Individual detector elements in opposing banks form coincidence pairs. During a scan, the tomograph is rotated about an axis parallel to the subject and is linearly translated along the same axis. In this manner, photons emitted over 360° in the plane of the detectors can be recorded. Using tomographic reconstruction techniques, the data can be used to map the distribution of the positron emitter in a slice through the subject. Time is required for rotation and translation of the tomograph and to acquire sufficient coincident events for adequate measurement statistics. As a result, PET spectra generally require scan times in the order of 10 to 15 minutes, during which millions of coincidence events will be collected, thus enabling the production of a full 3D reconstruction of the imaged object. Nowadays, PET tomographs achieve a spatial resolution of 1 to 2 mm.[7]

Application of PET to problems of industrial interest has only occurred recently.[8] PET has been shown to be capable of monitoring turbulent two-phase (liquid/gas) flows, using injected solutions of aqueous $Na^{18}F$ as radiotracer.[9] Hoff *et al.*[10] used PET for monitoring water diffusion in porous construction materials. This appears to be especially useful as an alternative for NMR measurements in building materials that have significant iron content.

Many dynamic processes occurring in industrial equipment, for example, mixing processes, occur on time-scales that are too short for complete measurement of a 3D image. However, flow pattern measurements of such rapid processes can be measured if one restricts the measurements to the tracking of a single, radio-labelled, tracer particle. This technique, known as Positron Emission Particle Tracking (PEPT),[8,11] has been used to measure physical processes such as powder mixing in a ploughshare batch mixer,[12] and particle motion within a fluidized bed.[13] Both PET and PEPT have been used in a study of axial diffusion of particles in rotating drums.[14] Ongoing studies on this subject proved to be very successful; slip of the bed at the walls was observed and axial dispersion coefficients were determined.[15,16]

The velocity and hold-up distribution of the solid phase in a liquid-solid riser has been studied with radioactive particle tracking and computed tomography (CT).[17] The goal of this research was the development of an understanding of the variables affecting the performance of liquid-solid risers, and of fundamentally-based scale-up rules. An improved PEPT system has recently been developed, capable of continuously following the 3D trajectory of a radiotracer particle (as small as $500\,\mu m$) moving at $0.1\,ms^{-1}$ with a resolution of 5 mm. The system has been used to measure *in situ* flow patterns of solids in a gas-solids Interconnected Fluidised Bed reactor.[18]

Single photon emission computed tomography (SPECT) was used by Kantzas *et al.*[19] to monitor chemical processes. In SPECT imaging, a gamma camera physically orbits the object of interest, taking a series of static images at different angles. The two-dimensional images are assimilated and combined into three-dimensional images by reconstruction algorithms. They used this technique for particle tracking studies in fluidised beds and for radioactive tracer studies of the water displacement of heavy oil.

Jonkers and co-workers[20,21] conducted the first study in which PET was applied to chemical reactions in reactors. These pilot experiments were performed at the State University of Gent, Belgium, using a commercial PET-camera designed for medical imaging, the NeuroECAT tomograph described above. The objective was to show that PET could be used to obtain images of a reaction occurring within a tubular, plug-flow reactor, operating under normal process conditions and that data could provide information for subsequent modelling of the reaction kinetics.

Since the early 1990s, a facility has been developed at the Eindhoven University of Technology (TU/e), dedicated to positron emission imaging of physical and chemical processes in catalytic reactors at practical operating conditions. A positron emission detector has been developed, that is specifically tailored to the measurement of activity distributions as a function of time along a single, axial direction, as measurement of concentration profiles in a single dimension is sufficient under axially-dispersed plug flow conditions (since concentration gradients in the radial direction are negligible). This detector[22] is called a Positron Emission Profiling (PEP) detector to distinguish it from its 3D parent.

1.4. *The TU/e PEP detector*

The Positron Emission Profiling (PEP) detector is shown in Fig. 2. It has been designed to be flexible so that it can be used with a variety of different sizes of reactors; measurements can be carried out on reactors having lengths between 4.0 cm and 50 cm and diameters of up to 25 cm. The detector consists of two banks, each containing an array of sixteen independent detection elements, and is mounted horizontally, with the reactor and furnace placed between the upper and lower banks. Each detection element comprised of a bismuth germanium oxide (BGO) scintillation crystal coupled to a photomultiplier. The detection elements are situated in a frame, which allows adjustment of the overall detector dimensions when required.

The detection principle is as follows. Each detector element in a bank can form a detection pair with each of the elements in the opposite bank. Coincident detection of the two 511 keV gamma photons, formed during positron-electron annihilation by a detection pair, is used to locate the position of the event along the cylindrical axis of the reactor at the point at which the chord joining the two elements intersects this axis. Temporal information is obtained by collecting data during a fixed sampling period. A minimum integration time of 0.1 s is required to obtain sufficient coincident events for reliable measurements. When the detector block is in its "close-packed configuration", i.e., when all of the detector elements are placed tightly together, the spatial resolution of the detector is 2.9 mm.[23]

In order to reduce errors resulting from the detection of Compton scattered photons (which can be as high as 35%[23]), which lead to anomalous positioning of annihilation events, energy selection of the photons is

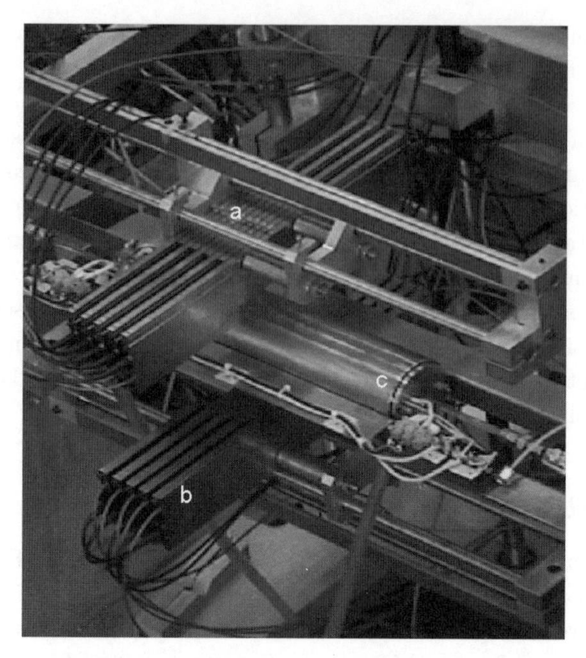

Fig. 2. Photo of the Positron Emission Profiling (PEP) detector showing the two banks of detector elements. Each element consists of a BGO crystal: (a) coupled to a photomultiplier; (b) In the middle, the reactor and furnace (c) are situated.

employed. However, the energy resolution of BGO scintillators is poor (only 28% FWHM), causing a large overlap of the measured energy of scattered and non-scattered gamma photons. Therefore, a dual energy window scatter correction technique[24,25] is applied. Gamma photons are now measured in an energy window around the photopeak energy (511 keV) and a second energy window in the Compton background. Adverse effects caused by scattering of the annihilation radiation can now be corrected for.

In nearly all detector systems, a minimum amount of time is required between consecutive events in order to be able to separate them. This amount of time is called the dead time, during which all other incoming events are lost. At high count rates, this loss of events becomes significant compared with the number of measured events, resulting in a non-linear response. At high count rate, the possibility also exists that two gamma-photons are detected as coincident that do not arise from the same annihilation event. These random coincidences give rise to incorrect position reconstructions. Since the number of random coincidences is proportional to the count rate, correction for their contribution is possible when the

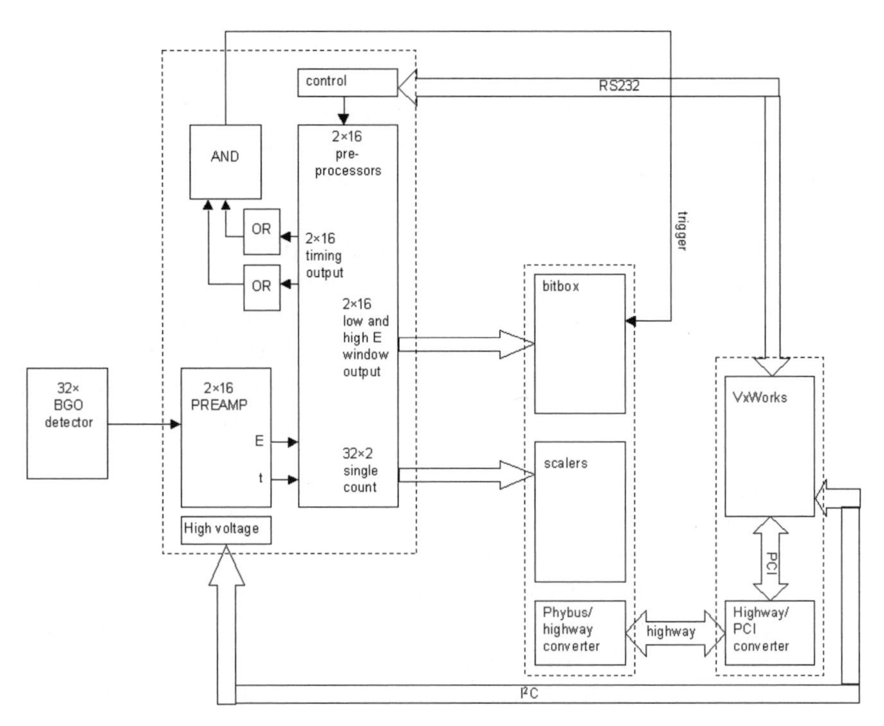

Fig. 3. Schematic overview of the data acquisition and control system of the new PEP detector set-up.

count rate of the individual detection elements is known. The detector provides the single count rates, both for the correction for dead time and for the chance coincidence.

Figure 3 presents an overview of the layout of all hardware components, and of the flow of data and control of the PEP detector. The heart of the PEP detector consists of two parallel banks of 16 gamma-ray detection elements. The charge signal of the photomultiplier tubes is converted into an energy signal and a timing signal by the preamplifiers (preamp). The energy signal is used for energy selection by the preprocessors. In the preprocessor, the energy of the detected gamma photon is determined, as well as whether this energy falls in the photopeak or Compton window. The timing signal is used by a coincidence circuit to determine whether a gamma photon was detected coincidentally in the upper and lower detection element. The coincidence circuit consists in two 16-fold OR-circuits (OR) for the two detector banks, whose two outputs are processed by an AND-circuit (AND).

The output of the coincidence unit is used to trigger the bitbox. This bitbox records the status of all preprocessors and reconstructs the annihilation position. By making a position histogram of a large number of these coincident events, a concentration profile is determined. This number of events is collected during each binning time, which can be programmed to be 0.1 s or longer.

Also, the number of events per binning time per detection element is recorded. Two scalers perform this task. These values are used to correct for dead time and chance coincidences. To measure the count rate in the photopeak window and the total count rate, the preprocessor gives two types of scaler pulses. The scalers count these pulses per binning time and the single count rate of the preprocessors can be determined. One scaler pulse is given for each event that was processed by the preprocessor, this pulse is used for dead time corrections. When the energy of the gamma-ray falls in the photopeak window, another scaler pulse is given, which is used for chance coincidence corrections.

The data acquisition is controlled by a server PC with the Vx/Works operating system, chosen because of its real time capabilities. The data that are collected by the Vx/Works system are sent to a user interface running on a separate PC. All functionality and control of the hardware through RS232, I^2C and PCI ports are implemented on the server.

2. Synthesis of Radio-labelled Molecules

The on-site preparation of proton rich radioisotopes is accomplished by irradiating an appropriate target material with energetic beams of protons or deuterons supplied by a cyclotron. During the irradiation process, the highly energetic particles impart large kinetic energies to the target molecules; these energies greatly exceed the bond dissociation energies. As a result, only very simple molecules survive as products. To produce more complex radio-labelled molecules containing positron emitters, post-irradiation chemical synthesis must be carried out to incorporate the smaller radio-labelled fragments into the larger framework. Strategies must be developed such that the desired molecule can be produced and separated from other reaction products within a few half-lives of the radioisotope. This normally means that one must use a precursor that is only one or perhaps two reaction steps from the desired target molecule.[6] This requirement has not prohibited the synthesis of radio-labelled analogues of complex molecules such as synthetic drugs and drug metabolites,[26] nor has the need for a dedicated cyclotron curtailed the use of this technique. In fact, it has led to the development of small, shielded, semi-automated cyclotrons, which are now commercially available.[6,27] A number of labelled molecules containing ^{11}C, ^{13}N, or ^{15}O, have been produced so far at the TU/e for use in PEP experiments of catalytic reactions,[28] among which are ^{11}C-labelled alkanes, ^{15}O-oxygen and ^{13}N-ammonia. The production and synthesis of ^{13}N-labelled ammonia is described below in further details. Naturally, speed and efficiency are necessary in the synthesis and purification processes, due to the short half-lives of the isotopes involved.

2.1. $^{13}NH_3$ production

The successful application of positron emission tomography in medical research caused an increase in the need of the production of $[^{13}N]$-NH_3. Nowadays, $[^{13}N]$-NH_3 is used as a tracer in animals and humans for heart or brain studies. A wide range of different production methods for $[^{13}N]$-NH_3 has been designed. When using a water target, the most applied nuclear reaction is $^{16}O(p,\alpha)^{13}N$. The indirect production methods like DeVarda's method[29–33] and $TiCl_3$ method[34] are still in use. In recent years, more research effort is focused on the direct ammonia production methods. The water target with reducing agents like H_2[35,36] and ethanol[34,36–38] has been intensively studied and these methods are in use nowadays. All these $[^{13}N]$-NH_3 production methods commonly lead to radioactive ammonia dissolved in water or a saline solution. These solutions can be directly used in positron emission tomography (PET). For the study of the gas phase ammonia oxidation reaction, ammonia dissolved in water is not a suitable end-product.

For the production of labelled and chemically pure gaseous ammonia concentrated in a small volume to be injected as a pulse into a gas stream of reactants, the following procedure has been developed.[39] Nitrogen-13 is produced by irradiating a water target with a proton beam with an energy of 16 MeV. During irradiation, the $^{16}O(p,\alpha)^{13}N$ nuclear reaction occurs. The ^{13}N-labelled species typically exist as nitrates (more than 85%) and nitrites[40] under the conditions in the aqueous target. These must be chemically reduced subsequently in order to produce ^{13}N-labelled NH_3. Production of $^{13}NH_3$ is performed by reducing the irradiated water (containing $^{13}NO_3$) over a mixture of DeVarda's alloy and NaOH. The reaction mixture is then flushed with a NH_3(3%)/He stream, to transport the liberated $^{13}NH_3$ and to prevent excessive loss of labelled ammonia on the walls of the gas tubing. Thus, labelled ammonia is not produced carrier free. The schematic setup is presented in Fig. 4. On a gas chromatograph, a 6-way heated pneumatic valve is installed with a sample coil of 5 ml

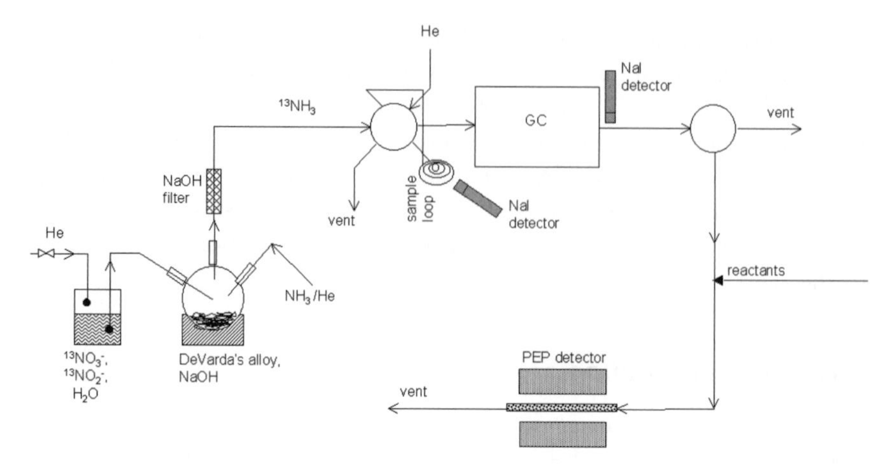

Fig. 4. Set-up for the production of $[^{13}N]$-NH_3.

(SS, 3×2 mm). A NaI detector continuously measures the passing activity through the sample loop. A GC run is remotely started at the moment that the sample loop is filled with $[^{13}N]$-NH_3. The GC is equipped with a Haysep P column (SS, mesh 80–100, O.D. 1/8). The GC run is started at 90°C, and after 1 minute, the temperature is increased with 10°C/min to 130°C. The products are analysed by a thermal conductivity detector (TCD). A NaI detector, directed at the TCD, monitors labelled products. Behind the TCD detector, a heated electrical 3-way valve selects a small part from the $[^{13}N]$-NH_3 peak. Depending on the experiment, a pulse time of 2–10 s is used to inject the labelled gaseous ammonia into the reactant stream. The required specific activity of the radioactive ammonia pulse is 0.1 MBq/ml minimum to meet the statistical requirements of the positron emission profiling experiments.

3. Positron Emission Profiling Study on the Catalytic Ammonia Oxidation

The PEP technique enables one to measure concentration profiles of labelled molecules (reactants and/or products) within reactors as a function of axial position and time. Once these profiles have been obtained, they provide substantial information on the reaction kinetics and involved intermediates. We will illustrate this with our research on the understanding of the mechanism of the ammonia oxidation on platinum based catalysts.

The study of the oxidation of ammonia to nitrogen and water at low temperature has become more and more important in recent years due to agricultural and industrial waste streams. Noble metals (Pt, Ir) are suitable for the selective, low temperature oxidation of ammonia to nitrogen and water. Alumina supported platinum catalysts are promising in the conversion of gaseous ammonia to N_2 and H_2O. They possess a high activity and selectivity in N_2 formation.[41] However, these catalysts are susceptible to rapid deactivation due to irreversible adsorption of reaction "intermediates" such as NH_x species.[41]

Gas phase ammonia oxidation on noble metal catalysts has been extensively studied with surface science techniques, like SIMS, AES, LEED, EELS, TPD and TPR.[42–45] The formation of NO, N_2 and water was observed and the selectivity towards nitrogen products was mainly dependent on the temperature and ammonia/oxygen ratio. Generally, it is believed that the surface is covered with reaction intermediates and adsorbed NO plays a key role in the product formation. Other reported species present on the surface are NH, O and OH. The reaction pathways change greatly with pressure. At ambient pressure and low temperature, the main products are N_2 and water with N_2O as a by-product. From approximately 573 K NO is formed.[46] IR experiments done by Matyshak[47] suggested that the platinum surface is mainly covered with nitrogen species. TPD, TPR and XPS studies showed that after the steady-state condition was reached, NH and OH species were the main intermediates present at the catalyst surface.[41] The role of $NO_{(ads)}$ is only attributed to the production of the by-product, N_2O. Ostermaier *et al.*[48,49] proposed that the initial deactivation of platinum supported on alumina is caused mainly by PtO formation. Still, the mechanism of selective ammonia oxidation and the fast initial deactivation of e.g., platinum catalysts are not well understood.

PEP experiments are performed using either $^{13}NH_3$ or $[^{15}O]$-O_2 to obtain further insight in the reaction mechanism and deactivation behaviour. Transient ammonia pulse experiments are performed to study the adsorption and dissociation of ammonia on pure platinum catalysts, followed by the focus on the deactivation of platinum. Finally, we will briefly discuss the influence of the alumina support.

3.1. *Activation of ammonia dissociation on Pt*

Evidence for the activation of the ammonia adsorption and dissociation promoted by oxygen on platinum is provided by PEP pulse experiments. The conversion and product formation due to [^{13}N]-NH_3 pulses on pre-oxidised platinum is determined as a function of temperature. The product selectivity is temperature dependent, and on a pre-oxidised platinum surface, part of the injected [^{13}N]-NH_3 remains. The nature of the nitrogen species, which remains after the ammonia pulse on pre-oxidised platinum, is further investigated with temperature programmed desorption (TPD) and temperature programmed oxidation (TPO), and by the removal of the adsorbed species by H_2, NH_3 and NO reaction experiments.

3.1.1. *Experimental details*

To exclude any influence of the catalyst carrier material, ammonia oxidation experiments are conducted on pure platinum sponge. The catalytic bed was 4 cm long and 4 mm in diameter, and contained 1.8 g of platinum sponge. The reactor effluent was analysed by a quadrupole mass spectrometer to determine the conversion and selectivity. Before each experiment, the platinum was reduced at 673 K for 2 hours, before the catalyst was flushed with helium for 20 minutes and cooled to the desired reaction temperature under helium flow. The platinum sponge is pre-oxidised by the following treatment: the reduced catalyst was pre-treated with 1 vol% O_2/He flow (48 cm^3/min) for 1 hr at 373 K, followed by flushing with He (48 cm^3/min) for 1 hr before the reaction temperature was set.

3.1.2. *Activation of ammonia adsorption by oxygen*

The reduced platinum catalyst was kept under a flow (48 cm^3/min) of 1.0 vol% ammonia/helium at 323 K. Figure 5(a) shows that an injected radiolabelled ammonia pulse travels through the platinum catalyst bed with a retention time of approximately 9 seconds, whereas the retention time of He is less than a second. At higher temperatures, the ammonia retention time decreases, giving a value of 5 s at 423 K. Judging from the constant broadness of the pulse, diffusion limitations of ammonia in the sponge material are considered negligible.[50] Figure 5(a) shows that the ammonia adsorption equilibrium is fast and indicates a weak molecular adsorption of ammonia.

In the case of injection of [^{13}N]-NH_3 into an ammonia-free He flow over the reduced platinum sponge [Fig. 5(b)], the injected ammonia also travels through the catalyst bed. Gas-phase analysis showed that only NH_3 is detected at the reactor outlet, indicating that ammonia does not dissociate under these conditions.[44,45,51] The PEP image shows the weak binding of ammonia on platinum. A relatively small amount of the labelled ammonia

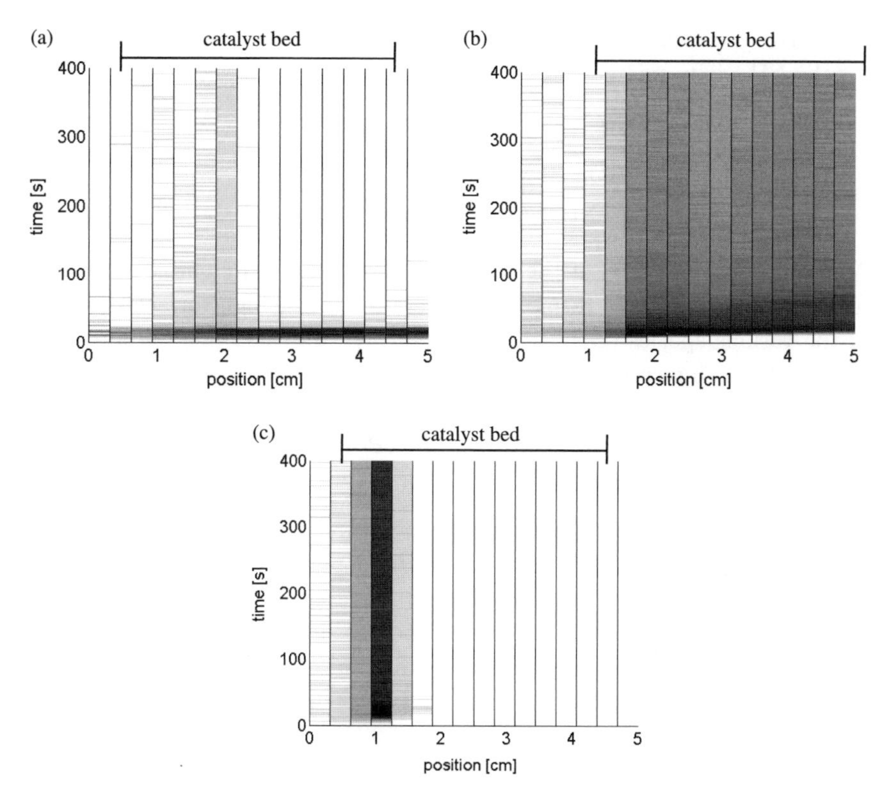

Fig. 5. PEP image of a pulse injection of [^{13}N]-NH_3 at 323 K: (a) in a 1 vol% NH_3/He flow (48 cm^3/min) on Pt sponge; (b) in He flow (48 cm^3/min) on Pt sponge; (c) in He flow (48 cm^3/min) on the pre-oxidised Pt sponge. The colour intensity represents the concentration of $^{13}NH_3$ (dark = high concentration).

species (15%) remains adsorbed at the platinum surface. This adsorption is most probably caused by incomplete reduction of the platinum surface. This is also observed when a pulse of [^{13}N]-NH_3 was injected into a hydrogen flow during the reduction of pre-oxidised platinum sponge at 373 K, suggesting that a residual presence of oxygen enhances the adsorption of ammonia.

In contrast to Fig. 5(b), the adsorption of ammonia on the pre-oxidised surface [Fig. 5(c)] is strong. The ^{13}N species are already adsorbed at the beginning of the catalyst bed and remained there during the experiment. This indicates that the presence of oxygen leads to enhanced dissociation of ammonia on Pt. This agrees with recent calculations performed by Fahmi and van Santen,[52] who calculated that adsorbed atomic oxygen activates the N-H bond cleavage. Accordingly, we propose the following mechanism for the NH_3 conversion:

$$NH_{3(a)} + O_{(a)} \rightarrow NH_{2(a)} + OH_{(a)} \quad (1)$$

$$NH_{2(a)} + O_{(a)} \rightarrow NH_{(a)} + OH_{(a)} \quad (2)$$

$$NH + O_{(a)} \rightarrow N_{(a)} + OH_{(a)} \quad (3)$$

As we did not detect N_2 at the relatively low temperature of 323 K, atomic nitrogen is probably not formed. However, due to the high surface coverage of the various species (OH, NH_2, NH, NO, N), diffusion may be impaired thereby hindering reconstruction to N_2.

3.1.3. *Influence of temperature on the ammonia dissociation on pre-oxidised Pt*

To investigate the influence of the temperature on the ammonia dissociation, we pulsed [^{13}N]-NH_3 over pre-oxidised platinum in an oxygen flow (1 vol% oxygen/helium) in the temperature range of 323–573 K. Figure 6 shows the PEP images, including the integrated ^{13}N concentration in the reactor, at different temperatures as a function of time. The PEP images of the radio-labelled ammonia pulses on pre-oxidised platinum change significantly with increasing temperature. At 323 K, almost no gaseous products are formed and the total concentration of ^{13}N remains at the catalyst [Figs. 6(a) and 6(d)]. Below 423 K, the PEP images are very similar and the ^{13}N concentration is observed at the beginning of the catalyst bed and the total radioactivity profile [Fig. 6(d)] shows a sharp decrease, as gaseous products are formed. The decrease of the ^{13}N concentration at 423 K is due to the formation of nitrogen and some nitrous oxide. Figure 7(a) displays the MS analysis of the formed products. N_2 and N_2O evolve simultaneously as a response to the $^{14}NH_3/^{13}NH_3$ pulse. Water is detected with an obvious delay, comparing to N_2 and N_2O evolution. This stems from the stronger interaction of water molecules with the catalytic surface. As can be observed, when comparing Figs. 7(a) and 7(b), this adsorption of water is less pronounced at higher temperatures. At a temperature of 573 K, the PEP image changes substantially [Fig. 6(c)]. At the beginning of the catalyst bed, a reaction zone is still observed, but in contrast to the observation at 423 K, not all the products leave the catalyst bed directly. This change is explained by the formation of NO, which readsorbs at the platinum surface throughout the catalyst bed. This is not the case for N_2 and N_2O, which have a very low adsorption constant. Consistently, Fig. 6(d) shows

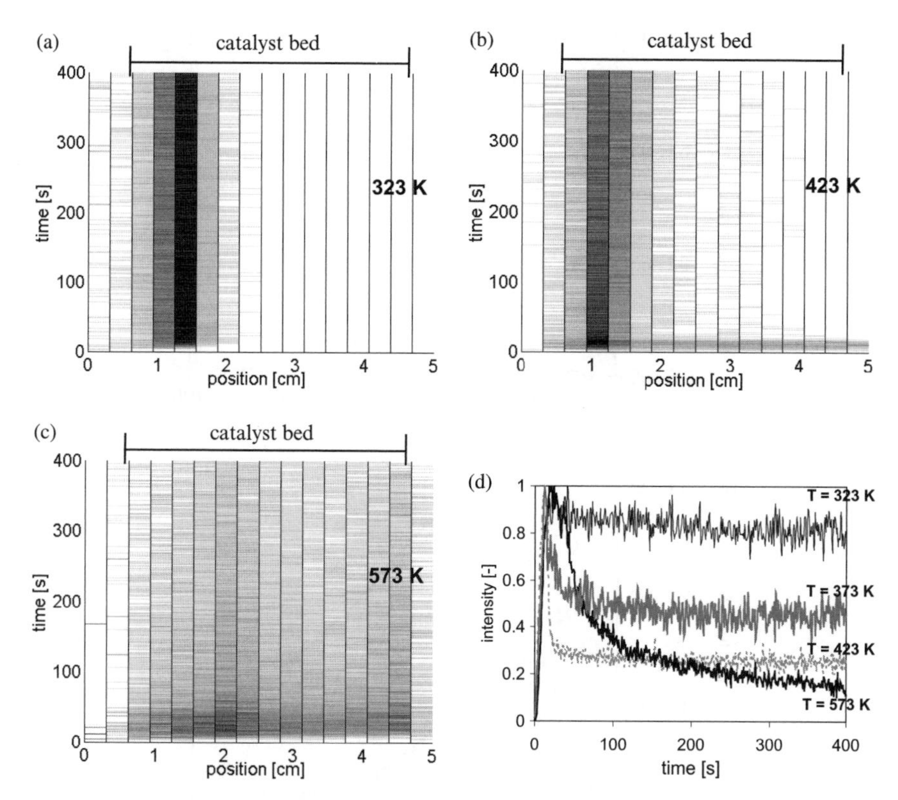

Fig. 6. PEP images of an pulse injection of [^{13}N]-NH_3 in 1.0 vol% O_2/He flow (48 cm^3/min) on pre-oxidised Pt sponge at: (a) 323 K; (b) 423 K; (c) 573 K; (d) integrated ^{13}N concentration as function of time for presented temperatures. The colour intensity represents the concentration of $^{13}NH_3$.

a relatively slow decrease of the ^{13}N concentration from the catalyst due to the NO readsorption. At this relatively high temperature, almost all injected ammonia reacts and forms gaseous products as shown in Fig. 6(d).

The above results indicate that a relatively high oxygen coverage on platinum promotes $NO_{(a)}$ formation. At low temperatures, $NO_{(a)}$ leads to N_2O via:

$$NO_{(a)} + N_{(a)} \rightarrow N_2O_{(g)} + 2^* \tag{4}$$

At temperatures higher than 573 K, the NO surface coverage is low due to desorption of NO, resulting in less N_2O formation. The dissociation of $NO_{(a)}$ at low temperature (<400 K) and also in the situation of a high adsorbates coverage is unfavourable.[45,53–55] The formation of nitrogen is

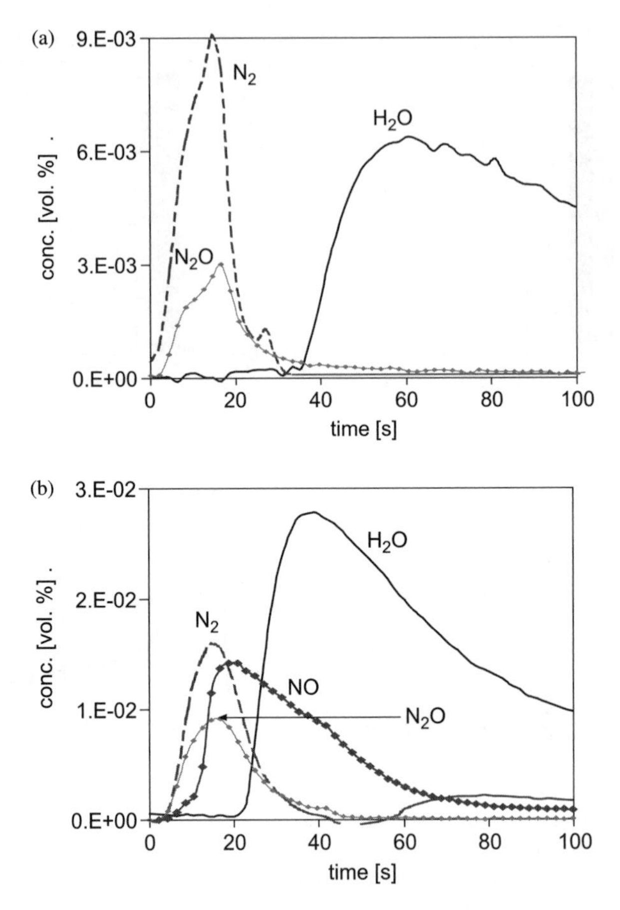

Fig. 7. Formation of N_2, N_2O, NO and H_2O measured by online mass spectrometry on a response of $[^{13}N]$-NH_3 injection in 1 vol% O_2/He flow (48 cm^3/min) on the pre-oxidised Pt sponge: (a) at 423; (b) at 573 K.

thus assumed to proceed via atomic nitrogen originating from NH_x species, rather than via the dissociation of $NO_{(a)}$.

3.1.4. *Determination of adsorbed species by temperature programmed desorption*

In this experiment, the temperature is raised from 373 to 573 K after injection of a $[^{13}N]$-NH_3 pulse on the catalyst (Fig. 8). The observed background of the MS signals of the products is relatively high since the catalyst could not be sufficiently flushed with He before starting a TPD experiment, due to

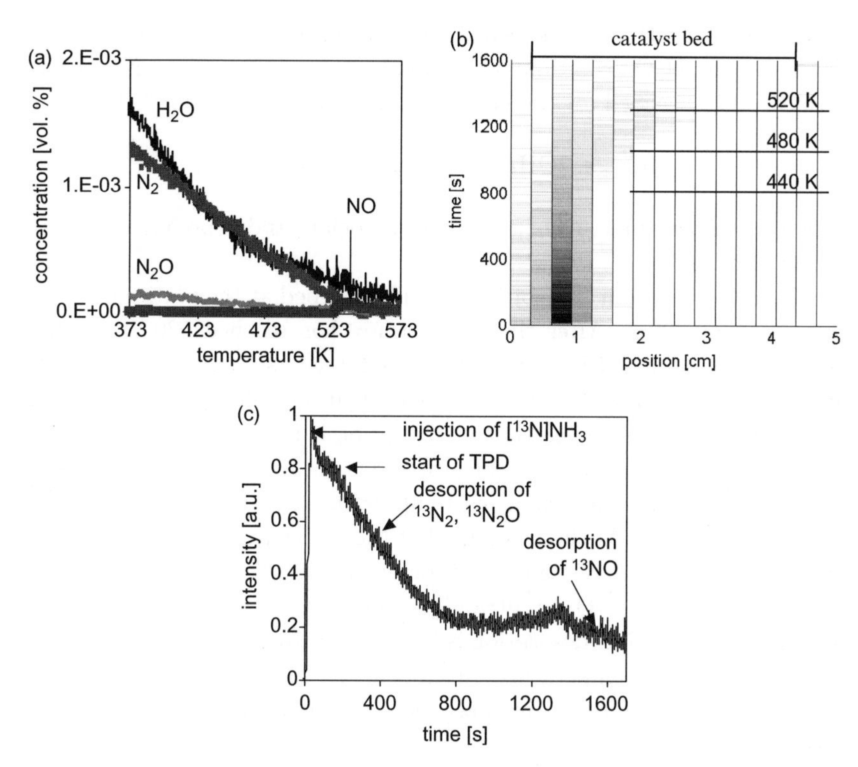

Fig. 8. A $[^{13}N]$-NH_3 pulse was adsorbed on the pre-oxidised platinum sponge kept under He flow ($40\,cm^3/min$) at 373 K followed after 170 s by a temperature programmed desorption experiment: (a) TPD spectrum; (b) PEP image, the colour intensity represents the concentration of $^{13}NH_3$; (c) normalised concentration as function of time (T = 373–573 K, 10 K/min, He flow of $40\,cm^3/min$).

a relatively fast ^{13}N decay. Desorption of the inert N_2 and N_2O molecules is qualitatively observed in the PEP images [Fig. 8(c)]. The signal of the integrated ^{13}N concentration decreases at the same time as N_2 and N_2O desorb. The desorption of N_2 and N_2O continues up to 480 K, because at that temperature, the total radioactivity profile [Fig. 8(c), from 800 s] shows no further decrease of the ^{13}N concentration. N_2 and N_2O have low adsorption constants,[56,57] thus surface reactions lead to their formation. In other words, N_2 and N_2O are not molecularly bound on the surface before the TPD experiment started. The formation of N_2O is assigned to reaction (4). Since radio-labelled ammonia was injected on a pre-oxidised surface, the oxygen coverage is relatively high and therefore the dissociation of NO on platinum is unfavourable. This leads to the assumption that $N_{(a)}$ originates from adsorbed NH_x species. Substantial formation of water accompanied by the N_2 and N_2O evolution also supports the presence of the NH_x on the surface. $NO_{(a)}$ needed for the N_2O formation originates from NH_x species

$$NH_{(a)} + 2OH_{(a)} \rightarrow NO_{(a)} + H_2O_{(a)} + H_{(a)} \quad (5)$$

At lower temperatures, $NO_{(a)}$ does not dissociate and is preferentially converted into N_2O.

Above 480 K, about 20% of ^{13}N is still adsorbed at the platinum surface (Fig. 8(c), at 800 s). After the temperature has reached 520 K, the total radioactivity profile [Fig. 8(c)] decreases for the second time. The PEP image [Fig. 8(b)] shows that the ^{13}N concentration is slowly moving through the catalyst bed, indicating that readsorption is taking place. The velocity of the ^{13}N moving through the catalyst bed is not linear because the desorption is enhanced by the increasing temperature, i.e., the ^{13}N desorption is accelerated in time. At the moment that the activity reaches the end of the catalyst bed ($T = 520$ K), only NO is detected with MS. The ^{13}N desorption is thus assigned to ^{13}NO. In ultra high vacuum experiments on single crystals, the energy for NO desorption is found to be around 150 kJ/mol[45,58,59] and NO desorbs at temperatures around 400 K. The PEP image shows that NO starts to desorb (moves to subsequent positions) around 440 K in these experiments.

3.1.5. *Determination of adsorbed species by temperature programmed oxidation*

In this experiment, a [^{13}N]-NH_3 pulse was adsorbed on pre-oxidised platinum sponge kept under He flow at 373 K. Then the temperature is raised from 373 to 573 K in 1 vol% O_2/He flow of 40 cm^3/min (Fig. 9). In the first minutes of this TPO experiment (up to 423 K), ^{13}N species desorb, which can be observed in Fig. 9(c) as the integrated ^{13}N concentration slowly decreases. This ^{13}N desorption is mainly assigned to nitrogen formation [Fig. 9(a)]. In contrast with the TPD experiment, only few N_2O are formed. Figure 9(c) shows that the nitrogen desorption stops at 400 s from the start of the experiment, which corresponds [Fig. 9(a)] to a temperature of 420 K. At that temperature, 30% of the injected [^{13}N]-NH_3 is still present at the surface. In the presence of gas phase oxygen, the major species desorbing at 540 K is NO. The actual desorption temperature of NO is 440 K, since at this temperature, the ^{13}N concentration starts to move in the catalyst bed. In comparison with the TPD experiment, NO desorbs at a similar

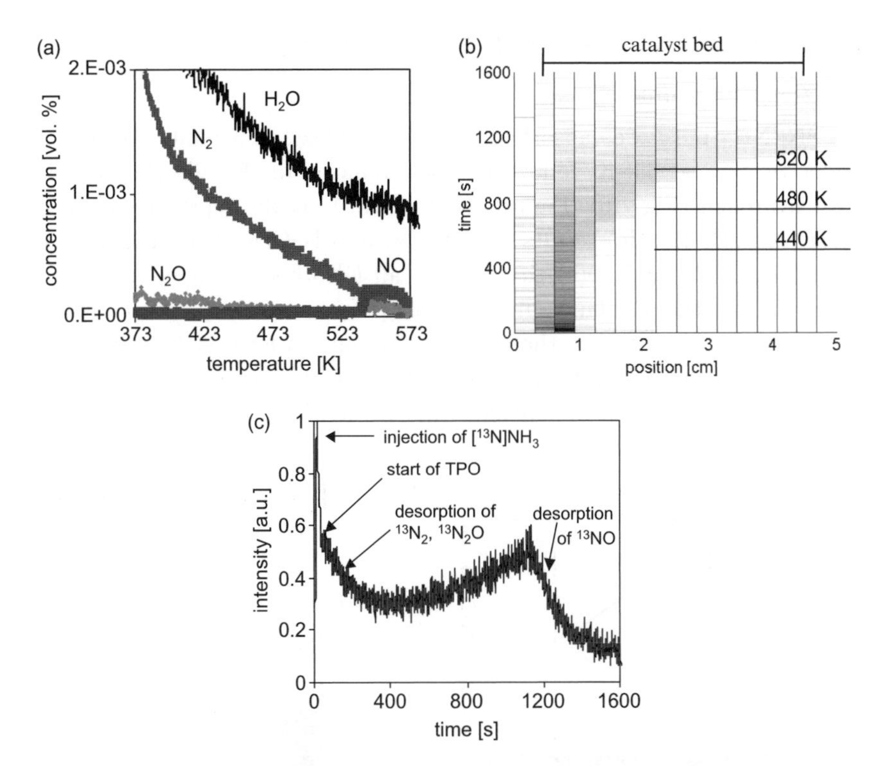

Fig. 9. A [^{13}N]-NH_3 pulse was adsorbed on the pre-oxidised platinum sponge kept under He flow (40 cm^3/min) at 373 K followed after 100 s by a temperature programmed oxidation experiment: (a) TPO spectrum; (b) PEP image, the colour intensity represents the concentration of $^{13}NH_3$; (c) normalised concentration as function of time (T = 373–573 K, 10 K/min, 1 vol% O_2/He flow of 40 cm^3/min).

temperature which is not surprising, because in both experiments, platinum was covered with oxygen. In correspondence with Fig. 8(b), Fig. 9(b) shows that equilibrium of ^{13}NO species is controlled by the readsorption on the platinum surface. As already mentioned, the NO transport is accelerated with increasing temperature. In this TPO experiment, all adsorbed ^{13}N species are removed.

Increasing the temperature to 400 K activates the bond cleavage of NH_x by atomic oxygen, according to reactions (2) and (3). Formation of nitrogen is assumed to proceed via the recombination of atomic nitrogen. At high oxygen surface coverage NO can be formed via

$$N_{(a)} + O_{(a)} \rightarrow NO_{(a)} +^* \tag{6}$$

$$NH_{(a)} + 2O_{(a)} \rightarrow NO_{(a)} + OH_{(a)} +^* \tag{7}$$

3.1.6. *Reaction of adsorbed N species with H_2, NH_3, and NO*

Reaction with hydrogen

A [^{13}N]-NH_3 pulse was adsorbed on a pre-oxidised platinum sponge kept under He flow at 348 K, followed by the removal of the adsorbed species with hydrogen (Fig. 10). Figure 10(a) shows that the replacement of He by H_2 (4 vol% H_2/He) results in N_2O, N_2 and H_2O formation. It is clear from Figs. 10(b) and 10(c) that almost all adsorbed nitrogen species are removed from the platinum surface in the presence of H_2. This means that the activation energies for the surface reactions to form N_2O and N_2 are relatively low, since all nitrogen species are already removed at 348 K. It is assumed that hydrogen first reacts with oxygen forming hydroxyl groups.

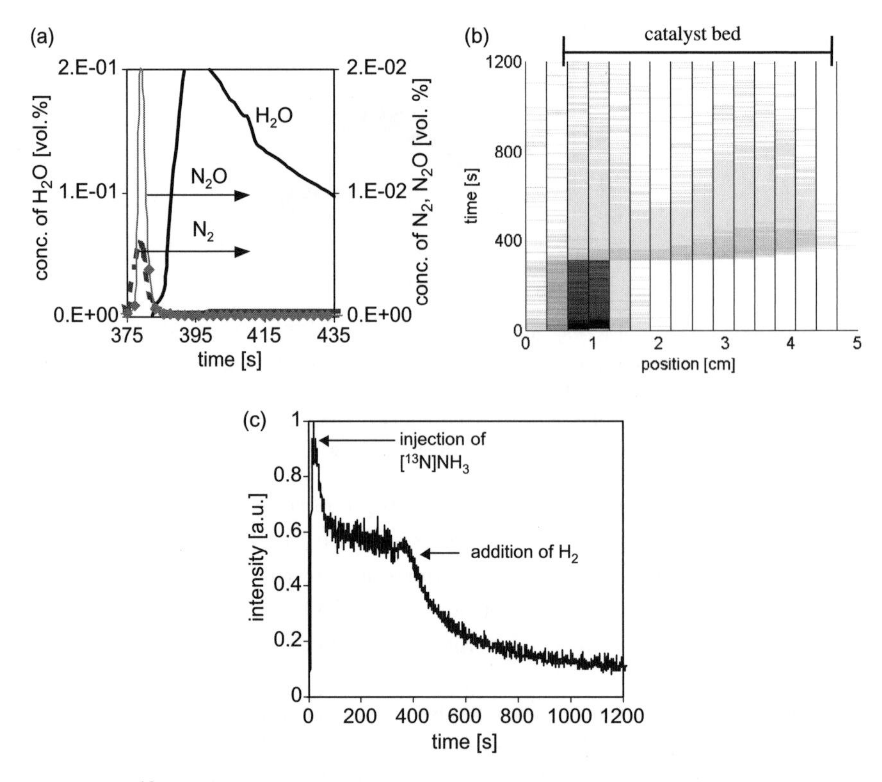

Fig. 10. A [^{13}N]-NH_3 pulse was adsorbed on a pre-oxidised platinum sponge kept under He flow (48 cm^3/min) followed after 380 s by a removal of N-species with hydrogen: (a) MS spectrum, shown from the moment of H_2 addition; (b) PEP image, the colour intensity represents the concentration of $^{13}NH_3$; (c) normalised concentration as function of time (T = 348 K, 10 vol% H_2/He flow of 48 cm^3/min).

OH groups do not only form water, but they also react with the adsorbed NH_x species to produce N_2

$$NH_{2(a)} + OH_{(a)} \rightarrow NH_{(a)} + H_2O_{(a)} \quad (8)$$

$$NH_{(a)} + OH_{(a)} \rightarrow N_{(a)} + H_2O_{(a)} \quad (9)$$

$$2N_{(a)} \rightarrow N_2 + 2^* \quad (10)$$

However, higher OH concentrations favour the formation of NO

$$NH_{(a)} + 2OH_{(a)} \rightarrow NO_{(a)} + H_2O_{(a)} + H_{(a)} \quad (11)$$

NO cannot desorb at this low temperature. Instead, it reacts with $N_{(a)}$ to form N_2O. NO can also be an intermediate leading to N_2 formation:

$$NH_{x(a)} + NO_{(a)} \rightarrow N_{2(a)} + OH_{x(a)} \quad (12)$$

However, this reaction can only occur at relatively low oxygen surface coverage, because the nitrogen/oxygen bond of NO needs to be broken. For that reason, this route to nitrogen formation can be activated after the initial desorption of N_2O and H_2O from the surface. Eventually, platinum is reduced and the oxygen surface coverage is lowered.

Reaction with ammonia

In this experiment, a $[^{13}N]$-NH_3 pulse was adsorbed on the pre-oxidised platinum sponge kept under He flow at 348 K, followed by a removal of the adsorbed species with ammonia (Fig. 11). Upon addition of ammonia (1 vol%) to the He flow, N_2, N_2O and H_2O are observed at the reactor outlet [Fig. 11(a)]. Figures 11(b) and 11(c) show that all ^{13}N species are instantaneously removed from the surface. Similar to the reaction with hydrogen, OH groups are produced, followed by desorption of water. These hydroxyl groups also react with $^{13}NH_x$ species to form adsorbed NO and eventually dinitrogen. The amount of N_2O formed is much lower than the amount of N_2, although it agrees well with the amount formed in the reaction with hydrogen. In contrast to nitrogen and water, N_2O is only formed in the first seconds after the replacement of He by the ammonia flow. This suggests that in the beginning of the ammonia addition, $NO_{(a)}$ is formed, which is similar to the observations with the H_2 flush experiment. Obviously, nitrogen and water are not only produced from the adsorbed ^{13}N species, but mainly originate from gas phase ammonia. The addition of ammonia results in a high concentration of the nitrogen containing species on the platinum surface. The high concentration of ammonia favours selective N_2 formation,

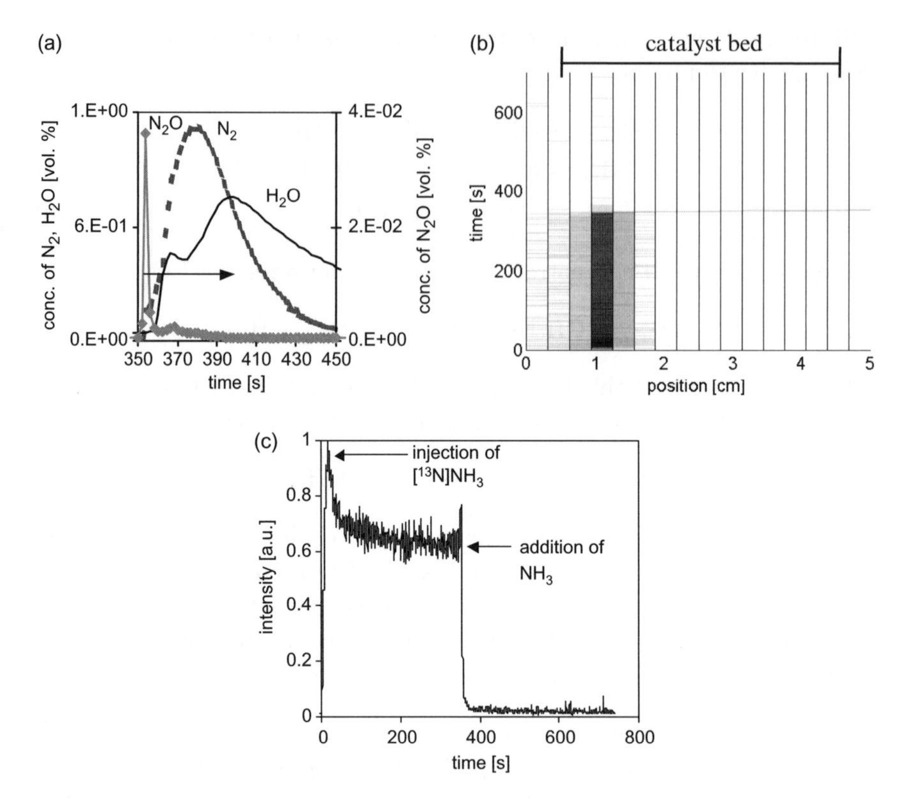

Fig. 11. A [^{13}N]-NH_3 pulse was adsorbed on the pre-oxidised platinum sponge kept under He flow (48 cm^3/min) followed after 350 s by a removal of N-species with ammonia: (a) MS spectrum, shown from the moment of NH_3 addition; (b) PEP image, the colour intensity represents the concentration of $^{13}NH_3$; (c) normalised concentration as function of time (T = 348 K, 1.0 vol% NH_3/He flow of 48 cm^3/min).

which is also observed in UHV experiments.[43,45] The production of N_2 and H_2O decreases in time because less oxygen is available for the reaction. This results in an oxygen-free surface, which is confirmed by measuring an additional PEP image after a [^{13}N]-NH_3 pulse in this ammonia flow (not shown). This image was identical to Fig. 5(a), showing the equilibrium of ammonia on platinum.

Reaction with NO

The role of NO as an reaction intermediate is further investigated in an experiment, in which a [^{13}N]-NH_3 pulse was adsorbed on pre-oxidised platinum sponge kept under He flow at 323 K, followed by the removal of the adsorbed species with nitric oxide (Fig. 12). Figures 12(b) and 12(c) show

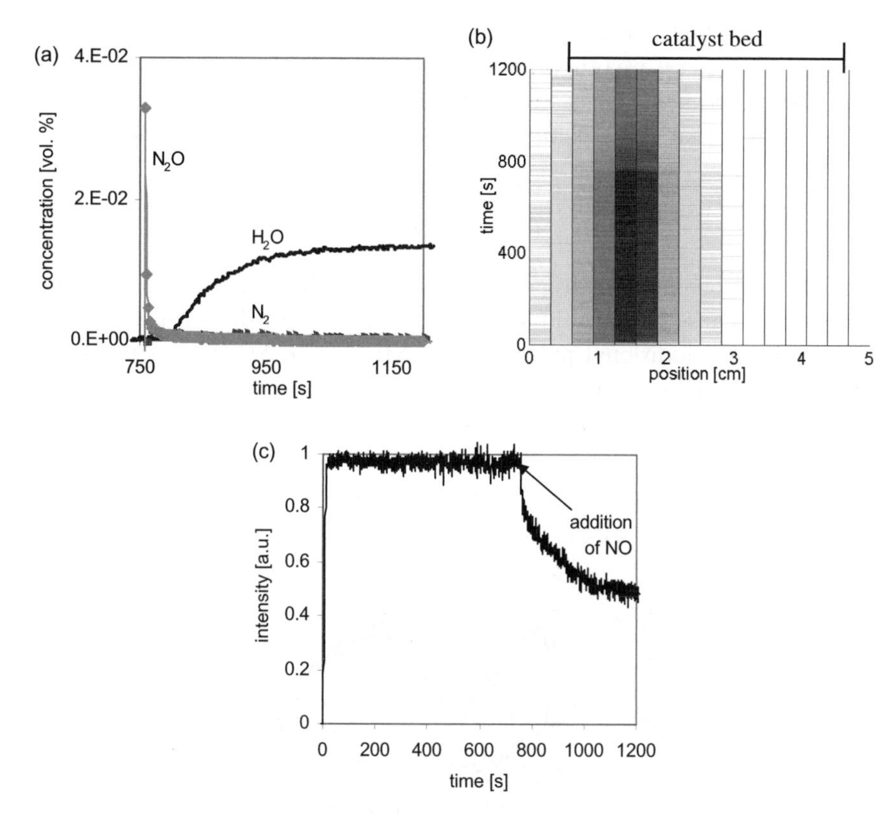

Fig. 12. A [^{13}N]-NH_3 pulse was adsorbed on the pre-oxidised platinum sponge kept under He flow (48 cm^3/min) followed after 750 s by a removal of N-species with nitric oxide: (a) MS spectrum, shown from the moment of NO addition; (b) PEP image, the colour intensity represents the concentration of $^{13}NH_3$; (c) normalised concentration as function of time (T = 323 K, 0.5 vol% NO/He flow of 48 cm^3/min).

that upon addition of an NO flow (0.5 vol% NO/He), about 50% of the ^{13}N species are instantaneously removed from the platinum surface. This partial conversion has two possible explanations. Firstly, due to inaccessibility or a high surface coverage, NO does not react with all of the adsorbed ^{13}N species. This explanation is in contrast with the experiments with hydrogen and ammonia, where all ^{13}N species were removed. Thus, the effect of the high surface coverage most probably can be excluded, even though NO preferably adsorbs on the same sites as oxygen (hollow sites), while ammonia adsorbs on the atop sites. The second explanation is that NO selectively reacts with one of the adsorbed ^{13}N species. Only N_2O is formed as a result of the NO flush [Fig. 12(a)]. High oxygen surface coverage, together with high partial pressure of NO, favours the formation of N_2O above N_2. Temperature programmed experiments, and ammonia and hydrogen flush experiments suggested that the adsorbed ^{13}N species are $^{13}NH_x$. It is plausible that NO selectively reacts with one of these species.

$$NH_{x(a)} + NO_{(a)} \rightarrow N_2O_{(a)} + xH_{(a)} \tag{13}$$

$$H_{(a)} + O_{(a)} \rightarrow OH_{(a)} \tag{14}$$

The abstraction of one hydrogen from $NH_{(a)}$ by NO giving $N_2O_{(a)}$ is more probable, because NH_2 would lead to H_2, which is not detected with MS.

Injecting an additional pulse of $[^{13}N]$-NH_3 in the NO flow at the same reaction conditions, 40 minutes after the first pulse (Fig. 13), it showed that on the platinum surface covered with $O_{(a)}$ and $NO_{(a)}$, the adsorption

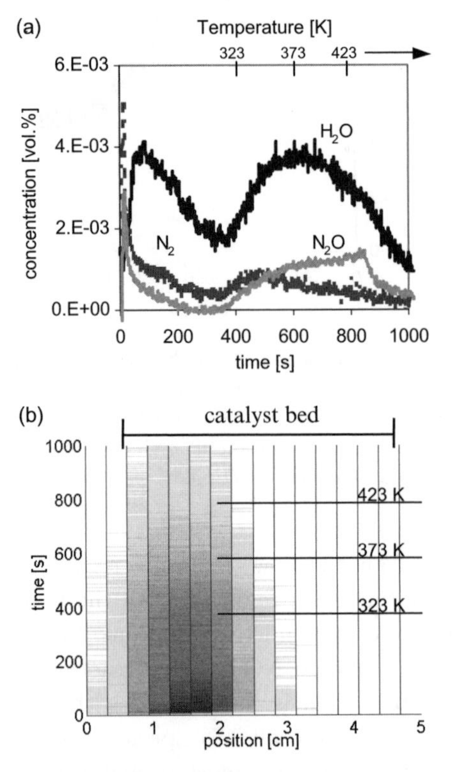

Fig. 13. A $[^{13}N]$-NH_3 pulse was injected on the pre-oxidised platinum sponge kept under NO/He flow at 323 K. After 400 s temperature was increased with 15 K/min: (a) MS spectrum; (b) PEP image, the colour intensity represents the concentration of $^{13}NH_3$ (0.5 vol% NO/He flow of 48 cm^3/min).

of ammonia is not prohibited. [^{13}N]-NH_3 directly reacts at the beginning of the catalyst bed, but in contrast to the first pulse (Fig. 12), where 80% of the injected ^{13}N is now retained at the surface. The second difference is the formation of N_2 in addition to N_2O. This [^{13}N]-NH_3 pulse also leads to water formation, and thus to the reduction of platinum. Since the oxygen surface coverage is lowered, dinitrogen formation becomes more favourable, which was also observed in the hydrogen and ammonia reaction experiments. By increasing the temperature, the remaining ^{13}N species desorb from the surface as N_2 and N_2O [Fig. 13(a)]. Desorption of nitrogen and nitrous oxide is again accompanied by water formation/desorption, indicating that NH_x species are mainly present at the surface. Both desorption peaks cease, what actually means that NO at this temperature does not decompose at the surface. The ^{13}N species are already removed from the surface at 573 K, thus ^{13}NO does not remain at the surface. These NO flush experiments suggest that ^{13}NO was not formed in large amounts at the surface and that the [^{13}N]-NH_3 pulse in the ^{14}NO flow at the pre-oxidised platinum surface leads to $^{13}NH_x$. Thus, the formation of N_2O and N_2 is due to the reaction of $NH_{x(a)}$ with $^{14}NO_{(a)}$, $O_{(a)}$ and $OH_{(a)}$.

3.1.7. *Conclusions*

The PEP results demonstrate that ammonia readily dissociates in the presence of co-adsorbed oxygen. At low temperatures, ammonia does not dissociate on the pure platinum sponge and the adsorption of ammonia is weak. Obviously, the presence of atomic oxygen decreases the activation barrier for the ammonia dissociation. Mechanistically, we envisage that oxygen atoms readily abstract hydrogen from co-adsorbed ammonia. Our results also suggest that oxygen and ammonia occupy different adsorption sites. Schematically

$$\text{Adsorption of NH}_3 \text{ on Pt}_{\text{reduced}} \rightarrow \text{weak molecular adsorption}$$
$$\text{Adsorption of NH}_3 \text{ on Pt-O} \rightarrow \text{strong dissociative adsorption}$$

The radio-labelled PEP experiments have shown that the pre-adsorbed oxygen favours the dissociation of ammonia, which leads to production of N_2, N_2O and NO. The product selectivity strongly depends on the temperature. Below 423 K, mainly nitrogen and nitrous oxide were formed and above this temperature, NO was formed. The PEP experiments indicate that all ammonia reacts at the beginning of the catalyst bed, and that [^{13}N]-NH_3 is partly converted into gaseous products and partly remains adsorbed at the surface in some dissociated form. Temperature programmed experiments indicate that the remaining adsorbed species at the surface are mainly NH_x species, because the formation of nitrogen and nitrous oxide is accompanied by the production of a large amount of water. The NO reaction experiment provides further indications that NH_x species remain at the surface. Thus, upon adsorption of ammonia, hydrogen atoms are stripped by adsorbed oxygen species. These exothermic reactions consecutively lead to the formation of atomic nitrogen and OH groups. Fahmi and van Santen[52] showed that NH bond scissioning is more costly than that of NH_2 and that the recombination of atomic nitrogen is not the rate-determining step. Our experiments are performed at relatively low temperature. The relatively high-surface coverages also contribute to the blocking of certain surface reaction pathways on the surface.

Part of [^{13}N]-NH_3 reacts towards N_2, N_2O and NO. A simplified model of the reaction mechanism includes formation of atomic nitrogen, which leads to nitrogen and nitrous oxide (below 423 K). OH groups form water and also promote the formation of $NO_{(a)}$ via the reaction with NH_x. At high oxygen surface coverage and below 350 K, the formation of nitrogen from NO dissociation is not favourable.[45] Therefore, it is concluded that atomic nitrogen is formed from the abstraction of hydrogen from the NH_x species. The NO formed, cannot desorb at these low temperatures and N_2O is selectively formed at these conditions. A known route for the formation of N_2O is the reaction between $N_{(a)}$ and $NO_{(a)}$. The reaction experiment with NO indicated another possibility; NO selectively reacts with $NH_{(a)}$ to form N_2O, even at 323 K. The formation of nitrogen becomes more favourable at lower oxygen surface coverage, which is shown in the hydrogen and ammonia reaction experiments. Above 440 K the TPD and TPO experiments show that NO desorbs from a pre-oxidised platinum surface. As NO desorption becomes possible at higher temperatures, the selectivity towards N_2O decreases in favour of the NO production. An interesting feature of the NO desorption from platinum is the readsorption which was distinctively observed in the PEP images.

3.2. *Low temperature ammonia oxidation on Pt*

The temperature dependence of the conversion and product formation in the ammonia oxidation over a platinum sponge catalyst is investigated. Positron emission profiling experiments demonstrate that below 413 K, the catalyst deactivates due to the poisoning of the catalyst surface, mainly by nitrogen species. The nature of the deactivating species is further investigated with temperature programmed desorption, oxidation and reaction. Experiments with pre-oxidised platinum sponge provide complementary information on the mechanism of the ammonia oxidation. A reaction mechanism of the ammonia oxidation on platinum at low temperatures is presented.

3.2.1. *Experimental details*

Again, a pure platinum sponge catalyst is used for the ammonia oxidation experiments to exclude any influence of the catalyst carrier material. The catalytic bed was 4 cm long and 4 mm in diameter, and contained 1.8 g of platinum sponge. The reactor effluent was analysed by a quadrupole mass spectrometer to determine the conversion and selectivity. Before each experiment, the platinum was reduced at 673 K for 2 hours. Then, the catalyst was flushed with helium for 20 minutes and cooled to the desired reaction temperature under helium flow. Pre-oxidised platinum sponge is obtained by the following treatment: the reduced catalyst was pre-treated with 1 vol% O_2/He flow (48 cm^3/min) for 1 hr at 373 K followed by flushing with He (48 cm^3/min) for 1 hr before the reaction temperature was set. For the ammonia oxidation, a reaction mixture was used consisting of 2.0 vol% of ammonia and 1.5 vol% of oxygen in helium with a total flow of 46.5 cm^3/min.

3.2.2. *Ammonia oxidation at 323–473 K*

Conversion and selectivity

Figure 14 shows the concentration of the formed products (N_2, N_2O and H_2O) in the ammonia oxidation at 323 and 373 K. NO formation was not observed. The detection of the water signal by the mass spectrometer is delayed due to the readsorption of water on platinum. Two regions can be distinguished in Figs. 14(a) and 14(b), in which the selectivity changes with time. The start of the ammonia oxidation shows selective formation of nitrogen and water. The second region in the N-product selectivity begins when N_2O evolves. In Fig. 14(a), this is after 40 seconds; subsequently, the N_2 production decreases and N_2O selectivity sharply increases to 16% and decreases in time to 7%. At 373 K, the high activity of the catalyst lasts longer and N_2O formation is observed after about 10 minutes. In the temperature range 323–473 K, the observed concentration profiles for N_2 and N_2O are similar. Initially, only N_2 is formed and the duration of

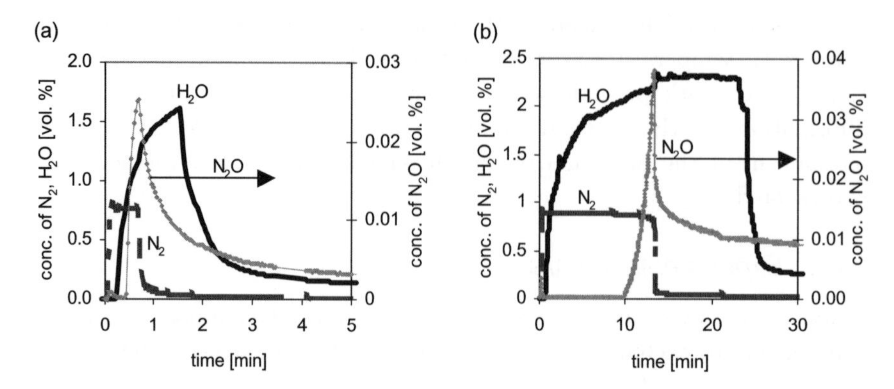

Fig. 14. Concentration of N_2, N_2O and H_2O versus time for the ammonia oxidation reaction at: (a) 323 K; (b) 373 K (GHSV $= 5600\,\mathrm{hr}^{-1}$, $NH_3/O_2 = 2/1.5$, flow $= 46.5\,\mathrm{cm}^3/\mathrm{min.}$).

this N_2 formation at a high conversion level is greatly extended at higher temperatures. It is thought that as the catalyst becomes covered with the reaction intermediates, N_2O evolves (low quantities) and the production of N_2 is strongly reduced. The time till deactivation changes with temperature, from 0.6 minutes at 323 K to 13 minutes at 373 K. At higher temperatures, the deactivation of the catalyst is further retarded. At 398 K, the catalyst is deactivated after 16.5 hours and at 423 K, not even after 24 hours time on-stream.

To investigate the influence of the temperature on the initial deactivation, a temperature programmed reaction experiment is performed (Fig. 15). After the catalyst is deactivated at 373 K, the activity starts to increase again at approximately 383 K when the temperature is raised. The N_2O formation shows an interesting development. First, the concentration of N_2O increases together with N_2 till 413 K, and then decreases very fast to a very low concentration. The reactivation behaviour of the catalyst is the reverse of the deactivation. Above 413 K, production of N_2 is favoured over N_2O. Finally, the catalytic activity of the platinum sponge catalyst is restored to its initial value and with similar high selectivity for N_2. Thus, the stable activity of the platinum catalyst above 413 K for a relatively long time suggests that adsorbed species, which deactivate the catalyst at lower temperature, are now not present at the surface.

Positron Emission Profiling

Figure 16(a) shows that upon injection of $[^{13}N]$-NH_3 in the reactant flow at the very beginning of the reaction, ^{13}N-labelled species are adsorbed in the

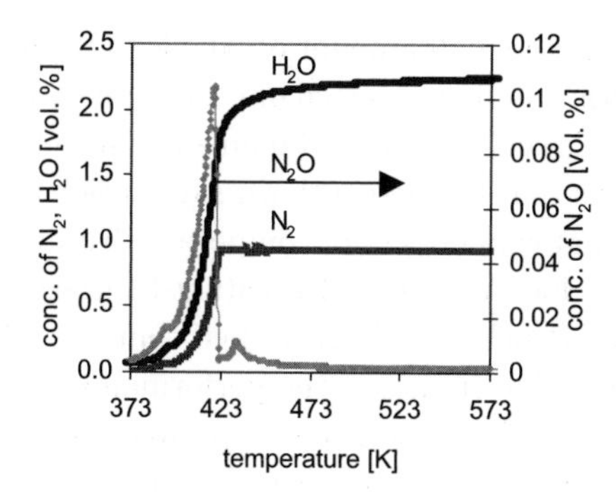

Fig. 15. Ammonia oxidation at 373 K until the catalyst is deactivated followed by a temperature programmed reaction. TP-reaction part is only shown (10 K/min, GHSV = 5000 hr^{-1}, $NH_3/O_2 = 2/1.5$, flow = 46.5 cm^3/min.).

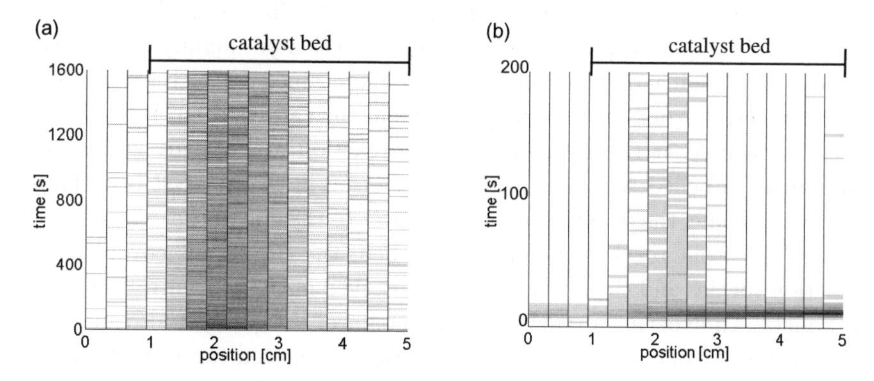

Fig. 16. PEP image of [^{13}N]-NH_3 and [^{15}O]-O_2 pulse injection into the reaction stream of NH_3/O_2/He in the first 2 s of the ammonia oxidation: (a) [^{13}N]-NH_3 PEP image; (b) [^{15}O]-O_2 PEP image, (T = 323 K, GHSV = 5600 hr^{-1}, $NH_3/O_2 = 2/1.5$, flow = 46.5 cm^3/min.). Colour intensity represents the concentration of [^{13}N]-NH_3 or [^{15}O]-O_2 (dark = high concentration).

front of the catalyst bed (catalyst bed starts at position 1.5 cm). A part of the [^{13}N]-NH_3 went through the total catalyst bed, observed as a thin line during the 10 seconds of the injection time, also at the last positions of the catalyst bed. Since ammonia is initially converted to N_2, the measured radioactivity is gaseous [^{13}N]-N_2 [Fig. 14(a)]. Figure 16(a) shows that part of the radio-labelled nitrogen species desorbs slowly from the catalyst surface, indicated by the slow intensity decrease. A substantial part of the labelled nitrogen species remains adsorbed at the catalyst surface and does not move through the catalyst bed during the measurement time of 30 minutes. Figure 16(b) shows that a labelled oxygen pulse passes through the catalyst bed, in contrast to Fig. 16(a), and only a small amount of labelled O-species is deposited at the surface. Oxygen is mainly converted into the gaseous products (water), as conversion of oxygen is very high at that moment. The retention time of water is short, which means that the desorption-readsorption equilibrium of water is fast. This means that water does not compete with ammonia for adsorption sites, and thus water does not poison the catalyst surface. This is in line with the results of van de Broek,[60] which showed that the addition of water to the reaction flow does not influence the performance of the catalyst. With increasing temperatures, less nitrogen is deposited on the platinum catalyst and the amount of oxygen deposition is rather stable up to 373 K. The PEP images do not differ too much up to 373 K. As already shown in Fig. 15, the catalyst is very active above 413 K and all ammonia and oxygen are converted to nitrogen and water.

The ammonia oxidation reaction proceeds in the first part of the catalyst bed [Fig. 16(a)]. This part is subsequently deactivated, mainly by nitrogen species. The high activity of the catalyst is maintained due to the movement of the reaction front to the next positions in the catalyst bed. When [^{13}N]-NH_3 is injected at the moment that the reaction was already 20 seconds on-stream, labelled N species adsorb further on in the catalyst bed. Thus, in time to come, the deactivation front moves to the end of the catalyst bed. When this front reaches the end of the bed, the catalyst is covered with reaction species and the deactivation is observed in the concentration of the products. An experiment with half an amount of the catalyst also supports this reaction front movement. This experiment showed the formation and concentration of the products in the same manner, however, the catalyst remained active for half the time of the normally applied catalyst bed. Thus, below 413 K, the catalyst remains initially active because the reaction zone moves to the next bed positions, after the previous positions became fully covered with the adsorbed reaction species. Injection of a [^{13}N]-NH_3 or [^{15}O]-O_2 pulse after the initial deactivation, confirmed that the platinum surface is fully covered and that conversion of ammonia and oxygen is low. No significant amount of nitrogen or oxygen species remains adsorbed at the catalyst surface.

3.2.3. *Characterisation of the adsorbed nitrogen and oxygen species*

Temperature programmed desorption

In the TPD spectrum, three products are observed: N_2, N_2O and H_2O. Figure 17 shows that N_2O desorbs first, already at 388 K. We presume that

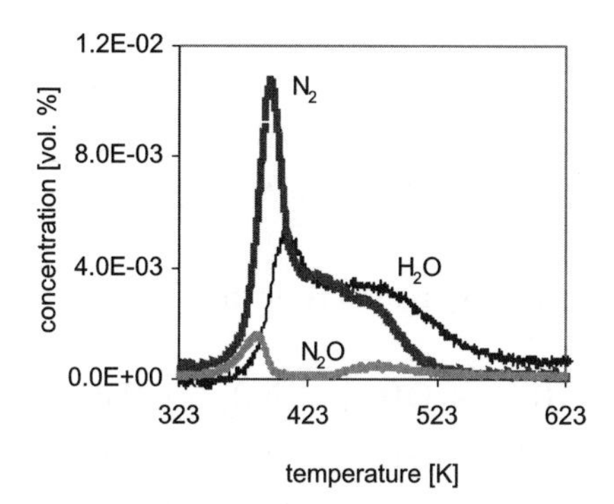

Fig. 17. Formation of N_2, N_2O, and H_2O measured by online mass spectrometry in a temperature programmed desorption experiment after ammonia oxidation at 323 K (10 K/min, He flow of 40 cm^3/min).

N_2 and N_2O are not molecularly bound on the surface at the start of the TPD experiment, because it is thermodynamically favourable of N_2 and N_2O to desorb from platinum.[56,57] The desorption of N_2O then indicates that the reaction of $NO_{(a)}$ with $N_{(a)}$ took place. The amount of produced N_2O is relatively low. Further, Fig. 17 shows that the peak of N_2 has two shoulders with peak maxima at 433 K and at 483 K. Again, water desorption is concomitant with nitrogen desorption. The mass spectrometer signal of water is somewhat delayed due to longer residence time of water on platinum caused by the readsorption of water, as will be shown below. This simultaneous desorption of nitrogen and water indicates that surface reactions of both oxygen and nitrogen containing species occur. The presence of atomic nitrogen on the surface is not likely, because recombination of atomic nitrogen would lead directly to N_2 without water formation. Absence of atomic nitrogen on the surface is also in agreement with XPS measurements.[61] At this relatively high surface coverage, some $O_{(a)}$ and OH is left on the surface, trapped in between NH_x species. For that reason, the production of water through the reaction of two hydroxyls is only partly responsible for the water formation

$$2OH_{(a)} \leftrightarrow H_2O + O_{(a)} + {}^* \qquad (15)$$

The atomic oxygen reacts further with NH_x abstracting hydrogen. Two possible routes for the OH involvement in the formation of N_2 are proposed, either from NH or NH_2. At low temperature, the reaction of NH with OH to form water is favoured with a reaction energy of -11 kcal/mol.[52] The reaction energy of the two step reaction of NH_2 (via NH) with

OH is $-26\,\text{kcal/mol}$.

$$NH_{(a)} + OH_{(a)} \rightarrow N_{(a)} + H_2O_{(a)} \tag{16}$$
$$2N_{(a)} \rightarrow N_2 + 2^* \tag{17}$$

At higher temperatures NH_2 reacts to form N_2

$$NH_{2(a)} + OH_{(a)} \rightarrow NH_{(a)} + H_2O_{(a)} \tag{18}$$
$$NH_{(a)} + OH_{(a)} \rightarrow N_{(a)} + H_2O_{(a)} \tag{19}$$
$$2N_{(a)} \rightarrow N_2 + 2^* \tag{20}$$

The third nitrogen peak with a maximum at 483 K most probably due to an $NO_{(a)}$ intermediate already formed at much lower temperatures. A relatively high surface coverage promotes the NO formation, but the desorption of NO is slow at these relatively low temperatures.[45,58,59]

Temperature programmed oxidation

The most important results from the TPO experiment are the evolution of one N_2 peak, together with one H_2O peak and additional NO evolution at higher temperature (Fig. 18). Nitrogen is formed at 383 K, followed by the water production with a peak maximum at 403 K. The production of N_2O is clearly much higher than in the TPD experiment. This can be explained by the reaction of NH with $O_{(a)}$ or $OH_{(a)}$, which gives atomic nitrogen and

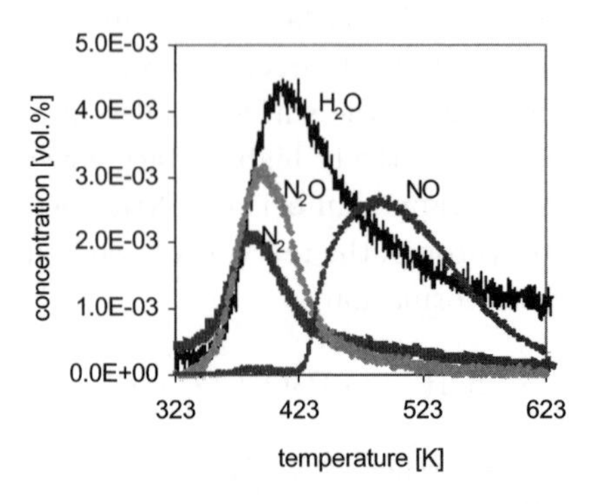

Fig. 18. Formation of N_2, N_2O, and H_2O measured by online mass spectrometry in a temperature programmed oxidation experiment after ammonia oxidation at 323 K (10 K/min, 1.0 vol% O_2/He flow of $40\,\text{cm}^3$/min).

water as that of the TPD experiment.

$$NH_{(a)} + OH_{(a)} \rightarrow N_{(a)} + H_2O_{(a)} \tag{21}$$

$$NH_{(a)} + O_{(a)} \rightarrow N_{(a)} + OH_{(a)} \tag{22}$$

$O_{2(g)}$ will adsorb and dissociate on vacant sites that becomes available. This will lead to production of $NO_{(a)}$

$$NH_{(a)} + 2O_{(a)} \rightarrow NO_{(a)} + OH_{(a)} + {}^{*} \tag{23}$$

$$N_{(a)} + O_{(a)} \rightarrow NO_{(a)} + {}^{*} \tag{24}$$

Thus, at the surface, there will be a competition between reactions towards N_2 or N_2O.

$$2N_{(a)} \rightarrow N_2 + 2^{*} \tag{25}$$

$$NO_{(a)} + N_{(a)} \rightarrow N_2O + 2^{*} \tag{26}$$

$$NO_{(a)} + NH_{(a)} \rightarrow N_2 + OH_{(a)} + {}^{*} \tag{27}$$

$$NO_{(a)} + NH_{(a)} \rightarrow N_2O + H_{(a)} + {}^{*} \tag{28}$$

The TPO spectrum shows that the maximum of the N_2 peak appears at lower temperature than the maximum of the N_2O peak. This indicates that the production of N_2O is dependent on the formation of $NO_{(a)}$ from $NH_{(a)}$ species or on the availability of free sites for oxygen adsorption and dissociation.

At higher temperature, an additional peak assigned to NO evolves with a maximum at 493 K. This suggests that at a certain moment, the sole species originating from NH_x on the platinum surface is $NO_{(a)}$. Apparently, oxygen strongly promotes the NO formation and blocks the N_2 and N_2O production.

Temperature programmed reaction with NO

Figure 19 shows the products of the reaction between the surface species, left after the ammonia oxidation, and NO from the gas phase. Nitrogen evolves at 398 K, followed by the water production with a peak maximum at 413 K. The N_2O production is much higher than in the TPD and TPO experiments. The maximum of the N_2O peak appears at 413 K, higher than the maximum of the N_2 peak. This was already observed in the TPO spectrum and indicates that the N_2O production depends on the formation of $N_{(a)}$ from $NH_{(a)}$ species, and the availability of free sites for NO adsorption. The increased N_2O production can be explained by the reaction of NH with

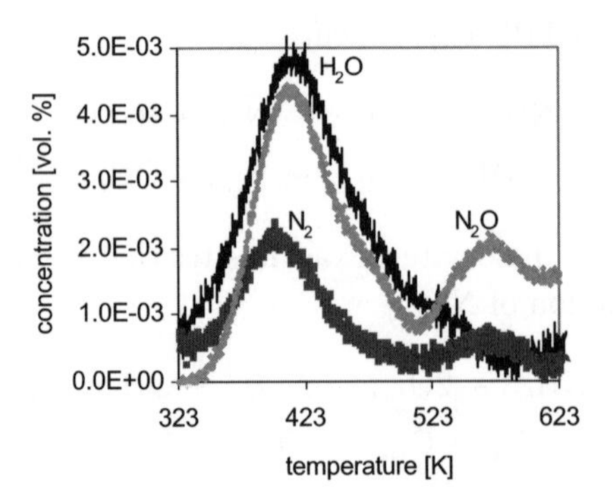

Fig. 19. Formation of N_2, N_2O, and H_2O measured by online mass spectrometry in temperature programmed NO experiment after ammonia oxidation at 323 K (10 K/min, 0.5 vol% NO/He flow of 40 cm^3/min).

$O_{(a)}$ or $OH_{(a)}$, which gave nitrogen and water in the TPD experiment. First, atomic nitrogen is produced

$$NH_{(a)} + OH_{(a)} \rightarrow N_{(a)} + H_2O_{(a)} \tag{29}$$

$$NH_{(a)} + O_{(a)} \rightarrow N_{(a)} + OH_{(a)} \tag{30}$$

N_2 starts to desorb and some vacant sites are created. $NO_{(g)}$ will adsorb on the available sites leading to the production of N_2O

$$N_{(a)} + NO_{(a)} \rightarrow N_2O_{(g)} + 2^* \tag{31}$$

Thus, as already stated, there is a competition between reactions towards N_2 or N_2O. The excess of $NO_{(a)}$ leads to a higher production of N_2O. The production of N_2O and N_2 seizes at 473 K. This indicates that NO dissociation does not take place. At 573 K, some $N_2O_{(g)}$ and $N_{2(g)}$ are formed again, which probably originate from the NH_2 species. The dissociation of the $NO_{(g)}$ starts above 623 K.

3.2.4. *Conversion and selectivity on pre-oxidised platinum*

Ammonia oxidation is conducted on a pre-oxidised platinum sponge catalyst. Figure 20 shows the conversion and selectivity at 373 K. The same selectivity characteristics as on the reduced platinum sponge catalyst are observed (Fig. 14). Thus, a high oxygen surface coverage does not favour initial nitrous oxide formation. The main difference with the reduced platinum sponge is the faster deactivation of the pre-oxidised catalyst below 413 K.

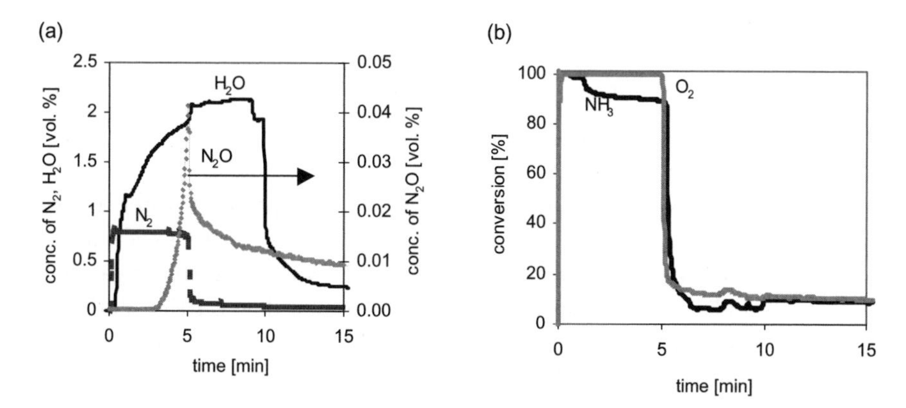

Fig. 20. Ammonia oxidation reaction performed at 373 K on a pre-oxidised platinum sponge catalyst: (a) concentration of N_2, N_2O and H_2O versus time for; (b) conversion of NH_3 and O_2 versus time (GHSV = $5600\,hr^{-1}$, $NH_3/O_2 = 2/1.5$, flow = $46.5\,cm^3/min.$).

However, above 413 K and also on the pre-oxidised catalyst, the high activity and selectivity towards nitrogen sustains. The presence of oxygen at the platinum surface apparently does not cause a permanent deactivation of the catalyst. Above 413 K, the catalyst is reduced by ammonia.

The deactivating effect of unreacted oxygen was confirmed by $[^{15}O]$-O_2 labelled PEP experiments. In these experiments, $[^{15}O]$-O_2 was first injected on reduced platinum sponge before the ammonia oxidation was started. In contrast to the $[^{15}O]$-O_2 pulse in the reaction flow stream, 10% of pulsed oxygen now remains at the catalyst surface. Thus, some oxygen species are formed at the surface, which are unreactive with respect to water or nitrous oxide. In a similar experiment, the adsorbed $[^{15}O]$-O_2 was subjected to a hydrogen flow which resulted in the formation of water, but a significant amount of (unreactive) oxygen also remained adsorbed at the surface. The formation of oxygen islands on platinum provides a possible explanation for this unreactive adsorbed oxygen phase.

3.2.5. *Desorption of water*

The PEP and TPD experiments indicated that mainly nitrogen containing species (NH_2, NH) cause the deactivation of the catalyst. The oxygen species, especially hydroxyls, are not causing the profound deactivation of the catalyst. This suggests that the production of water through the reaction of two hydroxyls is much faster than the endothermic reaction between

NH and OH

$$2OH_{(a)} \leftrightarrow H_2O_{(g)} + O_{(a)} + {}^* \quad \text{(fast)} \tag{32}$$

$$2NH_{(a)} + OH_{(a)} \rightarrow N_{2(g)} + H_2O_{(g)} + 3^* \quad \text{(slow)} \tag{33}$$

The formation of water via hydroxyls, and the desorption of water at low temperatures are investigated with PEP in $[^{15}O]$-O_2 pulse experiments injected in a hydrogen or ammonia flow. Figures 21(a) and 21(b) reveal that a $[^{15}O]$-O_2 pulse at 323 K in ammonia or hydrogen flow on the platinum sponge shows a distinct desorption profile, which is assigned to desorbing water. In this experiment, there were no nitrogen species adsorbed at the surface before oxygen dissociation took place, which is different comparing to the ammonia oxidation reaction.

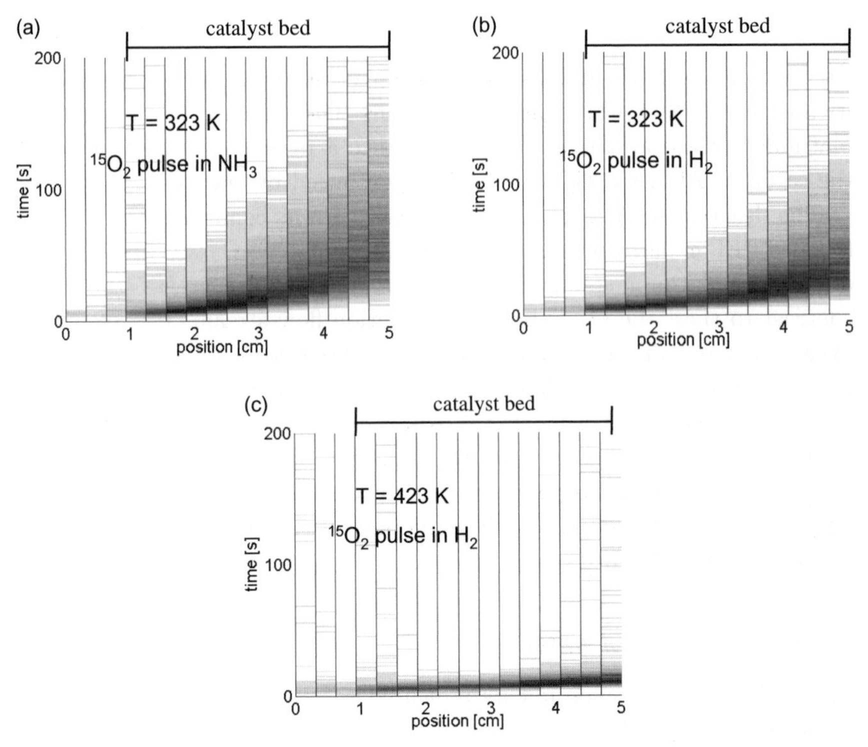

Fig. 21. PEP image of a pulse injection of $[^{15}O]$-O_2 on Pt sponge (flow $= 46.5\,cm^3/min.$): (a) in 1 vol% NH_3/He flow at 323 K; (b) in 4 vol% H_2/He flow at 323 K; (c) in 4 vol% H_2/He flow at 423 K.

As already noted, oxygen adsorbs dissociatively on platinum. The oxygen atoms react with ammonia or hydrogen towards water, the only oxygen-containing product. In both experiments, the water formation proceeds most probably via the reaction of hydroxyls. This indicates that water formation via hydroxyls is not the rate determining step for the ammonia oxidation at low temperatures. The desorbing water (low quantities) readsorbs on platinum and obviously with increasing temperature, water leaves the catalyst bed faster [Fig. 21(c)]. The recombination of OH also results in the formation of very reactive $O_{(a)}$.

3.2.6. *Reaction mechanism*

The dissociation of ammonia is greatly enhanced by the presence of atomic oxygen on the surface.[62]

$$O_{2(g)} + 2^* \rightarrow 2O_{(a)} \tag{34}$$

$$NH_{3(g)} + {}^* \leftrightarrow NH_{3(a)} \tag{35}$$

At the very beginning of the reaction, hydrogen is stripped from ammonia by dissociatively adsorbed oxygen forming NH_x and OH species. These reactions are exothermic, relatively fast and proceed at all used conditions.

$$NH_{3(a)} + O_{(a)} \rightarrow NH_{2(a)} + OH_{(a)} \tag{36}$$

$$NH_{2(a)} + O_{(a)} \rightarrow NH_{(a)} + OH_{(a)} \tag{37}$$

$$NH_{(a)} + O_{(a)} \rightarrow N_{(a)} + OH_{(a)} \tag{38}$$

Moreover, as shown above, the recombination of $OH_{(a)}$ towards water, and active $O_{(a)}$ takes place at all studied temperatures.

Temperatures below 398 K; active regime

The PEP and TPD experiments indicated that after the deactivation of the catalyst, the surface of platinum is fully covered with NH and NH_2 species. This means that the endothermic reactions between NH_x and OH are not proceeding very fast. The production of water through the reaction of two hydroxyls is much faster

$$2OH_{(a)} \leftrightarrow H_2O + O_{(a)} + {}^* \tag{39}$$

In this way, formed active oxygen reacts instantly with NH_x species. Atomic nitrogen is not observed in XPS measurements,[61] probably because the recombination of two $N_{(a)}$ forms N_2

$$2N_{(a)} \rightarrow N_2 + 2^* \tag{40}$$

The formation of atomic nitrogen proceeds via either $NH_{x(a)}$ or $NO_{(a)}$. The reaction of NH with atomic oxygen is exothermic and very fast. The reaction of NH_x with OH is less probable since for these reactions a higher energy barrier needs to be overcome. The reaction of NH with OH is also proposed by van den Broek[41] to be the rate determining step in the ammonia oxidation.

$$NH_{(a)} + O_{(a)} \rightarrow N_{(a)} + OH_{(a)} \tag{41}$$

$$NH_{(a)} + OH_{(a)} \rightarrow N_{(a)} + H_2O_{(a)} \tag{42}$$

The formation of nitrogen via an $NO_{(a)}$ intermediate is not favourable at low temperatures. At temperatures below 380 K, it has been reported on Pt(100) that the dissociation of NO is prohibited.[45,53–55] Both proposed reactions for the formation of $N_{(a)}$ via $NO_{(a)}$ require the dissociation of nitric oxide, and therefore the recombination of N ad-atoms is a more feasible option. Moreover, the formation of NO at these conditions is apparently not favourable, because the deactivated catalyst is mainly covered with NH_x species (NH and NH_2) instead of with NO. Thus, at low surface coverage, NO seems not to be the dominant species, probably due to the low oxygen surface coverage which disfavours the formation of NO.

Temperatures below 398 K; inactive regime

As the catalyst deactivates, the NH_x species are formed faster via ammonia adsorption than removed via nitrogen formation, the selectivity of the catalyst changes, and N_2O starts to be formed. First, $NO_{(a)}$ needs to be formed

$$NH_{(a)} + 2O_{(a)} \rightarrow NO_{(a)} + OH_{(a)} + {}^* \tag{43}$$

$$N_{(a)} + O_{(a)} \rightarrow NO_{(a)} + {}^* \tag{44}$$

$NO_{(a)}$ cannot desorb at these temperatures[44,45,51] and therefore reacts towards N_2O. N_2O can also be formed via the reaction of NO with NH_x species

$$N_{(a)} + NO_{(a)} \rightarrow N_2O_{(g)} + 2^* \tag{45}$$

$$NH_{x(a)} + NO_{(a)} \rightarrow N_2O_{(a)} + H_{x(a)} \tag{46}$$

The TPO experiment (Fig. 18) showed that NO desorbs from platinum from about 423 K, but only at high oxygen surface coverage. In Fig. 14, a drastic decrease of nitrogen and N_2O formation is observed, which can be explained in terms of the moving reaction front through the catalyst bed. As the reaction zone arrives at the last positions, N_2O cannot decompose anymore, since there is no fresh platinum surface left. As the last positions are deactivated, the catalyst's activity sharply decreases and the surface remains mainly covered with NH and NH_2. This is supported by XPS N(1s) measurement and indirectly by NO pulse experiments.[61]

A pre-oxidised catalyst deactivates much faster than reduced platinum sponge. Ammonia adsorption and dissociation are accelerated by the presence of oxygen. Thus, the NH_x species cover much faster the platinum surface. The concentration profiles for nitrogen and nitrous oxide do not change, which indicates that the reaction mechanism is not changed for the pre-oxidised catalyst.

Temperatures above 398 K

Deactivation of the catalyst is not observed at these temperatures and only N_2 and H_2O are formed. A different reaction mechanism depending on the surface coverage is suggested by the reactivation experiment (Fig. 15), in which N_2O formation rapidly decreases. This decrease cannot be explained only by the decomposition of N_2O at platinum; just low amounts of N_2O can be decomposed

$$N_2O_{(a)} \rightarrow N_2 + O_{(a)} \tag{47}$$

Above 398 K, the deposition of the NH_x species on platinum does not occur, as shown in the PEP experiments. Thus, the NH_x species are now much more reactive. The TPD, TPO and TP-NO experiments showed that above 423 K, NO is present at the surface leading to N_2, N_2O or NO, depending on the surface coverage. In conclusion, three reaction pathways are responsible for the nitrogen formation: via the exothermic reaction of NH_x species with $O_{(a)}$, endothermic reactions of NH_x with OH, or via the NO and NH_x reaction.

3.2.7. *Conclusion*

Below 388 K, two different product distributions in the ammonia oxidation are observed, depending on the surface coverage. At low coverage (catalyst is active), N_2 and H_2O are selectively produced via stripping of hydrogen from ammonia by oxygen atoms. Moreover, the recombination of $OH_{(a)}$

towards water and active $O_{(a)}$ takes place at all studied temperatures. With PEP, it observed that when the catalyst is still active, adsorbed nitrogen species are formed, which stay irreversibly adsorbed at the catalyst surface, and do not form any gaseous products. These nitrogen species mainly cause the poisoning of the platinum catalyst, thus retarding the adsorption of reactants. XPS[61] and temperature programmed experiments identified those as $NH_{2(a)}$ and $NH_{(a)}$. Oxygen species do not poison the catalyst, unless oxygen is pre-adsorbed on the platinum. The main reason for the deactivation of the platinum catalyst is the fact that the $NH_{x(a)} + OH_{(a)}$ reaction is much slower than the formation of water via hydroxyls. In this way, NH and NH_2 remain at the surface and block the active sites. Nitrogen formation proceeds via the recombination of atomic nitrogen, and not via an $NO_{(a)}$ intermediate. Low surface coverage does not favour $NO_{(a)}$ formation, and below 400 K, the dissociation of NO is prohibited. Moreover, addition of NO to the reaction flow causes faster deactivation of the platinum sponge. The pre-oxidised platinum sponge deactivates faster due to surface poisoning caused partly by the nitrogen species and partly by unreactive oxygen.

At high surface coverage, when the catalyst deactivates, the selectivity of the catalyst changes. Next to N_2, N_2O is also formed. The intermediate $NO_{(a)}$ seems to be mainly involved in the formation of very low amounts of N_2O. The initial deactivation of the platinum catalyst is obviously a case of self poisoning. The fact that the catalyst is regenerated after a reduction step with hydrogen to its initial activity supports this. Above 388 K, nitrogen and water are formed and the catalyst maintains its high initial activity.

3.3. *Ammonia adsorption on alumina*

Aluminas are commonly used as acid catalysts and as support material for catalysts. Here, we will investigate the influence of the alumina support on ammonia adsorption in the ammonia oxidation reaction over Pt/γ-alumina. Ammonia appears to be involved in a phenomenon called adsorption assisted desorption (AAD). Adsorption assisted desorption (AAD) has become well-known in catalysis.[63–68] We found that the rate of desorption of ammonia is increased by the partial ammonia pressure in the gas phase.

3.3.1. *Experimental details*

A Ketjen E-000 γ-alumina sample was used. The particle size of the samples was 250–425 μm. The surface area of 180 m^2/g and an average pore diameter of 110 Å were measured with the B.E.T. method. The length of the used catalyst bed was 4 cm, and the catalyst volume equal to 0.5 ml. Before the experiments, the pure γ-alumina was pre-treated with a 10 vol% hydrogen/helium flow of 40 cm^3/min (STP) at 673 K overnight.

3.3.2. *[^{13}N]-NH_3 equilibrium pulse experiments*

Figure 22 shows a typical PEP-image for a pulse of [^{13}N]-NH_3 on an ammonia pre-saturated γ-alumina in an ammonia/helium flow. This figure shows that in time, the labelled ammonia moves to positions further in the catalyst bed. After 800 seconds, there is no adsorbed $^{13}NH_3$ left on the alumina, suggesting that only reversible adsorption/desorption took place. Figure 22 also shows that the pulse almost does not broaden while leaving the catalyst bed, implying that diffusion limitations are negligible.

3.3.3. *[^{13}N]-NH_3 Adsorption Assisted Desorption pulse experiments*

Figure 23 shows a typical PEP image for the adsorption/desorption of a labelled pulse of ammonia on γ-alumina in a helium and ammonia/helium flow. In helium flow, [^{13}N]-NH_3 strongly adsorbs at the beginning of the catalyst bed, because the available amount of adsorption sites is much higher than the amount of pulsed labelled ammonia. [^{13}N]-NH_3 remains at the same position in the bed due to its very strong adsorption. In a similar experiment, injection of [^{13}N]-NH_3 in He flow over a clean γ-alumina surface, it was observed that $^{13}NH_3$ stays at the same bed positions for more than 2000 seconds. To observe [^{13}N]-NH_3 desorption and the consequent

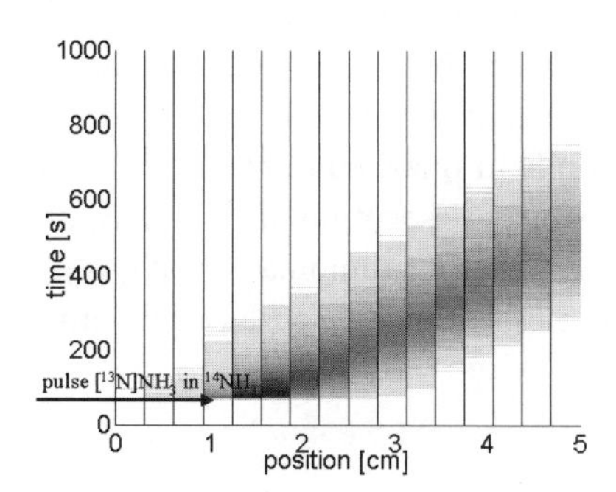

Fig. 22. Pulse of [^{13}N]-NH_3 in a 1 vol% NH_3/He-flow on γ-alumina (T = 423 K, F_t = 48 cm^3/min, GHSV = 5600 h^{-1}). Colour intensity represents the concentration of [^{13}N]-NH_3.

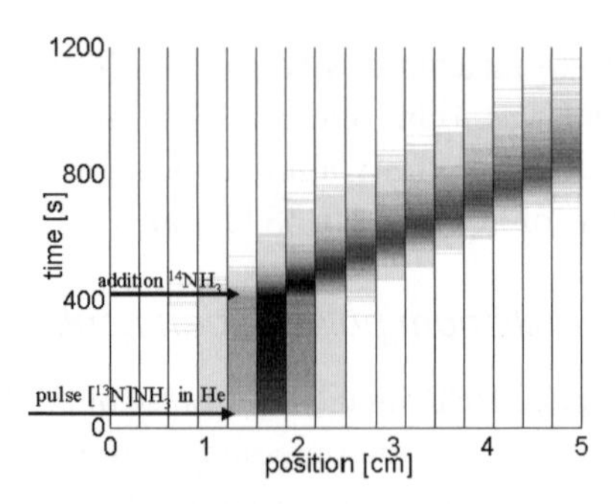

Fig. 23. Pulse of $[^{13}N]$-NH_3 in a He-flow on γ-alumina. At t = 400 seconds He-flow is changed to 1 vol% NH_3/He-flow (T = 423 K, F_t = 48 cm^3/min, GHSV = 5600 h^{-1}). Colour intensity represents the concentration of $[^{13}N]$-NH_3.

movement of labelled species through the bed, temperatures above 600 K were required. Even at these high temperatures, the retention time was over 200 minutes. This result is consistent with IR and TPD results,[69] $[^{13}N]$-NH_3 adsorbs preferentially at the strongest alumina acid sites and desorbs only at relatively high temperatures. This indicates that the adsorption of ammonia on γ-alumina at low temperatures is irreversible.

Figure 23 further shows that after changing the flow to NH_3/He, $[^{13}N]$-NH_3 desorbs and travels as a pulse through the reactor. This indicates that $[^{13}N]$-NH_3 exchanges rapidly with $[^{14}N]$-NH_3. At first sight, this exchange process is very similar to the experiment shown in Figure 22. However, in this case, radio-labelled ammonia is not in full equilibrium on γ-alumina. After switching to unlabelled ammonia, first of all, the available Lewis sites are saturated. The time to saturate the γ-alumina bed with ammonia, measured with the mass spectrometer at the outlet of the reactor, is equal to the retention time of radiolabelled ammonia in the catalyst bed. Thus, the radiolabelled ammonia moves with the saturation front, where ammonia adsorption/desorption is in quasi equilibrium. We conclude that gas phase ammonia clearly facilitates desorption of $[^{13}N]$-NH_3; it remains adsorbed at the same bed position without ammonia in the gas phase. This proves that Adsorption Assisted Desorption takes place for ammonia desorption from γ-alumina.

3.3.4. *Conclusions*

The $[^{13}N]$-NH_3/$[^{14}N]$-NH_3 exchange experiments on γ-alumina, showed that adsorption assisted desorption of ammonia takes place on γ-alumina.

We observed that a [^{13}N]-NH_3 pulse adsorbs irreversibly on γ-alumina in a He flow and only desorbs in the presence of ammonia in the gas phase.

3.4. *Low temperature ammonia oxidation on Pt/alumina*[70]

Alumina-supported Pt particles exhibit a lower activity in low temperature ammonia oxidation than Pt sponge materials. Moreover, deactivation is more severe for the supported catalysts. Below 400 K, Pt/γ-alumina was found to be almost inactive. Pre-exposure of the reduced catalyst to oxygen lowers the initial TOF, suggesting that adsorbed oxygen atoms accelerate the deactivation. An important drawback of the application of the γ-alumina support was the preferential adsorption of ammonia, resulting in a fast covering of the Pt surface by oxygen species. Concomitantly, the catalyst is further deactivated by strongly adsorbed NH_x species. Oxygen spillover to the alumina support appears to be rather small. Moreover, the alumina support itself is inactive for the oxidation of ammonia. At reaction temperature in excess of 400 K, Pt/γ-alumina shows a much higher initial activity, although typically below 520 K, deactivation is still observed. In this temperature range, deactivation is mainly caused by the inhibition by oxygen atoms. Due to the adsorption of ammonia on alumina, the initial reaction proceeds in an oxygen-rich environment. It is found that the catalyst exhibits a stable activity at reaction temperatures in excess of 520 K, presumably due to the removal of adsorbed oxygen and NH_x species.

References

[1] Krestel E, *Medical Imaging in Nuclear Medicine*, Siemens Aktiengesellschaft, Berlin, 1990.

[2] Mangnus AVG, *A Detection System for Positron Emission Profiling*, PhD Thesis, Eindhoven University of Technology, Eindhoven, The Netherlands, 2000.

[3] Ferrieri RA, Wolf AP, *J Phys Chem* **88**: 2256, 1984.

[4] Ferrieri RA, Wolf AP, *J Phys Chem* **88**: 5456, 1984.

[5] Baltensperger U, Ammann M, Bochert UK, Eichler B, Gäggeler HW, Jost DT, Kovacs JA, Türler A, Sherer UW, Baiker A, *J Phys Chem* **97**: 12325, 1993.

[6] Feliu AL, *J Chem Ed* **65**: 655, 1988.

[7] Lewis JS, Achilefu S, Garbow JR, Laforest R, Welch MJ, *Eur J Cancer* **38**: 2173, 2002.

[8] Hawkesworth MR, Parker DJ, Fowles P, Crilly JF, Jefferies NL, Jonkers G, *Nucl Instrum Methods* **A310**: 423, 1991.

[9] Hensel F (Institut Sicherheitsforschung, Forschungszentrum Rossendorf, Dresden D-01314, Germany). FZR, FZR-152, 12, 1996.
[10] Hoff WD, Wilson MA, Benton DM, Hawksworth MR, Parker DJ, Fowles P, *J Mater Sci Lett* **15**: 1101, 1996.
[11] Hawkesworth MR, Parker DJ, *Process Tomography: Principles, Techniques and Applications*, in Williams RA, Parker DJ (eds.), Butterworth-Heinemann Ltd., Oxford, pp. 199, 1995.
[12] Broadbent CJ, Bridgwater J, Parker DJ, *Chem Eng J* **56**: 119, 1995.
[13] Bemrose CR, Fowles P, Hawkesworth MR, O'Dwyer MA, *Nucl Instrum Methods A* **273**: 874, 1988.
[14] Parker DJ, Hawkesworth MR, Broadbent CJ, Fowles P, Fryer TD, McNeil PA, *Nucl Instrum Methods A* **348**: 583, 1994.
[15] Parker DJ, Dijktra AE, Martin TW, Seville JPK, *Chem Eng Sci* **52**: 2011, 1997.
[16] Parker DJ, Allen DA, Benton DM, Fowles P, McNeil PA, Tan M, Beynon TD, *Nucl Instrum Methods A* **392**: 421, 1997.
[17] Roy S, Chen J, Kumar SB, Al-Dahhan MH, Duduković MP, *Ind Eng Chem Res* **37**: 4666, 1997.
[18] Stellema CS, Vlek J, Mudde RF, de Goeij JJM, van den Bleek CM, *Nucl Instrum Methods A* **404**: 334, 1998.
[19] Kantzas A, Hammilton K, Zarabi T, Bhargava A, Wright I, Brook G, Chen J, *Chem Eng J* **77**: 19, 2000.
[20] Jonkers G, Vonkeman KA, van der Wal SWA, van Santen RA, *Nature* **355**: 63, 1992.
[21] Jonkers G, Vonkeman KA, van der Waal SWA, *Precision Process Technology*, Weijnen MPC, Drinkenburg AAH (eds.), Kluwer Academic Publishers, The Netherlands, pp. 533, 1993.
[22] Mangnus AVG, van IJzendoorn LJ, de Goeij JJM, Cunningham RH, van Santen RA, de Voigt MJA, *Nucl Instrum Methods B* **99**: 649, 1995.
[23] Haast MAM, *Performance of the EUT Positron Emission Profiling Detector*, Thesis MSc, Eindhoven University of Technology, Eindhoven, The Netherlands, 1995.
[24] Jaszcak RJ, Greer KL, Floyd CE, Harris CC, Coleman RE, *J Nucl Med* **25**: 893, 1984.
[25] Grootoonk S, Spinks TJ, Jones T, Michel C, Bol A, in the *Proceedings of the Nuclear Science Symposium and Medical Imaging Conference*, IEEE New York, NY, USA, pp. 1569, 1992.
[26] Dannals RF, Ravert HT, Wilson AA, *Nuclear Imaging in Drug Discovery, Development and Approval*, in Burns HD, Gibson R, Dannals R, Siegel P (eds.), Birkhauser, Boston pp. 55, 1993.
[27] Wolf AP, Schlyer DJ, *Nuclear Imaging in Drug Discovery, Development and Approval*, in Burns HD, Gibson R, Dannals R, Siegel P (eds.), Birkhauser, Boston pp. 33, 1993.
[28] Cunningham RH, Mangnus AVG, van Grondelle J, van Santen RA, *J Mol Catal A* **107**: 153, 1996.

[29] Vaalburg W, Kamphuis WAA, Beerling van der Molen HD, Reiffers S, Rijskamp A, Woldring MG, *Int J Appl Radiat Isot* **26**: 316, 1975.
[30] MacDonald JS, Cook JS, Birdsall RL, MacConnel LJ, *Transl Am Nucl Soc* **33**: 927, 1979.
[31] Slegers G, Vandecasteele C, Sambre J, *J Radioan Chem* **59**, 585: 1980.
[32] Gatley SJ, Shea C, *Appl Radiat Isot* **42**: 793, 1991.
[33] Suzuki K, Shikano N, Kubodera A, Label J, *Comp Radiopharm* **37**: 644, 1995.
[34] Suzuki K, Yoshida Y, *Appl Radiat Isot* **50**: 497, 1999.
[35] Mulholland GK, Kilbourn MR, Moskwa JJ, *Appl Radiat Isot* **41**: 1193, 1990.
[36] Berridge MS, Landmeier BJ, *Appl Radiat Isot* **44**: 1433, 1993.
[37] Wieland W, Bida G, Padget H, Hendry G, Zippi E, Kabalka G, Morelle JL, Verbruggen R, Ghyoot M, *Appl Radiat Isot* **42**: 1095, 1991.
[38] Helmeke HJ, Harms T, Matzke KH, Meyer GJ, Hundeshagen H, *Appl Radiat Isot* **45**: 274, 1994.
[39] Sobczyk DP, van Grondelle J, de Jong AM, de Voigt MJA, van Santen RA, *Appl Rad Isotop* **57**: 201, 2002.
[40] Krizek H, Lembares N, Dinwoodie R, Gloria R, Lathrop KA, Harper PV, *J Nucl Med* **14**: 629, 1973.
[41] van den Broek ACM, van Grondelle J, van Santen RA, *J Catal* **185**: 297, 1999.
[42] Fogel YM, Nadykto BT, Rybalko VF, Shvachko VI, Korobchanskaya IE, *Kinetika I Kataliz* **5**: 431, 1964.
[43] Gland JL, Korchak VN, *J Catal* **53**: 9, 1978.
[44] Mieher WD, Ho W, *Surf Sci* **322**: 151, 1995.
[45] Bradley JM, Hopkinson A, King DA, *J Phys Chem* **99**: 17032, 1995.
[46] Il'chenko NI, *Russ Chem Rev* **45**: 1119, 1976.
[47] Matyshak VA, Lefler E, Schnabel K Kh, *Kinetika I Kataliz* **28**: 1199, 1987.
[48] Ostermaier JJ, Katzer JR, Manogue WH, *J Catal* **33**: 457, 1973.
[49] Ostermaier JJ, Katzer JR, Manogue WH, *J Catal* **41**: 277, 1976.
[50] Ruthven DM, *Principles of Adsorption and Adsorption Processes*, Wiley Interscience, 1984.
[51] Vajo JJ, Tsai W, Weinberg WH, *J Phys Chem* **89**: 3243, 1985.
[52] Fahmi A, van Santen RA, *Z Phys Chem* **197**: 203, 1996.
[53] Lombardo SJ, Esch F, Imbihl R, *Surf Sci Lett* **271**: L367, 1992.
[54] Lombardo SJ, Slinko M, Fink T, Löher T, Madden HH, Esch F, Imbihl R, Ertl G, *Surf Sci* **269/270**: 481, 1992.
[55] Veser G, Esch F, Imbihl R, *Catal Lett* **13**: 371, 1992.
[56] Avery NR, *Surf Sci* **131**: 501, 1983.
[57] Campbell JH, Bater C, Durrer WG, Craig JH Jr, *Surf Sci* **380**: 17, 1997.
[58] Gorte RJ, Gland JL, *Surf Sci* **102**: 348, 1981.
[59] Burgess D Jr, King DS, Cavangh RR, *J Vac Sci Technol A* **5**: 2959, 1987.
[60] van den Broek ACM, *Low Temperature Oxidation of Ammonia over Platinum and Iridium Catalysts*, Universiteitsdrukkerij, Eindhoven, 1998.

[61] Sobczyk DP, van Grondelle J, Thuene PC, Kieft IE, de Jong AM, van Santen RA, *J Catal* **225**: 466, 2004.
[62] Sobczyk DP, de Jong AM, Hensen EJM, van Santen RA, *J Catal* **219**: 156, 2003.
[63] Yamada T, Onishi T, Tamaru K, *Surf Sci* **133**: 533, 1983.
[64] Yamada T, Tamaru K, *Surf Sci* **138**: L155, 1984.
[65] Yamada T, Tamaru K, *Surf Sci* **146**: 341, 1984.
[66] Tamaru K, *Colloids Surf* **38**: 125, 1989.
[67] Tamaru K, *Appl Catal A* **151**: 167, 1997.
[68] Xu M, Iglesia E, *J Phys Chem B* **102**: 961, 1998.
[69] Sobczyk DP, Hesen JJG, van Grondelle J, Schuring D, de Jong AM, van Santen RA, *Catal Lett* **94**: 37, 2004.
[70] Sobczyk DP, Hensen EJM, de Jong AM, van Santen RA, *Top Catal* **23**: 109, 2003.

Index